T0215374

Geographical Gerontology

Understanding where ageing occurs, how it is experienced by different people in different places, and in what ways it is transforming our communities, economies and societies at all levels has become crucial for the development of informed research, policy and programmes.

This book focuses on the interdisciplinary field of study – geographical gerontology – that addresses these issues. With contributions from more than 30 leading geographers and gerontologists, the book examines the scope and depth of geographical perspectives, concepts and approaches applied to the study of ageing, old age and older populations. The book features 25 chapters organized into five parts that cover the field's theoretical traditions and intellectual evolution; the contributions of key disciplinary perspectives from population geography, social and cultural geography, health geography, urban planning and environmental studies; the scales of inquiry within geographical gerontology from the global to the embodied; the thematic breadth of contemporary issues of interest that define the field (places, spaces and landscapes of ageing); and a discussion about challenges, opportunities and agendas for future developments in geography and gerontology.

This book provides the first comprehensive foundation of knowledge about the state of the art of geographical gerontology that will be of interest to scholars of ageing around the world.

Mark W. Skinner, Ph.D., is Professor of Geography, Canada Research Chair in Rural Aging, Health and Social Care, and Director, Trent Centre for Aging & Society, Trent University, Canada.

Gavin J. Andrews, Ph.D., is Professor and founding Chair, Department of Health, Aging and Society, McMaster University, Canada.

Malcolm P. Cutchin, Ph.D., is Professor, Department of Health Care Sciences, Wayne State University, USA.

Routledge Studies in Human Geography

For a full list of titles in this series, please visit www.routledge.com/series/SE0514

This series provides a forum for innovative, vibrant, and critical debate within Human Geography. Titles will reflect the wealth of research which is taking place in this diverse and ever-expanding field. Contributions will be drawn from the main sub-disciplines and from innovative areas of work which have no particular sub-disciplinary allegiances.

Geographical Gerontology

Perspectives, Concepts, Approaches

**Edited by Mark W. Skinner,
Gavin J. Andrews, and
Malcolm P. Cutchin**

Routledge
Taylor & Francis Group

LONDON AND NEW YORK

First published 2018 by Routledge

2 Park Square, Milton Park, Abingdon, Oxfordshire OX14 4RN

52 Vanderbilt Avenue, New York, NY 10017

Routledge is an imprint of the Taylor & Francis Group, an informa business

First issued in paperback 2019

British Library Cataloguing-in-Publication Data
A catalogue record for this book is available from the British Library

Library of Congress Cataloging-in-Publication Data
Names: Skinner, Mark William, 1975– editor. | Andrews, Gavin J., 1970– editor. | Cutchin, Malcolm P., editor.
Title: Geographical gerontology : perspectives, concepts, approaches / [edited by] Mark Skinner, Gavin Andrews, Malcolm Cutchin.
Other titles: Routledge studies in human geography ; 74.
Description: Abingdon, Oxon ; New York : Routledge, 2018. | Series: Routledge studies in human geography ; 74
Identifiers: LCCN 2017037397 | ISBN 9781138241152 (hardback) | ISBN 9781315281216 (ebook)
Subjects: | MESH: Aging | Geography, Medical | Population Dynamics | Sociological Factors
Classification: LCC RC952 | NLM WT 104 | DDC 618.97—dc23
LC record available at https://lccn.loc.gov/2017037397

ISBN: 978-1-138-24115-2 (hbk)
ISBN: 978-0-367-88556-4 (pbk)

Typeset in Bembo
by Apex CoVantage, LLC

Contents

Figures

Contributors

Gavin J. Andrews is a professor and inaugural chair (2006–2011) of the Department of Health, Aging and Society, McMaster University, Canada. A geographer, he has wide-ranging interests that include the dynamics between space/place and ageing, holistic medicine, health care education and work, specific phobias, fitness cultures, health histories of places and popular music. Much of his work is positional and considers the development, state-of-the-art and future of health geography. In recent years he has become interested in the potential of non-representational theory, specifically the idea of affect, in conveying the liveliness and 'taking place' of health and well-being. Dr. Andrews has led the publication of authoritative reviews of geographical gerontology (in *Ageing & Society*, *Geography Compass*, *Social Science & Medicine*), is on the editorial board of the international journal *Health & Place* and edited the pioneering Routledge book *Ageing and Place: Perspectives, Policy, Practice* (2005).

Tine Buffel is a research fellow in the School of Social Sciences at the University of Manchester, United Kingdom. Dr. Buffel is funded through the Economic and Social Research Council *Future Research Leaders* scheme on a project which involves a comparative study of social exclusion among older people living in urban areas in three European Union nation states. She has published extensively on a range of social and urban policy issues. Before coming to the University of Manchester, Dr. Buffel was a PhD student, Research Associate and Lecturer in Educational Sciences at the Free University of Brussels (VUB), where she continues to hold an honorary position. She was closely involved with the development of the Belgian Ageing Studies.

Denise S. Cloutier is a professor in the Department of Geography and the Institute on Aging and Lifelong Health at the University of Victoria, Canada. As a health and social geographer, she studies models of health service delivery and the continuum of care for older adults. Her research has focused on the care of populations who are living in rural environments, socially isolated, stroke-affected, at the end of life and clients of home care and institutional long-term care. Her research supports the use of mixed methods (quantitative and qualitative) and has been funded by the Canadian Institutes of Health Research, Social Sciences and Humanities Research Council of Canada, Michael Smith Foundation for Health Research and the World Health Organization. She publishes in leading interdisciplinary and geographical journals such as the *Journal of Gerontology*, *The Gerontologist*, *Progress in Human Geography*, *Social and Cultural Geography, Health & Place* and *Social Sciences & Medicine*.

Tara M. Coleman is a professional teaching fellow in the School of Social Sciences, University of Auckland, New Zealand. Her research interests encompass ageing in place, young people and sexuality, feminist geographies, phenomenology, health policy and qualitative research methods.

Malcolm P. Cutchin is a professor in the Department of Health Care Sciences at Wayne State University, United States. His research has been funded by the National Institutes of Health and his work generally falls into the category of human-place relationships and well-being. Gerontological project foci have included assisted living residences, ethnic populations living near petrochemical plants, preventive home visits for at-risk older adults, and neighbourhood stress and the daily activities of older African Americans. He uses both qualitative and quantitative methods in research and has an abiding interest in philosophy and theory – the writings of John Dewey and those who study Dewey figure prominently in his thinking and writing. He has been actively involved in groups within the Association of American Geographers, the Southern Gerontological Society and the Society for the Study of Occupation: USA. From 2007 to 2011, he was the editor in chief of the *Journal of Applied Gerontology*, and he is a fellow of the Gerontological Society of America.

Keith Diaz Moore is the dean and a professor in the College of Architecture and Planning at the University of Utah, United States. He is the author of over 60 publications on the role the physical setting plays in the ageing experience, particularly for those experiencing dementia, and, as an architect, has consulted on more than 100 projects across the globe. His research ranges across environmental scales from product to urban design issues associated with ageing and draws heavily upon theories of human development. He is lead author of the award-winning book *Designing a better day* (Johns Hopkins University Press, 2006) and co-author, with Andrew Scharlach, of the chapter on "Aging-in-Place" in the *Handbook of theories of aging* (3rd edn). He has been elected both Chair of the Environmental Design Research Association and President of the Architectural Research Centers Consortium.

Susan J. Elliott is a professor of geography and public health at the University of Waterloo, Canada. She has worked as a faculty member at three Canadian research-intensive universities and has more than 200 peer-reviewed publications related to environment and (public) health, focused on Canada and the developing world. For the past eight years, she has been a programme lead for the AllerGen NCE (gene/environment interactions as they relate to allergic disease) for the theme 'Patients, Policy and Public Health'. She is a principal investigator (PI) for the development of a National Food Allergy Strategy for Canada and the PI of GLOWING – the development of a Global Index of Wellbeing, based on the successful Canadian Index of Wellbeing.

Zhixin Feng is a senior research fellow in the Centre for Research on Ageing at the University of Southampton, United Kingdom. He obtained his PhD in Human Geography at University of Bristol. His research interests are population health, health geography, health inequality and ageing in both the United Kingdom and China. He has published widely on these areas in journals, including *Social Science and Medicine, Health & Place* and *Population, Space and Place.*

Anthony C. Gatrell is a professor of the geography of health at Lancaster University (UK) and former dean of both the Faculty of Arts & Social Sciences and the Faculty of Health and Medicine. He graduated with First Class Honours from Bristol University and obtained a PhD from Pennsylvania State University before returning to the United Kingdom to lecture at Salford, and then Lancaster, University. His books include *Mobilities and Health* (Ashgate, 2012); *Geographies of Health: An Introduction* (Wiley, third edition 2014, co-authored with Susan J. Elliott) and *Interactive Spatial Data Analysis* (Longman, 1995, co-authored with Trevor J. Bailey).

Stephen M. Golant, a gerontologist and geographer, is a professor in the Department of Geography, University of Florida (Gainesville), United States. Previously, he was a faculty member in the Committee on Human Development (Department of Behavioral Sciences) and in the Department of Geography at the University of Chicago. He is a fellow of the Gerontological Society of America and a Fulbright Senior Scholar award recipient. He is the recipient of the Richard M. Kalish award from the Gerontological Society of America in recognition of his insightful and innovative publication on ageing and life development in the behavioral and social sciences. This was for his article: "The Quest for Residential Normalcy by Older Adults: Relocation but One Pathway" published in *The Journal of Aging Studies* (25(3), 193–205). His latest book is *Aging in the Right Place* (Health Professions Press, 2015).

Neil Hanlon is a professor of geography at the University of Northern British Columbia, Canada. His research interests include rural and remote health care delivery, age- and able-minded community development and the changing role of voluntarism in health and social care. With Mark Skinner, he co-edited *Ageing Resource Communities: New Frontiers of Rural Population Change, Community Development, and Voluntarism* (Routledge, 2016).

His work appears in leading journals, such as *Health & Place, Social Science & Medicine* and *Health & Social Care in the Community*.

Rachel Herron is an assistant professor in the Department of Geography at Brandon University (Canada), where she specializes in health geography, rural mental health, and ageing. She obtained her Ph.D. in Geography from Queen's University. Her current research examines the vulnerability and complexity of care relationships, social inclusion and meaningful engagement for people with dementia, and the diversity of lived experiences of rural mental health. Her research has been published in peer-reviewed journals such as *Health & Place, Social Science & Medicine, Gender, Place & Culture*, and *Environment and Planning* as well as in edited volumes by international academic publishers.

Alun E. Joseph is a university professor emeritus in the Department of Geography at the University of Guelph in Canada. Over the last three decades, Alun has published extensively on a range of topics related to ageing and to the geography of healthcare. His work has engaged with methodological issues and the policy implications of processes of change. Alun has contributed to large multidisciplinary research programmes in Canada and New Zealand. Recent collaborative research in Canada (with Mark Skinner and others) has focused on voluntarism as a mechanism of adjustment or resistance to processes of change in ageing rural communities.

Gillian M. Joseph is the director of Clear Pane Research Services, Canada. Previously she was a research associate at the Centre for Families, Work & Well-Being at the University of Guelph and a lecturer at that university and at the University of Guelph Humber in Toronto. She has served as the research chair for the Ontario Caregiver Coalition and on numerous boards and committees. Gillian has published on topics associated with eldercare, caregiving policy, weather and long-term care work, seniors and the environment and temporary foreign caregivers. She has also recently edited a book on international issues in ageing with leading gerontologists, *Diverse Perspectives on Aging in a Changing World* (Routledge, 2016).

Robin Kearns is a professor of geography in the School of Environment, University of Auckland, New Zealand. His research interests range across the life course, with his most recent book being *Children's Health and Wellbeing in Urban Environments* (Routledge, 2017; with Christina Ergler and Karen Witten). He is an editor of the journal *Health & Place* and is currently involved in researching the implications of housing tenure on well-being for older people.

Sarah A. Lovell is a senior lecturer at the University of Canterbury, New Zealand. Her research interests focus on health sector responses to social inequality. In recent years, this has included examining issues of access to healthcare, the use and efficacy of community capacity building strategies in health promotion, and health sector responses to climate change. Dr. Lovell has published in journals such as *Social Science & Medicine, British Medical Journal, Critical Public Health* and *Journal of Primary Heath Care*.

Anne Martin-Matthews is a professor of sociology at the University of British Columbia, Canada. From 2004 to 2011, she was the scientific director of the national Institute of Aging, Canadian Institutes of Health Research, and championed the launch of the Canadian Longitudinal Study of Aging in 2008. Anne has published two books, three edited volumes, and more than 200 papers on ageing, social care, social support, later-life transitions and ageing environments. Past editor in chief of *Canadian Journal on Aging*, she is editorial board member for *Ageing & Society* and the *Journal of Aging Studies*. She is a member of the Research Council of the Canadian Institute for Advanced Research, and Fellow of the Canadian Academy of Health Sciences, and Gerontological Society of America. Anne serves on scientific advisory committees for the Cohorts and Longitudinal Studies Enhancement Research initiative (UK), the AGEWELL NCE (Canada) and the European project, EXTEND: Social Inequalities in extending working lives of an ageing workforce.

Christine Milligan is a professor of health and social geography and the director of the Centre for Ageing Research at Lancaster University, United Kingdom. Her research interests focus particularly around active ageing and the provision of care and support for older people – including care at home, ageing in place and new care technologies to support ageing in place. Christine has published more than 100 books, refereed journal articles, book chapters and reports around these topics, is an editor of *Health and Place*, an international journal. She also sits on the research committees of several major national and international funding bodies. Funding for her own research has come from the Europen Union, research councils, the National Health Service and various charitable bodies.

Shelia Peace is an emeritus professor of social gerontology at The Open University, United Kingdom. A human geographer by first discipline, she gained her PhD in the area of environment and ageing in 1977 before moving into social gerontology. A long research career has concerned inclusive design and quality-of-life issues for older people including those with special needs living in care homes, supportive and mainstream housing, as well as intergenerational social interaction within the community utilising mixed and participatory methodologies. A fellow of the British Academy of Social Sciences, Sheila was the president of the British Society of Gerontology from 2014 to 2016. Widely published, she co-authored the seminal study in care homes *Private Lives in Public Places*, 1987; *Environment and Identity in Later Life*, 2006; and co-edited *Ageing in Society: European Perspectives in Gerontology*, 2007. Her new book *Environment and Ageing: Space, Place and Materiality* will be published by Policy Press.

David R. Phillips is the chair professor of social policy at Lingnan University, Hong Kong. He has research and teaching interests in social gerontology, ageing, global health and social epidemiology, with special reference to ageing in China and the Asia-Pacific region. He has published widely in these areas, and his book *Global Health* recently appeared in a second edition (K. McCracken and D. R. Phillips, Routledge, 2017). On several occasions, he has been an advisor and resource person to the World Health Organization and other international agencies.

Judith E. Phillips is a professor of gerontology and deputy principal (Research) at the University of Stirling, Scotland, United Kingdom. Her research interests are in the social and environmental aspects of ageing. Following a geography degree at the University of Wales, Aberystwyth she went to study at Stockholm University, Jesus College Oxford and the University of East Anglia before joining the Centre for Social Gerontology at Keele University and then establishing the Interdisciplinary Research Centre in Ageing at Swansea University in Wales. Her book publications include *Care: Key Concepts*, (Polity Press, 2007), *Key Concepts in Social Gerontology* (with Ajrpuch and Hillcoat-Nalletamby; Sage, 2010) and *Critical Issues in Social Work with Older People* (with Bernard and Ray; Palgrave, 2008).

Chris Phillipson is a professor of sociology and social gerontology at the University of Manchester, United Kingdom. He was the director of the Manchester Institute for Collaborative Research into Ageing (MICRA; 2013–2016). He is a past-president of the British Society of Gerontology. His areas of specialist interest include social theory and ageing, globalisation and social policy, work and retirement and age-friendly cities. He has published extensively on sociological aspects of ageing societies.

Andrea Rishworth is a PhD candidate in the Department of Geography and Environmental Management at the University of Waterloo, Canada. Her research interests lie in health geography with specific focus on health inequalities and environmental health linkages. Her current work addresses issues of health and well-being among ageing populations in Uganda. She has also conducted research in Ghana on maternal health and health policy.

Mark W. Rosenberg is a professor in the Department of Geography and Planning and cross-appointed as a professor in the Department of Public Health Sciences at Queen's University in Kingston, Canada. He is the Tier 1 Canada Research Chair in Development Studies. Over the course of his career, Professor Rosenberg has carried out a wide range of studies on access to health care and its utilisation with a particular emphasis on understanding how barriers to access affect utilisation of health services among vulnerable populations, especially older populations, women in need of diagnostic services or reproductive health services and Aboriginal Peoples.

Graham D. Rowles is a professor of gerontology at the University of Kentucky, United States. He earned his PhD at Clark University (Massachusetts) and spent nine years on the faculty at West Virginia University before moving to the University of Kentucky in 1985. An environmental gerontologist, he focuses his research on the lived experience of ageing. A central theme of this work is exploration – employing qualitative methodologies – of the changing relationship between older adults and their environments and the implications of this relationship for health and well-being. He has conducted in-depth ethnographic research with elderly populations in urban (inner-city), rural (Appalachian), and nursing facility environments. Recent research has focused on long-term care and the meaning of place in old age. Dr. Rowles has served as the president of the Southern Gerontological Society and of the Association for Gerontology in Higher Education. He is currently the chair of the Commonwealth of Kentucky Institute on Aging.

Mark W. Skinner is a professor of geography at Trent University (Canada), where he holds the Canada Research Chair in Rural Aging, Health and Social Care, and is the founding director of the Trent Centre for Aging & Society. Mark's research examines how rural people and places are responding to the challenges and opportunities of population ageing, particularly the evolving role of the voluntary sector and volunteers in supporting older people and sustaining rural communities. His previous book (with Dr. Neil Hanlon), *Ageing Resource Communities: New Frontiers of Rural Population Change, Community Development and Voluntarism*, is published by Routledge (2016). He is appointed to the Royal Society of Canada's College of New Scholars, Artists and Scientists.

Anna Tarrant is a lecturer in sociology at the University of Lincoln, United Kingdom. Her PhD in Social Geography at Lancaster University explored the influence of intergenerational relations on the construction and performance of contemporary grandfathering identities. She is currently funded by the Leverhulme Trust to conduct a three-year research project titled 'Men, Poverty and Lifetimes of Care' examining men's care responsibilities within evolving configurations of care in low-income families. The study consolidates her linked research interests in men and masculinities, family life and care and poverty and qualitative longitudinal methods. Throughout her research career, Anna has published in a number of interdisciplinary journals including *Sociology, Men & Masculinities, Social & Cultural Geography* and *Gender, Place & Culture*.

Kieran Walsh is the acting director at the Irish Centre for Social Gerontology and the director of Project Lifecourse at the National University of Ireland, Galway. With interests in the role of place in the construction of later-life exclusionary and inclusionary experiences, he has been responsible for conducting a series of studies on ageing in different environmental contexts. A significant portion of this work explores the intersection of place and ageing through the life course, and interrogates such concepts as home, belonging, and placed-based identity for older people. Dr. Walsh also coordinated the 3-Cities Project, as a part of Project Lifecourse, which explores the shared and differential experiences of participation in city communities amongst older people, people with disabilities and children and youth.

Janine L. Wiles is an associate professor in population health at the University of Auckland, New Zealand. She is a geographer and gerontologist and uses a critical positive ageing framework. Her research encompasses three disciplinary areas – social/health geographies, critical social gerontologies and community health – and links three themes – care, place and ageing. She has published numerous articles, book chapters and reports on topics including older people's resilience, well-being and identity and strengths-based approaches to ageing, homes and communities as sites or landscapes of care, the contextualised experiences of family or 'lay' caregivers at home and in communities – in light of social and political change and health-care access for marginalised groups.

Kathi Wilson is a professor in the Department of Geography and Program in Environment at the University of Toronto (Mississauga), Canada. Professor Wilson has carried out a wide range of studies on Indigenous health and well-being and Indigenous Peoples' migrations and rights to the city. More broadly, Professor Wilson studies the challenges that vulnerable populations face in accessing health care and social services as well as global migration, settlement and health.

Rachel Winterton is a research fellow at the John Richards Initiative, La Trobe University, Australia. Her research focuses on how rural environments are responding to challenges posed by population ageing through examining systems of governance, health and social infrastructure. Rachel's research interests include rural voluntarism and civic engagement, rural retirement migration, organisational capacity to facilitate rural age-friendly communities and the role of rural systems and structures in facilitating wellness for rural ageing populations.

Preface

Population ageing has become a defining attribute of the twenty-first century. Understanding where ageing occurs, what the diverse experiences of older people in different places are, and the ways in which demographic change is transforming our societies have become crucial for the development of informed research, policy and programmes. This book focuses on the field of study – *geographical gerontology* – that addresses these problems. The aim is to contribute to the growing interdisciplinary interest in the spatial dimensions of ageing, old age and older populations by establishing, for the first time, a comprehensive foundation of knowledge about geographical perspectives, concepts and approaches applicable to gerontology. Bringing together contributions from leading geographers and geographically oriented gerontologists, this book provides a robust compendium on the importance of place, space, scale, landscape, mobility, community and environment among other ideas, for scholars of ageing around the world.

In an important way, *Geographical Gerontology: Perspectives, Concepts, Approaches* is decades overdue. Any contemporary effort to advance the field must acknowledge the pioneering work of scholars who initiated the interdisciplinary conversation between geography and gerontology more than forty years ago. Chapters in this book illuminate the paper trail of scholarship and acknowledge prominent geographers who have led lines of inquiry within the field along the way (many of whom are contributors themselves to this volume, notably Stephen Golant, Alun Joseph, Christine Milligan, David Phillips and Mark Rosenberg). Of particular significance in prefacing this book are the foundational work of Graham Rowles (University of Kentucky) and Anthony Warnes (University of Sheffield), and their continued inquiries into the geographical contributions to the study of ageing. The historiography of the field begins, arguably, with Rowles's (1978) groundbreaking and still-influential book *Prisoners of Space? Exploring the Geographic Experience of Older People* (Westview) and Warnes's (1982) edited volume and the first dedicated to the field of inquiry *Geographical Perspectives on the Elderly* (Wiley). Many of the very same geographical questions that are asked of gerontology today were first raised in these now-classic publications. Indeed, dog-eared copies are still being handed out to (post-)graduate students as foundational reading, evidence of a legacy that the present volume hopes to take forward.

The genesis of this book was a conversation among the editors about the potential for an authoritative volume that would bring coherence to the various perspectives, concepts and approaches applied to geographical gerontology. We were struck by the growing interest in the field among geographers as evident in the popularity of Gavin Andrews' and David Phillips' (2005) *Ageing and Place: Perspectives, Policy, Practice* (Routledge) and the frequency of invited sessions organised for the annual conferences of the Association of American Geographers (AAG), Canadian Association of Geographers (CAG), and Royal Society of Geographers-Institute of British Geographers (RSG-IBG) in the 2010s. We were also struck by the growing interest among gerontologists as evident in the publication of special issues on geographical aspects of ageing in leading interdisciplinary journals (e.g., *Journal of Rural Studies*, *The Gerontologist*, etc.), not to mention the invitation for geographers to headline the plenary addresses at national gerontology association conferences in the same period (e.g., Malcolm Cutchin at the 2014 Canadian Association of Gerontology meeting in Niagara Falls; Mark Skinner at the 2015 Australian Association on Gerontology meeting in Alice Springs). Our own involvement in previous efforts to define the constitution of the field (Gavin Andrews), to delineate its disciplinary scope and depth (Mark Skinner), and to debate its importance to gerontology (Malcolm Cutchin) led us to recognise the value of showcasing the intellectual breadth of the field. We view *Geographical Gerontology: Perspectives, Concepts, Approaches* as a foundational and timely contribution that will inform further developments in the field.

As editors, we are grateful to the contributing authors for their dedication to this book project and to the *Routledge Studies in Human Geography Series* for supporting the advancement of geographical gerontology. We owe a specific thank you to Susan Lawrence (Wayne State University) for her editorial assistance without which this book would not have come to fruition. This book was undertaken, in part, thanks to funding from the Canada Research Chairs program.

Mark W. Skinner
Gavin J. Andrews
Malcolm P. Cutchin
2018

Part I
Introduction

1 Introducing geographical gerontology

Mark W. Skinner, Gavin J. Andrews and Malcolm P. Cutchin

Introduction

Ageing is a key narrative of the twenty-first century. Around the world, individuals, households, communities, organizations, businesses, industries, governments and civil societies are facing the opportunities and challenges of unprecedented numbers of older people. Understanding where ageing occurs, how it is experienced by different people in different places, and in what ways it is transforming our societies, economies and environments at all levels – in other words, he *geography of ageing* – has become crucial for the development of informed research, policy and programmes. Whilst geographical questions are not new to gerontology, the implications of population ageing (for families, health care systems, public finance, etc.) in an era of global economic, environmental and technological change have made them more important than ever. The aim of this book is to bring greater attention to, and build a foundational understanding of, the field of study that addresses these problems.

Geographical gerontology is a burgeoning body of interdisciplinary scholarship encompassing the application of geographical perspectives, concepts and approaches to the study of ageing, old age and older populations. Notwithstanding the growing interest in geographical dimensions of ageing, not to mention the popularity of recent efforts to articulate this interest (see Andrews and Phillips, 2005b; Andrews *et al.*, 2007, 2009; Skinner, Cloutier and Andrews, 2015), what has been missing is an authoritative volume that describes, explains and expands the depth of knowledge about geographical gerontology by elucidating the theoretical, analytical and thematic scope of the field. In taking up this challenge, this book is a distinctive addition to the geographical literature on ageing and the interdisciplinary gerontology literature in that it seeks to establish a comprehensive foundation for understanding geographical gerontology as a contemporary field of study. The emphasis on perspectives, concepts and approaches that are sensitive to the importance of place, space, scale, landscape and territory, among other geographical dimensions, as they relate to gerontological questions lays the groundwork for future developments in the field.

Geographical gerontology

Our starting point is the need to bring the perspectives, concepts and approaches that guide geographical gerontology into focus. The current field emerged at the confluence of a long-standing disciplinary tradition within 'geographies of ageing' (Rowles, 1978, 1986; Warnes, 1981, 1982, 1990; Golant, 1984; Harper and Laws, 1995; Hodge, 2008; Hopkins and Pain, 2007; Tarrant, 2010; Davies and James, 2011; Skinner *et al.*, 2015) and a more recent 'spatial turn' within social gerontology (Andrews and Phillips, 2005b; Wiles, 2005; Andrews *et al.*, 2007, 2009; Andrews, Evans and Wiles, 2013; Cutchin, 2009). The result is a growing body of interdisciplinary scholarship encompassing the application of geographical perspectives, concepts and approaches to the study of ageing, old age and older populations by geographers, gerontologists and allied scholars from across the health and social sciences. The contributions to the landmark volume *Ageing and Place: Perspectives, Policy, Practice*, edited by Andrews and Phillips (2005a), as well as to the special issues on ageing edited by Hardill (2009), Schwanen, Hardill and Lucas (2012) and Rosenberg (2015), exemplify this progression. Subsequent inquiries into the constitution of the field in the early 2000s led to the mapping of its historical, epistemological, thematic and professional terrain (Andrews *et al.*, 2007, 2009). As observed in a recent plenary address by Cutchin (2014), however, whilst the foundations of the contemporary field of study have been established, the full depth and scope of geographical gerontology have yet to be examined.

What is missing from the literature on geographical gerontology, and what lies at the intellectual core of this book, is a comprehensive understanding of the perspectives, concepts and approaches that distinguish the field of study in its most recent era. That is, this book aims to set out an understanding of what is distinctly 'geographical' about geographical gerontology (in contrast, for instance, to what distinguishes environmental gerontology, cultural gerontology and so on). Underlying the evolution of the field, both within geography and gerontology, is a long-standing and increasingly nuanced appreciation for the reflexive, transactional and mutually constitutive relationships between older people and the spaces and places in which and through which age and ageing occur (Andrews *et al.*, 2013). Indeed, the importance geographical constructs – place, space, scale, landscape, territory and so on – for gerontological thinking and research has been advocated for more than forty years by leading geographers such as Stephen Golant, Alun Joseph, Glenda Laws, Christine Milligan, David Phillips, Mark Rosenberg, Graham Rowles and Anthony Warnes as well as by leading gerontologists such as Sarah Harper, Sheila Peace and Judith Phillips, among others.

Notwithstanding the explicit reference to geographical aspects of older populations (e.g., locations, distributions, movements, etc.) within gerontology since its formalisation in the mid-twentieth century, the relationship between ageing and physical/social environments primarily has been the domain of environmental gerontology. The latter is a field informed for the most part by psychology (Scheidt and Schwartz, 2013; Wahl and Weisman, 2003), although the connection to place as an environmental theme is long-standing (Kendig, 2003), as is evident in recent authoritative volumes in environmental gerontology (see Rowles and Bernard, 2013; Peace, 2019). It is increasingly clear, however, that the breadth of geographical perspectives, concepts and approaches applicable to ageing research lie well beyond the prevailing psychosocial considerations of environmental gerontology (Cutchin, 2009, 2014). The recent emergence of geographical gerontology as a coherent domain of interest has broadened both the conceptual and empirical scope of gerontological research to include the wider array of geographical questions.

To elaborate, as first documented by Andrews *et al.* (2007, 2009) the foci of contemporary geographical gerontology range, in general terms, from the spatial patterns of demographic ageing, movement and migration of older populations, health care services and infrastructure, and living arrangements and environments of older people, to the place-embedded experiences of ageing, health, care and well-being, and, more recently, to the embodiment and representations of such experiences, each of which is actively under-way in the field today (Skinner *et al.*, 2015). As an illustration, Figure 1.1 encapsulates the thematic scope of geographical gerontology in reference to the most germane branches of geography and gerontology that have influenced the development of the field. Underlying this body of work, is an increasingly concerted effort to examine and explain how geographi-cal approaches (both long-standing and emergent) relate to understanding gerontological issues such as the ways in which theoretical developments in health geography, social and cultural geography, and population geography have been applied to the study of older people (e.g., therapeutic land-scapes, non-representational theory, post-colonial discourse) (Skinner *et al.*, 2015). By elucidating the full breadth and scope of perspectives, concepts and approaches that underlie the field of study, this book seeks to make a

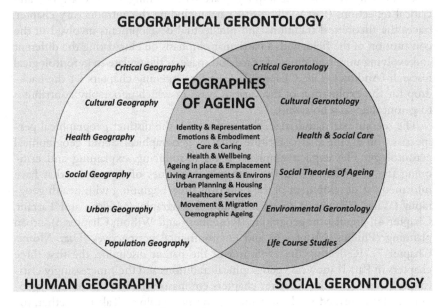

Figure 1.1 Scope of geographical gerontology

Source: adapted from Skinner *et al.* (2015: 778)

distinct and timely contribution to the advancement of geographical gerontology and the broader field of gerontology more generally.

Geographical perspectives, concepts and approaches

The core argument of this book is that geographical perspectives, concepts and
approaches provide constructive, distinct and essential contributions to the interdisciplinary study of ageing, old age and older populations, contributions that
emphasize the importance of scale, place and space in building our understanding of the processes and outcomes, representations and experiences of ageing
around the world. To advance our knowledge about geographical gerontology,
the book draws together leading researchers in the field to build a conceptual,
analytical and methodological foundation that is international in scope. Fea-
turing contributions from more than 30 established and emerging experts at
universities in Australia, Canada, England, Hong Kong, Ireland, New Zealand,
Scotland and the United States, the 24 chapters that comprise the remainder of
this book are illuminating in their focus on the diversity of geographical questions that emerge within gerontology and, perhaps most important, in their
collective emphasis on establishing the foundations for future empirical work in
the field. The contributors are leading scholars who have been examining and
advancing geographical gerontology for the past several decades including some
of the most important leaders in the early development of the field.

To achieve the objectives of this book, the chapters are organized into five
major parts emphasising theoretical foundations (Part I), key geographical perspectives (Part II), scales of inquiry (Part III), defining issues (Part IV) and
critical reflections (Part V). The first part, including this introductory chapter,
traces the theoretical traditions and intellectual developments involved in the
constitution of the field, with a particular emphasis on elucidating the different
and evolving understandings of place and space as they relate to gerontological
research (Andrews *et al.*, Chapter 2). The two opening chapters set the backdrop for the exploration of the various ways in which geography contributes
to gerontological scholarship.

The second part familiarizes the reader with the distinct geographical perspectives on ageing that constitute what is 'geographical' about geographical
gerontology. The explicit emphasis is on introducing, explaining and critiquing the contributions of the core sub-disciplines of geography that have
informed the development of the field over time, beginning with health geography (Wiles, Chapter 3), social and cultural geography (Milligan and Tarrant,
Chapter 4), population geography (Rosenberg and Wilson, Chapter 5), urban
planning (Phillips, Chapter 6) and environmental perspectives (Diaz Moore,
Chapter 7). Reflecting the evolution of the parent discipline, the first three
chapters in Part II showcase geographical traditions and their increasingly critical perspectives, whilst the latter chapters emphasize the links to broader, geographically informed fields in planning and gerontology. Taken together, the
five chapters in Part II set out the (inter)disciplinary foundation for understanding geographical concepts and approaches.

The third part explores the different scales at which inquiry within geographical gerontology occurs, from the macro to the micro and in between. Attention to scale is a uniquely geographical pursuit, and the emphasis here is on building an understanding of how geographical perspectives, concepts and approaches inform our understanding of ageing as a global phenomenon (Phillips and Feng, Chapter 8) and, internationally, as a particular challenge for low- and middle-income countries (LMICs) of the global South (Rishworth and Elliott, Chapter 9). The focus then shifts to geographies of ageing in cities (Buffel and Phillipson, Chapter 10) and in rural regions (Skinner and Winterton, Chapter 11) as well as how ageing issues manifest in communities (Lovell, Chapter 12), households (Martin-Matthews and Cloutier, Chapter 13) and, at the most immediate scale, the body (Herron, Chapter 14). The seven chapters in Part III provide a comprehensive view across the geographical scales at which ageing occurs.

The fourth part examines the thematic breadth of geographical research on ageing. The emphasis is on key issues that best define the scope of interest within the field; organized around the core geographical concepts of place, space and landscape. The first four chapters explore the issues of ageing in place (Golant, Chapter 15), place attachment (Rowles, Chapter 16), people-place transactions (Cutchin, Chapter 17), and health care and place (Hanlon, Chapter 18). These are followed by three chapters that examine older people's movement and mobilities (Gatrell, Chapter 19), domains of inclusion and exclusion that affect their lives (Walsh, Chapter 20), as well as spaces of resistance to the imperatives of ageing and care (Joseph and Joseph, Chapter 21). Closing out Part IV are two chapters that explore landscapes as they are constructed and experienced in the everyday lives of older people (Kearns and Coleman, Chapter 22) and the particular therapeutic landscapes of ageing (Winterton, Chapter 23). The nine chapters in Part IV capture the most salient issues central to contemporary geographical gerontology.

The fifth (and final) part features a discussion about future research in geographical gerontology. The focus is on summarizing the overarching themes of this book and identifying opportunities to further expand the theoretical, analytical and empirical scope of geographically informed gerontological scholarship. The discussion begins with a view from outside of the field by a leading international gerontologist who has observed the evolution of geographical aspects of gerontology over several decades and offers a constructive commentary on the key challenges it faces in the twenty-first century (Peace, Chapter 24). Reflecting on the key contributions of this book, the concluding chapter considers the need to expand further our understanding of the aspirations and applications of geographical perspectives, concepts and approaches to the study of ageing (Cutchin *et al.* Chapter 25). The closing chapter further cements the foundation for advancing geographical scholarship in gerontology by highlighting emergent issues that remain neglected within the literature and by offering a critically informed agenda for future research that speaks to the need for reinforcing the importance of geographical dimensions (and imaginations) of ageing, old age and older populations.

Concluding comments

> The study of geographical dimensions of aging has never reached its full poten-
> tial . . . only a fraction of the depth and scope of the collected theories, concepts
> and methods of geography has been applied to gerontological thinking and research.
>
> Cutchin (2009: 440)

Taking up the challenge implied in Cutchin's (2009) view of the field, *Geo-graphical Gerontology: Perspectives, Concepts, Approaches* makes several important and timely contributions to contemporary scholarship in geography and ger-ontology. As outlined in this introductory chapter, the two dozen chapters featured in the book inform our understanding of the importance of studying the geographical dimensions of ageing and, more important, the significance of geographical perspectives, concepts and approaches for informing gerontologi-cal research, policy and practice. They advance our theoretical and analytical framing of the complex ways in which place, space, scale, landscape, move-ment and mobility, community, identity, environment, housing, health, care and embodiment are key elements of understanding the processes, outcomes, expe-riences and representations of ageing. They do so with an eye to distinguishing the field from environmental gerontology at the same time recognizing the complementarity of geographical and environmental insights into many of the same gerontological questions. Taken together, the contributions to this book expand the recent waves of inquiry into the defining parameters of geographi-cal gerontology by establishing a foundation of knowledge about the perspec-tives, concepts and approaches that are brought to bear in the field.

Whilst incomplete in coverage of the full range of topics of interest to schol-ars in the field, the foundational emphasis of the book is designed to be a useful point of departure for subsequent endeavours to uncover the empirical and applied scope of geographical gerontology. Indeed, as illustrated in the recent proliferation of special issues and symposia related to geographical ger-ontology in academic journals and conferences organized by geographers and gerontologists alike,[1] the extent and potential impact of the burgeoning field is coming into focus. Underlying the interdisciplinary interest in geographical gerontology is a shared concern for articulating its active and potential applica-tion in under-researched experiences and contexts of ageing such as indigenous peoples, LGBTQ communities and populations of the global South. With this is mind, and notwithstanding the tradition of excellent regional volumes by Phillips (2004) among others (e.g., Moore, Rosenberg and McGuinness, 1997; Troisi and von Kondratowitz, 2013), there is a need to articulate the contextual, empirical and applied breadth of geographical gerontology more comprehen-sively. A subsequent edited book that addresses the latter will be a welcome accompaniment to the present volume.

Looking forward, the foundational aspirations of this book include advo-cating for greater attention to creating opportunities for the further applica-tion of geographical thinking within gerontology. The spatial turn already underway within gerontology bodes well for future engagement with – and co-development among – allied sub-disciplines; for instance, in social geography and political geography, where advancements in research on intersecting spaces of activism, racialization and sexuality of older populations have great potential to contribute to an increasingly critical gerontology (see del Casino Jr., 2009). Ultimately, in setting out the perspectives, concepts and approaches that define contemporary geographical gerontology, this book and its contributors have established a robust intellectual foundation from which a diversity of new geographically informed questions can be asked about ageing, old age and older populations. Geographies of ageing – the places, spaces and landscapes of older people – are ever more important in a century being defined by demographic change.

Note

1 See contributions to special issues on ageing in *The Professional Geographer* (edited by Hardill, 2009), *Journal of Rural Studies* (edited by Milbourne, 2012), *Geoforum* (edited Schwanen, Hardill and Lucas, 2012) and *Progress in Geography* (edited by Mark Rosenberg, 2015), as well as to the recent sessions on ageing organized for the 2016 annual meetings of the Association of American Geographers (AAG), Canadian Association of Geographers (CAG), and Royal Society of Geographers – Institute of British Geographers (RSG-IBG).

References

Andrews, G. J., Cutchin, M., McCracken, K., Phillips, D. R. and Wiles, J. (2007). Geographical gerontology: The constitution of a discipline. *Social Science & Medicine*, 65(1), 151–168. http://dx.doi.org/10.1016/j.socscimed.2007.02.047

Andrews, G. J., Evans, J. and Wiles, J. (2013). Re-spacing and re-placing gerontology: Relationality and affect. *Ageing & Society*, 33(8), 1339–1373. http://dx.doi.org/10.1017/S0144686X12000621

Andrews, G. J., Milligan, C., Phillips, D. R. and Skinner, M. W. (2009). Geographical gerontology: Mapping a disciplinary intersection. *Geography Compass*, 3(5), 1641–1659. http://dx.doi.org/10.1111/j.1749-8198.2009.00270.x

Andrews, G. J. and Phillips, D. R. (eds). (2005a). *Ageing and place: Perspectives, policy, practice*. New York: Routledge.

Andrews, G. J. and Phillips, D. R. (2005b). Geographical studies in ageing: Progress and connections to social gerontology, in G. J. Andrews and D. R. Phillips (eds), *Ageing and place: Perspectives, policy, practice* (pp. 7–12). New York: Routledge.

Cutchin, M. P. (2009). Geographical gerontology: New contributions and spaces for development. *The Gerontologist*, 49(3), 440–444. http://dx.doi.org/10.1093/geront/gnp095

Cutchin, M. P. (2014). *Beyond environmental gerontology: The many geographies of aging*. Invited plenary address to the 2014 Annual Scientific and Educational Meeting of the Canadian Association on Gerontology, McMaster University. Available from: mpc@wayne.edu.

Davies, A. and James, A. (2011). *Geographies of ageing: Social processes and the spatial unevenness of population ageing*. Aldershot: Ashgate.

Del Casino Jr., V. (2009). Ageing and the 'new' social geographies of older people, in V. Del Casino Jr. (ed.), *Social geography: A critical introduction* (pp. 238–263). Chichester: Wiley-Blackwell.

Golant, S. M. (1984). The geographical literature on aging and old age: An introduction. *Urban Geography*, 5, 262–272.

Hardill, I. (2009). Introduction: Geographies of aging. *The Professional Geographer*, 61(1), 1–3. http://dx.doi.org/10.1080/00330120802577509

Harper, L. and Laws, G. (1995). Rethinking the geography of ageing. *Progress in Human Geography*, 19(2), 199–221. http://dx.doi.org/10.1177/030913259501900203

Hodge, G. (2008). *Geography of aging: Preparing communities for the surge in seniors*. Montreal and Kingston: McGill-Queen's University Press.

Hopkins, P. E. and Pain, R. (2007). Geographies of age: Thinking relationally. *Area*, 39, 287–294.

Kendig, H. (2003). Directions in environmental gerontology: A multidisciplinary field. *The Gerontologist*, 43, 611–614.

Milbourne, P. (2012). Growing old in rural places. *Journal of Rural Studies*, 28(4), 315–317. http://dx.doi.org/10.1016/j.jrurstud.2012.10.001

Moore, E. G., Rosenberg, M. W. and McGuinness, D. (1997). *Growing old in Canada*. Toronto: ITP Nelson.

Peace, S. (2019). *Environment and ageing: Space, place and materiality*. Bristol: Policy Press.

Phillips, D. R. (ed.). (2004). *Ageing in the Asia-Pacific region: Issues, policies and future trends*. London: Routledge.

Rosenberg, M. W. (2015). Special issue on geographies of ageing. *Progress in Geography*, 34(12), 1479–1479.

Rowles, G. D. (1978). *Prisoners of space? Exploring the geographic experience of older people*. Boulder, CO: Westview.

Rowles, G. D. (1986). The geography of ageing and the aged: Towards an integrated perspective. *Progress in Human Geography*, 10, 511–539.

Rowles, G. D. and Bernard, M. (eds). (2013). *Environmental gerontology: Making meaningful places in old age*. Dordrecht, the Netherlands: Springer.

Scheidt, R. J. and Schwartz, B. (eds). (2013). *Environmental gerontology: What now?* London: Routledge.

Schwanen, T., Hardill, I. and Lucas, S. (2012). Spatialities of ageing: The co-construction and co-evolution of old age and space. *Geoforum*, 43(6), 1291–1295. http://dx.doi.org/10.1016/jgeoforum.2012.07.002

Skinner, M. W., Cloutier, D. and Andrews, G. J. (2015). Geographies of ageing: Progress and possibilities after two decades of change. *Progress in Human Geography*, 39(6), 776–799. http://dx.doi.org/10.1177/0309132514558444

Tarrant, A. (2010). 'Maturing' a sub-discipline: The intersectional geographies of masculinities and old age. *Geography Compass*, 4, 1580–1591.

Troisi, J. and von Kondratowitz, H-J. (eds). (2013). *Ageing in the Mediterranean*. Bristol: The Policy Press.

Wahl, H. W. and Weisman, J. (2003). Environmental gerontology at the beginning of the new millennium: Reflections on its historical, empirical, and theoretical development. *The Gerontologist*, 43(5), 616–627. http://dx.doi.org10.1093/geront/43.5.616

Warnes, A. M. (1981). Toward a geographical contribution to gerontology. *Progress in Human Geography*, 5(3), 317–341. http://dx.doi.org/10.1177/030913258100500301

Warnes, A. M. (ed.). (1982). *Geographical perspectives on the elderly*. Chichester: Wiley.

Warnes, A. M. (1990). Geographical questions in gerontology: Needed directions for research. *Progress in Human Geography*, 14, 24–56.

Wiles, J. (2005). Conceptualising place in the care of older people: The contributions of geographical gerontology. *International Journal of Older People Nursing*, 14, 121–129.

2 Space and place in geographical gerontology

Theoretical traditions, formations of hope

Gavin J. Andrews, Malcolm P. Cutchin and Mark W. Skinner

Introduction

How and why space and place matter are important questions for many scholars who study ageing, particularly those involved in the academic enterprise and field of geographical gerontology. Together these researchers have produced an array of answers based largely on specific sets of empirical observations they have made of older peoples' lives and services, and most of those answers are important from health, well-being and welfare standpoints. Such answers include, for example, that space and place matter because older people can no longer navigate their neighbourhoods as quickly and easily as they once could, because they continue to value and enjoy their homes, because they want to migrate in their retirement years, because they appreciate nature and wish to participate in outdoor leisure activities, because the small town that they have lived in for many years remains important to them, because they have established close bonds in a retirement community, because their emotional and physical needs are or are not being met in residential care settings, or because they enjoy the acceleration, wind and laughter while riding fairground rides with their grandchildren (Andrews *et al.*, 2007, 2009; Skinner, Cloutier and Andrews, 2015). Underlying almost all of these quite specific answers is the implicit and often understated realisation amongst scholars that space and place matter to older people – and those concerned with their happiness – because they are active phenomena that create particular outcomes. Indeed, being 'things' (i.e., nouns) that fundamentally possess and/or facilitate and/or reflect human and non-human agency, spaces and places help create, realise or dash hopes in some way. Complicating this geographical facilitation of hope is the fact that space and place have themselves been articulated in quite different ways across successive generations of geographical gerontology. Acknowledging this fluidity, the current chapter traces the ways in which hope emanates from different understandings of space and place as illustrated by six theoretical traditions in the sub-discipline. Of the following, the first four are well-established, while the remaining two emerged more recently: (1) spatial and population science, (2) political economy, (3) humanism, (4) social constructionism/

constructivism, (5) relational thinking and theory and (6) non-representational theory. Expanding on the introduction to geographical gerontology in Chapter 1 (Skinner, Andrews and Cutchin), the aim of this chapter is to review scope and progress, setting the theoretical scenes for the remainder of the book.

Spatial and population science: hope in realising volumes, distances and demands

Drawing largely on a positivistic tradition that first emerged in human geography in the 1950s and 1960s, one way that space has been understood in geographical gerontology is in a Newtonian sense as 'absolute space'. Space as something in itself, independent of and yet facilitative of what happens in it; a macro-scale, featureless, neutral underlying template for the unfolding of agency. However, it is also understood that when 'things' (such as humans, objects, or the events they create) are located at particular points in space, then it begins to emerge – to academics and officialdom, for example – as a substantial and important feature of human existence. This is because at this juncture, space becomes mathematically distinguishable and divisible. Indeed, *at* these points, rates, volumes and other localised measures become visible and calculable, and *between* these points, times, distances and differences also become visible and calculable. Thus, based on such understandings and approaches, certain geographical gerontology has been concerned with aerial differentiation and the quantitative calculation and representation of distributive trends related to ageing across space. The belief is that it is possible to find regular spatial patterns in ageing and age-related phenomenon – some of their fundamental underlying 'geometries'.

Using these understandings, geographical gerontologists have developed two concerns. The first of these is to trace how older people concentrate in space and distribute collectively across it over time (see Golant, 1989). This is an approach taken, for example, in numerous demographic and migration studies focused on international (Kinsella and Velkoff, 2001; Knodel, 1999; Phillips, 2000), national (Moore et al., 1997; Warnes, 1987; Warnes and Law, 1984) and subnational scales (Golant, 1990; Rosenberg, Moore and Ball, 1989; Watkins, 1990), as well as across these scales (Haberkorn, 2002; Heleniak, 2003; Moore and Pacey, 2004). Notably, in this research, not only are older people themselves placed and mapped, but so too are their quantifiable attributes and characteristics (e.g., health status and life expectancy; Pickle *et al.*, 1996; Warnes, 1999), and their health behaviours, incidence rates and events (e.g., injuries and falls (Yiannakoulias *et al.*, 2003). The second concern, albeit less commonly investigated, has been to trace the distributive features of services provided specifically for and used solely by older people and/or general services used by older people across space (Hiltner, Smith and Sullivan, 1986; Meyer, 1981, 1990).

This research is not, however, just an exercise in mapping but often attempts to explain trends related to ageing populations. Indeed, certain studies articulate, for example, why spatial and other patterns are not always regular, and/or they focus on the consequences of patterns and/or attempt to critically understand the individual, social, economic and political processes that underpin them (Bell and Rees, 2006; Davies and James, 2011; Everitt and Gfellner, 1996; Moore and Rosenberg, 2001; Moore, Rosenberg and McGuiness, 1997; Speare and Meyer, 1988) and of services provided for and used by them (Curtis, 1989; Hamnett and Mullings, 1992; Phillips and Vincent, 1986; Rosenberg and Everitt, 2001). The common 'hope' underlying this spatial and population science is that by researchers revealing the dynamics of older populations and related services, policy-makers and those in command of health and welfare systems might pick up this information and subsequently be prepared to meet future challenges and be better able to accurately target policies, programmes and other resources to areas of greatest need. Hence, academic hope is the prime reason for pursuing this kind of geographical gerontology (rather than, for example, anything felt or experienced by older people). It is the underlying push and motivation to undertake scholarly inquiry based on the perceived expertise and primacy of academics.

Political economy: hope for equitable distributions of limited resources

Emerging fifteen years after the introduction of spatial science, but often deployed in unison with it, was the political economy tradition in geographical gerontology. This is concerned primarily with production and exchange systems and the influence of commercial and political forces in relation to age (Harper and Laws, 1995). Specifically then, political economy in geographical gerontology is concerned with how health and welfare systems and policies pertaining to older people play out spatially or have spatial implications, including how resources and services are distributed – either by public-sector decision-making or by market forces (Rosenberg and Everitt, 2001; Cartier, 2003) – often in relation to utilisation patterns (Nemet and Bailey, 2000). Specific empirical interests include residential care provision for older people (Phillips and Vincent, 1986; Ford and Smith, 1995; Smith and Ford, 1998; Andrews and Phillips, 2002; Ford and Smith, 2008), retirement and labour relations (Harper and Thane, 1989; Harper, 1989, 1990) and meeting the needs of older people in remote rural communities (Hanlon and Halseth, 2005; Hanlon *et al.*, 2014; Skinner and Hanlon, 2016; Skinner *et al.*, 2014).

In terms of theoretical underpinnings, Marxist theory has informed much early political economy in geographical gerontology, the argument being that unequal allocation of resources and services are the realisation of unequal social and economic relations. In more recent years, however, Marxist theory has been supplemented with post-structural ideas and thinking. Broadly speaking, post-structural political economy provides a critique of neoliberalism – and often globalisation – by challenging universal definitions and by highlighting their local impacts, guises and embeddedness. This includes how categories such as 'age' and 'ageing environments' are actively reproduced and used by political and market forces. In addition to focusing on some of the topics mentioned already, post-structuralist inquiries are also focused thematically, for example, on the visions, biases, restructurings and 'offloadings' of both localised and central governance (Chalmers and Joseph, 1998; Hanlon, Halseth *et al.*, 2007; Hanlon, Rosenberg and Clasby, 2007; Milligan, 2000; Skinner and Rosenberg, 2005; Waldbrook *et al.*, 2013).

Although often understated and somewhat idealistic, a utilitarian ethical belief and hope for the future, often underlies the political economy tradition in geographical gerontology: that where older people live should not seriously compromise their health, welfare or longevity and that systems should work towards this reality through the optimal and most efficient spatial allocations of resources. More practical, however, is the way in which hope underpins much of this research in terms of the potential impact of studies. Researchers hope that through their production of knowledge and its translation to key decision makers, the spatial design of health and social care systems might be improved to better meet the needs of older populations. Indeed, in explaining distributive features in the supply of services for older people – and often their relation-ships to health needs, patterns and outcomes – political economy constitutes a theoretical perspective that is thought to contribute directly and critically to debates on rationing, efficiency and equity in service planning and provision.

Humanism: hope or despair felt through places

The humanistic tradition in geographical gerontology emphasises the value of human experience, values, subjectivity and agency and, importantly, their realisation through 'places'. Indeed, in celebrating the tradition, Harper and Laws (1995) suggest that, in contrast to positivistic paradigms, the humanistic tradition requires geographical gerontologists to listen to older people's voices and their concerns, because they are, by far, best placed to reason and articulate their own circumstances and lives. How places are qualitatively experienced by older people starts with the idea of 'embedded knowledge' which, based on Heidegger's thinking (although he often tried to distance himself from human-ism), posits that humans can only relate to and beyond themselves through their situation, their literal 'being-in-the-world' and their relational consciousness of other humans and non-humans in the world. In Husserl(ian) terms, being-in-the-world and forms of encounter occurring in it allow the 'essences' of places to be apprehended. Indeed, just as objects possess essences (i.e., their facets influence what humans feel emotionally *about* them), so, too, do places (i.e., their qualities influence what humans feel emotionally *about* and *in* them). This is how 'sense of place' emerges. Places can evoke a broad range of emotions, ranging from the basic to complex and from the positive to negative. Sense of place thereby becomes a central concept deployed throughout the tradition (Relph, 1976).

As Harper and Laws (1995) suggest, the early work of Rowles (1978, 1983a, 1983b, 1988) on urban and rural older lives stands out as foundational in human-istic geographical gerontology, closely followed by Rollinson's (1990a, 1990b) research on older tenants of rooming hotels and their navigation of poverty, and Harper's (1987a, 1987b, 1987c) own research on place and kinship networks. More recently, since this pathbreaking work, humanistic studies fall loosely into the following four categories. Category one includes studies that address the nature and challenges of urbanicity, rurality and ageing, particularly in relation to social isolation and social exclusion (Chalmers and Joseph, 1998; Smith, 2009). The second category contains studies that unpack the dynamics between place – particularly homes and other community settings – and experiences and practices of life and care (Martin et al., 2005; Milligan, 2003; 2005; Mowl, Pain and Talbot, 2000; Walsh and O'Shea, 2008; Wiles, 2003a). The third includes new 'mobilities' studies that critically unpack the motivations, meanings and implications of varied movement forms in later life (Schwanen and Paez, 2010; Schwanen and Ziegler, 2011) including older peoples' tourism mobilities, their leisure, lifestyle and fitness activities and transport (Davey, 2007; Griffin and Phoenix, 2014; McConatha and Volkwein-Caplan, 2012). The fourth includes more conceptual studies have pursued the idea of 'social space'. Whilst on one level this speaks to the process and experience of space being immediately used, navigated, divided, owned and contested by older people and others, on another, akin to the lay notion of 'personal space', it speaks to the intimate spatial proximities among people that impact upon narrative and emotional interpersonal proximities (Andrews and Peter, 2006; Andrews et al., 2005; Dyck et al., 2005).

In sum, both on its own, and in more recent years informing a range of allied theoretical positions and traditions, the humanistic perspective has certainly brought space and place 'to life' in geographical gerontology, making them far less abstract concepts. Hope in this tradition is found primarily in older peoples' potential or actual experiences of spaces and places with the future figuring in two distinct ways. On one hand, in the realm of preferences for particular encounters, an older person might *'hope for'* a place to provide them with particular experiences (e.g., greater mobility, independence, or social contact) and indeed they might *'hope to'* experience these things. On the other hand, following particular encounters, these experiences themselves, then, might provide an older person with a greater emotional *'feeling of hope'*; a positive attitude and sense about the likelihood of better futures. Otherwise, hope also arises as researchers hope that they might provide in-depth person–centred, sensitive insights into older peoples' experiences, preferences and less-than-ideal situations. The ultimate goal is that policy-makers and service providers are able to use those insights to directly inform designs and interventions (Rowles and Bernard, 2013).

Social constructionism and social constructivism: hope or despair represented in places

Being closely aligned and complementary, both social constructionism and social constructivism understand the social world to be a subjective rather than objective reality produced through group interactions. The main difference between the two theories is that whilst the former is concerned more with the actual constructions produced, the latter is concerned more with the individual learning and perception which is part of the overall process. Although numerous human and non-human phenomena are socially constructed, geographers are concerned more than many other social scientists with the additional involvement/construction of place. Specifically, an understanding has developed in geography that, at one level, because of people and structural features in situ, places possess basic agency; they '*do*' (e.g., nursing homes provide/ do residence and care, and so on). At another level, however, it is recognised that underlying this agency, social and psychological processes are at work. Indeed, it is posited that through human presence, interactions, perception and judgement, places become '*about*' certain things (i.e., about what humans do in them). The argument follows that such feelings, when experienced or over time, are an important part of the overall identities and attachments humans ascribe to, and develop with, places (e.g., nursing homes become about dependency, decline and closeness to death).

Informed by both social constructionism and social constructivism, a strong tradition in geographical gerontology is concerned with how places related to ageing are complex social constructions. Here, place is more than objective functions and locations on maps; places represent and make older people, and older people represent and make them. As Harper and Laws (1995), suggest, this tradition dates back more than thirty years (Cole, 1992; Laws, 1995a; Townsend, 1981; Ward, 1984). Such work, they argue, is crucial to geographical gerontologists' understandings of ageing processes and the ageing of societies. Just over ten years ago, Wiles (2005) summarised how, consistent with social constructionist understandings, places have been conceptualised in geographical gerontology in four ways (a typology which remains accurate). First, place is a process and an essential part of meaningful social relations (including family life and caring relationships). Second, place is subject to ongoing negotiation (such as when older people move into or out of place, or their needs change whilst *in situ*). Third, place is fought over and contested (such as between the different groups that occupy a home where caring takes place). Fourth, place expresses power relations (such as professional hierarchies in care settings, or gender inequalities in unpaid home care). Across these four categories, geographical gerontology has investigated many forms of socially constructed places of ageing including the seaside (Blaikie, 1997), social clubs (Pain, Mowl and Talbot, 2000), neighbourhoods (Wiles *et al.*, 2009), different types of housing (Cutchin, 2003; Hockey, Penhale and Sibley, 2001; Swenson, 1998), sites of seasonal migration (Gustafson, 2001; McHugh and Mings, 1996) and retirement communities (Katz and McHugh, 2010; Laws, 1995b; McHugh, 2000; McHugh and Larson-Keagy, 2005). Notably whilst a common theme running through this empirical literature is the role of these places in supporting self-determination and identity in older age, these latter two groups of studies in particular help challenge stereotypes that position older people and certain places as both being 'sick' and 'sad', by showcasing places associated with older peoples' preferences: independence, resilience and well-being.

In terms of theoretical advancement, recently both humanist and social constructionist/constructivist perspectives have led to post-structuralist scholarship. Inquiries drawing on the continental philosophies of Foucault, (Andrews *et al.*, 2005), Bourdieu (Antoninetti and Garrett, 2012) and other traditions (e.g., Feminism) (Laws, 1995a) to critically unpack the power and bodily processes behind ageing and place constructions and experiences. In all of the aforementioned constructionist/constructivist research, hope relates to whether these place constructions and identities provide positive experiences and futures for older people or, in the worst cases, whether they contribute to their marginalisation and unhappiness. Moreover, hope also exists in researchers' minds: hope that their studies might assist the modification or representation of places to improve experiences, constructions and identities. Ultimately then, geographical gerontologists harbour hope that ageism, stereotyping and marginalisation can be combatted.

Relational thinking and theory: hope emerging between places and across spaces

One criticism of the aforementioned understandings of space and place is that, although they recognise their rich and complex composition, they generally portray space and place as somewhat discrete and static. Space tends to be conceptualised as an isolated area (such as a country, region or block of census tracts), and place tends to be conceptualised as a fixed, parochial centre of meaning resulting from social inscription (such as a house, retirement community or nursing home). In response to this limitation, simply put, 'thinking space and place relationally' implies a twist in how they are theorised. Relational thinking evokes an image of spaces and places emerging not only 'in situ' but also through their connections within networks of interactions (in other words, spaces and places are thought to be highly related to, and produced by, many other spaces and places at multiple scales). Relational thinking thus complicates the conventional assumption that 'intrinsic' qualities of single spaces and single places exist that affect and reflect people's lives. Notably, relational accounts also place emphasis on temporality as they consider spaces and places to be ever-developing and changing over time. Indeed, as 'bundles of interrelations', spaces and places are considered only temporary accomplishments which are forever coming into being (Darling, 2009) Thus, as Jones (2009) suggests, the research agenda for relational thinkers consequently centres on ideas and theories of space and place that portray them as encountered, performed and fluid.

Although it might be a little premature to claim the emergence or existence of a fully blown 'relational geography of ageing', perhaps the first significant signpost towards substantive relational thinking in geographical gerontology has come in the form of six groups of studies that consider at least the 'interrelatedness' of place (Andrews, Evans and Wiles, 2013; Wiles, 2005). The first of these groups is theoretically oriented and centred on actor-network theory. Indeed, actor-network theory is employed as an approach to help explain the networking, relationality and equal importance of human bodies and non-human objects in events pertaining to older age (Cutchin, 2005; Tatnall and Lepa, 2003). The second group of studies is centred on home. Rowles's (1978, 1981) early work which developed the concept of the 'surveillance zone' was pathbreaking in this regard. The home, being thought of as a base that facilitates the process of monitoring events outside, hosts the emergence of watchful reciprocal social networks and provides the foundation for a sense of personal identity. Most recently, extending these lines of thought, Wiles *et al.* (2009) illustrated older people's physical, imaginative, emotional and symbolic experiences of, and connections to, place across time. Places are viewed as varied and relationships to them elastic, stretching from home to other settings, neighbourhoods and cities to other spaces beyond (see also Cutchin, 2001, 2007, 2013; Peace, Holland and Kellaher, 2006; Rosel, 2003). The third group of studies, meanwhile, is centred on care. It articulates the multiple and connected places and scales through which care operates (Milligan and Wiles, 2010; Wiles, 2003b; Wiles *et al.*, 2012). For example, Dyck *et al.* (2005) and Angus *et al.* (2005) expose the superimposition of 'external' medical on domestic fields as part of formal home care, and Zhou (2012) shows the body and care are experienced across transnational settings and scales. The fourth group of studies is centred on life course and its mobility. It shows the multiple places frequented, experienced and attached to by people both sequentially and simultaneously throughout their lives (Gustafson, 2001; McHugh and Mings, 1996; Schwanen, Hardill and Lucas, 2012). The fifth group of studies is focused on intergenerationality, on how older generations combine, collide and interrelate with younger generations in place (Hopkins and Pain, 2007; Vanderbeck, 2007), an emerging focus of this group being on grandparenthood and grandparenting (Tarrant, 2010, 2013, 2016). Finally, the sixth group of studies considers the relationalities between the past and present of both older lives and the places they have frequented. Notably the famous precedent here is the oral history research by Peter Townsend (1957, 1962) in *The Family Life of Older People* and the *Katherine Buildings Collection*, which recollects social change in East London, relating past and present experiences of that particular historical locale. This type of inquiry has come to fuller fruition in more recent oral histories of the British Homefront (Andrews *et al.*, 2006; Andrews, 2017) and analysis of biographical road cinema (McHugh, 2009).

As a whole, relational research reflects hope as complexly constructed for older people, originating from multiple geographical sources, ebbing and flowing and subject to enhancement or degradation over time. Indeed, hope emerges through networks and within relational assemblages that change. Researchers themselves hope to make visible these relations so that complex pasts and lives might be realised and current circumstances improved.

Non-representational theory: hope emerging within the forward movement of space and time

Whilst the aforementioned relational forms of understanding are insightful, a new paradigm of post-structuralist relational thinking known as non-representational theory (NRT) helps geographers think about how (relationally constructed) space and place is itself relationally performed (Anderson and Harrison, 2010). NRT is not, however, simply an orientation of relational thinking. It has arisen partly as a reaction to two 'representational' mainstays of contemporary social science (found, for example, throughout the spatial science, political economy, constructivism and humanism reviewed earlier). On one hand, the tendency to take a predominantly retrospective perspective, 'stopping the world' it studies and thinking about and articulating events as if they were discrete and complete. On the other hand, heavily theorising and interpreting events; endlessly peeling off layers of words and actions to find their underlying mechanisms, consequences and meanings (Thrift, 2008). For NRT's proponents, these mainstays of much inquiry have had the effect of embalming the subjects studied, for they have neglected the moment, immediate and active, the detail of what is 'taking place' in space and time (Thrift, 2008). As a solution then, NRT places the basic raw performance of the world at the centre of inquiries. On one level it communicates the many subtle, wordless, automatic and accidental everyday practices in life. On another level it communicates how these practices are registered, sensed and acted upon by as well as affect humans (for overviews in human geography, see Anderson and Harrison, 2010; Thrift, 2008; particularly for overviews in ageing and health-related contexts see Andrews, 2014; Andrews *et al.*, 2014; Andrews and Grenier, 2015). In research terms, as Skinner *et al.* (2015) describe, a number of key interests – two of which are discussed below – together create the overarching style of NRT as a new materialist, post-humanist way of researching.

First, NRT presents the 'onflow' of life; the moving frontier of existence as it rolls out creating new space and time. It recognises that, for an infinitely small moment, all life is brand new, physical and less-than-fully-consciously acted and experienced. Moreover, that it is this infinitely small moment that humans occupy – a moment happening all the time and everywhere. As Andrews and Grenier (2015) argue, the idea of onflow resonates with certain accounts and ideas in critical gerontology. Geographer Kevin McHugh has, however, directly advanced ideas on onflow in ageing contexts. Indeed, studying the relationship between the motion of older bodies, memory and landscape, McHugh (2009) argues that the passage through time and place is not a succession of states. Instead, older peoples' experience is that of an unbroken duration, of a dynamic continuation of movement and sensation. In fact, he argues,

> [f]or [the philosopher] Bergson time is real, that is, qualitative difference at each instant, not quantitative difference in degree that we know as measured clock-time. Real time, Bergson's duration, is duality in movement, . . . That is, at each "instant" the present endures, simultaneously dilated toward the past and contracted toward the future. This is consciousness, this is life, incessant creation, Bergson's élan vital.
>
> (McHugh, 2009: 210–211)

Second, NRT sets out to focus on practice and performance (as part of this onflow and movement), the expressive and felt engagement of the body, bodies together and bodies and objects together, including their spacings and timings. These practices and performances might be habitual or non-habitual, scripted or unscripted, remarkable or unremarkable, yet they underlie much of life. As Andrews and Grenier (2015) note, although not strictly NRT, perhaps the most well-aligned trend in gerontology with regard to practice and performance is a focus on the relationships between the performance of older bodies in places and cultural discourses such as decline, management, reconstruction or reconfiguration (Tulle, 2008; Wahidin, 2002). Some of this research is specifically concerned, for example, with the empirical phenomenon of older athletes (Phoenix and Grant, 2009; Tulle, 2008; Wainwright and Turner, 2004). Also aligned are gerontological studies that convey the performative embodied rhythms that are part of the processional nature of caring practices and places older people encounter as their health status and social contexts change (Antoninetti and Garrett, 2012; Rowles, 2000; Rowles and Watkins, 2003).

Whilst the traditional understanding of hope in research – and society at large – is of it as a prior, variably founded belief in or desire for particular outcomes (many of which are often hypothetical or utopian), NRT instead understands hope as the belief or desire that arises *because of* the unfinished nature of situations (because there is always some positive potential in future life). Moreover, it understands hope as a belief or desire that often emerges during, *and is part of*, the actual actions, movements and creations occurring in the journeys taken towards particular outcomes. Consequently, because NRT is also itself a radical practice with this view of hope, a range of methodological initiatives are intended to 'change and boost' the world, to 'build new life' in the world and to assist it in 'speaking back'. Practically, these have involved arts-based approaches to insert new movements, affect and messages into the world. These approaches are communicating in different, and often more direct and powerful ways, as more far-reaching forms of knowledge translation (Collective Disruption, 2017; Dupuis *et al.*, 2016). Other examples include a growing commitment to an activist and public scholarship, particularly where the researcher aligns with, and participates in, the performance of reactionary counter-politics on behalf of, or with, older people (Putney, Alley and Bengtson, 2005).

Conclusion

To summarise the discussion and points thus far, each theoretical orientation in geographical gerontology promotes different ideas about space and place that, in turn, facilitate different forms and foundations of hope. In spatial and population science, researchers hope to deal with ageing populations through realising volumes, distances and demands and then articulating them to officialdom. In political economy, researchers hope that by articulating inequalities and inefficiencies, more equitable distributions of limited resources might be reached. In humanism the hope or despair felt by older people in places is brought to light, whilst researchers harbour hope for change by articulating these feelings. In social constructionism and constructivism the hope or despair represented by the places experienced by older people is illuminated, whilst researchers hope that, through their scholarship, these identities and experiences might be modified. In relational studies, the hopes of older people are thought to be networked – shaped and changed between places and across spaces. Finally, NRT helps animate the hope emerging or retracting in the moment 'taking place' within the actual forward motion of space and time.

This list of theoretical orientations is certainly not exhaustive. Others exist, tweaking ideas about space and place in their own ways, yet the ones described showcase the variability within the geographic gerontology. As these theoretical orientations advance and develop in this field, each might continue to tell us something more about hope. And, as commentators have pointed out (Andrews *et al.*, 2013; Andrews and Grenier, 2015; Skinner *et al.*, 2015), it is perhaps the latter two – relational thinking and NRT – that are least explored and, hence, hold the greatest potential to break new empirical and theoretical ground. In sum, older people have hopes that are contingent on space and place, and geographical gerontologists trust that these hopes can be met by offering geographical knowledge that might inform particular solutions and ways forward. The chapters presented in this book reveal these hopes in their many conceptual, analytical and empirical guises.

References

Anderson, B. and Harrison, P. (2010). *Taking place: Non-representational theories and geography*. Farnham, Surrey: Ashgate Publishing.

Andrews, G. J. (2014). Co-creating health's lively, moving frontiers: Brief observations on the facets and possibilities of non-representational theory. *Health & Place*, 30, 165–170. http://dx.doi.org/10.1016/j.healthplace.2014.09.002

Andrews, G. J. (2017). The relational making of people and place: The case of the Teignmouth World War II homefront. *Ageing & Society*, 37(4), 725–752. http://dx.doi.org/10.1017/S0144686X15001415

Andrews, G. J., Chen, S. and Myers, S. (2014). The 'taking place' of health and wellbeing: Towards non-representational theory. *Social Science & Medicine*, 108, 210–222. http://dx.doi.org/10.1016/j.socscimed.2014.02.037

Andrews, G. J., Cutchin, M., McCracken, K., Phillips, D. R. and Wiles, J. (2007). Geographical gerontology: The constitution of a discipline. *Social Science & Medicine*, 65(1), 151–168. http://dx.doi.org/10.1016/j.socscimed.2007.02.047

Andrews, G. J., Evans, J. and Wiles, J. (2013). Re-spacing and re-placing gerontology: Relationality and affect. *Ageing & Society*, 33(8), 1339–1373. http://dx.doi.org10.1017/S0144686X12000621

Andrews, G. J. and Grenier, A. M. (2015). Ageing movement as space-time: Introducing non-representational theory to the geography of ageing. *Progress in Geography*, 34(12), 1512–1534. http://dx.doi.org/10.18306/dlkxjz.2015.12.003

Andrews, G. J., Holmes, D., Poland, B., Leheux, P., Miller, K-L., Pringle, D. and McGilton, K. S. (2005). 'Airplanes are flying nursing homes': Geographies in the concepts and locales of gerontological nursing. *Journal of Clinical Nursing*, 14(S2), 109–120. http://dx.doi.org/10.1111/j.1365-2702.2005.01276.x

Andrews, G. J., Kearns, R., Kontos, P. and Wilson, V. (2006). 'Their finest hour': Older people, oral histories, and the historical geography of social life. *Social and Cultural Geography*, 7(2), 153–177. http://dx.doi.org/10.1080/14649360600600338

Andrews, G. J., Milligan, C., Phillips, D. R. and Skinner, M. (2009). Geographical gerontology: Mapping a disciplinary intersection. *Geography Compass*, 3(5), 1641–1659. http://dx.doi.org/10.1111/j.1749-8198.2009.00270.x

Andrews, G. J. and Peter, E. (2006). Moral geographies of restraint in nursing homes. *Worldviews on Evidence-Based Nursing*, 3(1), 2–7. http://dx.doi.org/10.1111/j.1741-6787.2006.00044.x

Andrews, G. J. and Phillips, D. R. (2002). Changing local geographies of private residential care for older people 1983–1999: Lessons for social policy in England and Wales. *Social Science & Medicine*, 55(1), 63–78. http://dx.doi.org/10.1016/S0277-9536(01)00207-6

Angus, J., Kontos, P., Dyck, I., McKeever, P. and Poland, B. (2005). The personal significance of home: Habitus and the experience of receiving long-term home care. *Sociology of Health & Illness*, 27(2), 161–187. http://dx.doi.org/10.1111/j.1467-9566.2005.00438.x

Antoninetti, M. and Garrett, M. (2012). Body capital and the geography of aging. *Area*, 44(3), 364–370. http://dx.doi.org/10.1111/j.1475-4762.2012.01089.x

Bell, M. and Rees, P. (2006). Comparing migration in Britain and Austrialia: Harmonisation through use of age – time plans. *Environment and Planning A*, 38, 959–988. http://dx.doi.org/10.1068/a35245

Blaikie, A. (1997). Beside the sea: Visual imagery, ageing and heritage. *Ageing & Society*, 17(6), 629–648. Available from: www.cambridge.org/core/services/aop-cambridge-core/content/view/6B787C86CBC37D926DF344F0EDB5DA55/S0144686X9700670Xa.pdf/beside_the_sea_visual_imagery_ageing_and_heritage.pdf

Cartier, C. (2003). From home to hospital and back again: Economic restructuring, end of life and the gendered problems of place-switching health services. *Social Science & Medicine*, 56(11), 2289–2301. http://dx.doi.org/10.1016/S0277-9536(02)00228-9

Chalmers, A. I. and Joseph, A. E. (1998). Rural change and the elderly in rural places: Commentaries from New Zealand. *Journal of Rural Studies*, 14(2), 155–165. http://dx.doi.org/10.1016/S0743-0167(97)00050-8

Cole, T. (1992). *The journey of life: A cultural history of aging in America*. New York: Cambridge University Press.

Collective Disruption. (2017). Cracked: New light on Dementia, in J. Gray (ed.), *ReView: A collection of plays committed to social justice* (pp. 67–138). Rotterdam: Sense Publishers.

Curtis, S. (1989). *The geography of public welfare provision*. London: Routledge.

Cutchin, M. P. (2001). Deweyan integration: Moving beyond place attachment in elderly migration theory. *International Journal of Aging and Human Development*, 52, 29–44. http://dx.doi.org/10.2190/AF2D-A0T4-Q14C-1RTW

Cutchin, M. P. (2003). The process of mediated aging-in-place: A theoretically and empirically based model. *Social Science & Medicine*, 57(6), 1077–1090. http://dx.doi.org/10.1016/S0277-9536(02)00486-0

Cutchin, M. P. (2005). Spaces for inquiry into the role of place for older people's care. *Journal of Clinical Nursing*, 14(s2), 121–129. http://dx.doi.org/10.1111/j.1365-2702-2005.01280.x

Cutchin, M. P. (2007). From society to self (and back) through place: Habit in transactional context. *OTJR: Occupation, Participation, and Health*, 27, 50S–59S. http://dx.doi.org/10.1177/15394492070270S107

Cutchin, M. P. (2013). The complex process of becoming at-home in assisted living, in G. D. Rowles and M. Bernard (eds), *From knowledge to practice in environmental gerontology: Making meaningful places in old age* (pp. 105–124). New York: Springer Publishing Co.

Darling, J. (2009). Thinking beyond place: The responsibilities of a relational spatial politics. *Geography Compass*, 3(5), 1938–1954. http://dx.doi.org/10.1111/j.1749-8198.2009.00262.x

Davey, J. A. (2007). Older people and transport: Coping without a car. *Ageing & Society*, 27(1), 49–65. http://dx.doi.org/10.1017/S014686X06005332

Davies, A. and James, A. (2011). *Geographies of ageing: Social processes and the spatial unevenness of population ageing*. Farnham, Surrey: Ashgate Publishing.

Dupuis, S. L., Kontos, P., Mitchell, G., Jonas-Simpson, C. and Gray, J. (2016). Re-claiming citizenship through the arts. *Dementia*, 15(3), 358–380. http://dx.doi.org/10.1177/1471301216637206

Dyck, I., Kontos, P., Angus, J., McKeever, P. and Poland, B. (2005). The home as a site of long-term care: Meanings and management of bodies and spaces. *Health & Place*, 11(2), 173–185. http://dx.doi.org/10.1016/j.healthplace.2004.06.001

Everitt, J. and Gfellner, B. (1996). Elderly mobility in a rural area: The example of south-west Manitoba. *Canadian Geographer*, 40(4), 338–351. http://dx.doi.org/10.1111/j.1541.0064.1996.tb00461.x

Ford, R. G. and Smith, G. C. (1995). Spatial and structural change in institutional care for the elderly in south-east England, 1987–1990. *Environment and Planning A*, 27(2), 225–248. http://dx.doi.org/10.1068/a270225

Ford, R. G. and Smith, G. C. (2008). Geographical and structural change in nursing care provision for older people in England 1993–2001. *Geoforum*, 39(1), 489–498. http://dx.doi.org/10.1016/j.geoforum.2007.10.001

Golant, S. M. (1989). The residential moves, housing locations, and travel behavior of older people: Inquiries by geographers. *Urban Geography*, 10(1), 100–108. http://dx.doi.org/10.2747/0272-3638.10.1.100

Golant, S. M. (1990). Post-1980 regional migration patterns of the U.S. elderly population. *Journal of Gerontology*, 45(4), S135–S140. http://dx.doi.org/10.1093/geronj/45.4.S135

Griffin, M. and Phoenix, C. (2014). Learning to run from narrative foreclosure: One woman's story of aging and physical activity. *Journal of Aging and Physical Activity*, 22(3), 393–404. http://dx.doi.org/10.1123/JAPA.2012-0300

Gustafson, P. (2001). Retirement migration and transnational lifestyles. *Ageing & Society*, 21(4), 371–394. http://dx.doi.org/10.1017/S0144686X01008327

Haberkorn, G. (2002). Ageing in rural and regional Australia. *Rural America*, 17(3), 32–39.

Hamnett, C. and Mullings, B. (1992). The distribution of public and private residential homes for elderly persons in England and Wales. *Area*, 24(2), 130–144. Available from: www.jstor.org/stable/pdf/20003074.pdf

Hanlon, N. and Halseth, G. (2005). The greying of resource communities in northern British Columbia: Implications for health care delivery in already-underserviced communities. *Canadian Geographer*, 49(1), 1–24. http://dx.doi.org/10.1111/j.0008-3658.2005.00077.x

Hanlon, N., Halseth, G., Clasby, R. and Pow, V. (2007). The place embeddedness of social care: Restructuring work and welfare in Mackenzie, BC. *Health & Place*, 13(2), 466–481. http://dx.doi.org/10.1016/j.healthplace.2006.05.006

Hanlon, N., Rosenberg, M. and Clasby, R. (2007). Offloading social care responsibilities: Recent experiences of local voluntary organisations in a remote urban centre in British Columbia, Canada. *Health & Social Care in the Community*, 15(4), 343–351. http://dx.doi.org/10.1111/j.1365-2524.2007.00693.x

Hanlon, N., Skinner, M. W., Joseph, A. E., Ryser, L. and Halseth, G. (2014). Place integration through efforts to support healthy aging in resource frontier communities: The role of voluntary sector leadership. *Health & Place*, 29, 132–139. http://dx.doi.org/10.1016/j.healthplace.2014.07.003

Harper, S. (1987a). The kinship network of the rural aged: A comparison of the indigenous elderly and the retired immigrant. *Ageing & Society*, 7(3), 303–327. http://dx.doi.org/10.1017/S0144686X00012836

Harper, S. (1987b). The rural-urban interface in England: A framework of analysis. *Transactions of the Institute of British Geographers*, 12(3), 284–302. http://dx.doi.org/10.2307/622406

Harper, S. (1987c). A humanistic approach to the study of rural populations. *Journal of Rural Studies*, 3(4), 309–319. http://dx.doi.org/10.1016/0743-0167(87)90050-7

Harper, S. (1989). The impact of the retirement debate on post-war retirement trends, in A. Gorst, L. Johnman and W. S. Lucas (eds), *Post-war Britain, 1945–1964: Themes and perspectives* (pp. 95–108). London: Pinter.

Harper, S. (1990). The emergence and consolidation of the retirement tradition in post-war Britain, in M. Bury and J. McNicol (eds), *Aspects of ageing: Essays on social policy and old age* (pp. 12–29). Surrey: Royal Holloway Bedford New College (Social Policy Papers).

Harper, S. and Laws, G. (1995). Rethinking the geography of ageing. *Progress in Human Geography*, 19(2), 199–221. http://dx.doi.org/10.1177/030913259501900203

Harper, S. and Thane, P. (1989). The consolidation of 'Old Age' as a phase of life, 1964–65, in M. Jeffreys (ed). *Growing old in the twentieth century* (pp. 43–61). London: Routledge.

Heleniak, T. (2003). Geographic aspects of population ageing in the Russian Federation. *Eurasian Geography and Economics*, 44(5), 325–347. http://dx.doi.org/10.2747/1538-7216.44.5.325

Hiltner, J., Smith, B. and Sullivan, J. (1986). The utilization of social and recreational services by the elderly: A case study of northwestern Ohio. *Economic Geography*, 62(3), 232–240. http://dx.doi.org/10.2307/144007

Hockey, J., Penhale, B. and Sibley, D. (2001). Landscapes of loss: Spaces of memory, times of bereavement. *Ageing & Society*, 21(6), 739–757. http://dx.doi.org/10.1017/S0144686X01008480

Hopkins, P. and Pain, R. (2007). Geographies of age: Thinking relationally. *Area*, 39(3), 287–294. http://dx.doi.org/10.1111/j.1475-4762.2007.00750.x

Jones, M. (2009). Phase space: Geography, relational thinking, and beyond. *Progress in Human Geography*, 33(4), 487–506. http://dx.doi.org/10.1177/0309132508101599

Katz, S. and McHugh, K. E. (2010). Age, meaning, and place: Cultural narratives and retirement communities, in T. R. Cole, R. Ray and R. Kastenbaum (eds), *A guide to humanistic studies in aging* (pp. 271–292). Baltimore, MD: Johns Hopkins University Press.

Kinsella, K. and Velkoff, V. A. (2001). *An aging world: 2001*. US Census Bureau, Series P95/01-1. Washington, DC: U.S. Government Printing Office. Available from: www.census.gov/prod/2001pubs/p95-01-1.pdf

Knodel, J. (1999). The demography of Asian ageing: Past accomplishments and future challenges. *Asia-Pacific Population Journal*, 14(4), 39–56.

Laws, G. (1995a). Understanding ageism: Lessons from postmodernism and feminism. *The Gerontologist*, 35(1), 112–118. http://dx.doi.org/10.1093/geront/35.1.112

Laws, G. (1995b). Embodiment and emplacement: Identities, representation and landscape in Sun City retirement communities. *International Journal of Aging and Human Development*, 40(4), 253–280. http://dx.doi.org/10.2190/BR6H-GQGF-RDN8-XA31

Martin, G. P., Nancarrow, S. A., Parker, H., Phelps, K. and Regen, E. (2005). Place, policy and practitioners: On rehabilitation, independence and the therapeutic landscapes in the changing geography of care provision to older people in the UK. *Social Science & Medicine*, 61(9), 1893–1904. http://dx.doi.org/10.1016/j.socscimed.2005.04.001

McConatha, J. T. and Volkwein-Caplan, K. (2012). *The Social geography of healthy aging: The importance of place and space.* Maidenhead: Meyer & Meyer Sport (UK) Ltd.

McHugh, K. E. (2000). The "ageless self"? Emplacement of identities in sun belt retirement communities. *Journal of Aging Studies,* 14(1), 103–115. http://dx.doi.org/10.1016/S0890-4065(00)80018-3

McHugh, K. E. (2009). Movement, memory, landscape: An excursion in non-representational thought. *GeoJournal,* 74(3), 209–218. http://dx.doi.org/10.1007/s10708-008-9222-0

McHugh, K. E. and Larson-Keagy, E. M. (2005). These white walls: The dialectic of retirement communities. *Journal of Aging Studies,* 19(2), 241–256. http://dx.doi.org/10.1016/j.jaging.2004.07.004

McHugh, K. E. and Mings, R. C. (1996). The circle of migration: Attachment to place in aging. *Annals, Association of American Geographers,* 86(3), 530–550. http://dx.doi.org/10.1111/j.1467-8306.1996.tb01765.x

Meyer, J. W. (1981). Equitable nutrition services for the elderly in Connecticut. *The Geographical Review,* 71(3), 311–323. Available from: www.jstor.org/stable/214703?seq=2#page_scan_tab_contents

Meyer, J. W. (1990). Research on services for the elderly. *Urban Geography,* 11(4), 394–401. http://dx.doi.org/10.2747/0272-3638.11.4.394

Milligan, C. (2000). 'Bearing the burden': Towards a restructured geography of caring. *Area,* 32(1), 49–58. http://dx.doi.org/10.1111/j.1475-4762.2000.tb00114.x

Milligan, C. (2003). Location or dis-location: From community to long term care – the caring experience. *Social and Cultural Geography,* 4(4), 455–470. http://dx.doi.org/10.1080/146493603200013790

Milligan, C. (2005). From home to 'home': Situating emotions within the caregiving experience. *Environment and Planning A,* 37(12), 2105–2120. http://dx.doi.org/10.1068/a37419

Milligan, C. and Wiles, J. (2010). Landscapes of care. *Progress in Human Geography,* 34(6), 736–754. http://dx.doi.org/10.1177/0309132510364556

Moore, E. G. and Pacey, M. A. (2004). Geographic dimensions of aging in Canada, 1991–2001. *Canadian Journal on Aging,* 23(Supplement), S5–S21. http://dx.doi.org/10.1353/cja.2005.0037

Moore, E. G. and Rosenberg, M. W. (2001). Canada's elderly population: The challenges of diversity. *The Canadian Geographer,* 45(1), 145–150. http://dx.doi.org/10.1111/j.1541-0064.2001.tb01179.x

Moore, E. G., Rosenberg, M. W. and McGuiness, D. (1997). *Growing old in Canada: Demographic and geographic perspectives.* Ottawa and Toronto: Statistics Canada and ITP Nelson.

Mowl, G., Pain, R. and Talbot, C. (2000). The ageing body and the homespace. *Area,* 32(2), 189–197. http://dx.doi.org/10.1111/j.1475-4762.2000.tb00129.x

Nemet, G. F. and Bailey, A. J. (2000). Distance and health care utilization among the rural elderly. *Social Science & Medicine,* 50(9), 1197–1208. http://dx.doi.org/10.1016/S0277-9536(99)00365-2

Pain, R., Mowl, G. and Talbot, C. (2000). Difference and the negotiation of 'old age'. *Environment and Planning D: Society and Space,* 18(3), 377–393. http://dx.doi.org/10.1068/d31j

Peace, S., Holland, C. and Kellaher, L. (2006). *Environment and identity in later life.* Maidenhead: Open University Press.

Phillips, D. R. (ed.). (2000). *Ageing in the Asia-Pacific region: Issues, policies and future trends.* London: Routledge.

Phillips, D. R. and Vincent, J. A. (1986). Private residential accommodation for the elderly: Geographical aspects of developments in Devon. *Transactions of the Institute of British Geographers,* 11(2), 155–173. http://dx.doi.org/10.2307/622003

Phoenix, C. and Grant, B. (2009). Expanding the agenda for research on the physically active aging body. *Journal of Aging and Physical Activity*, 17(3), 362–379. http://dx.doi.org/10.1123/japa.17.3.362

Pickle, L. W., Mungiole, M., Jones, G. K. and White, A. A. (1996). *Atlas of United States mortality*, Centers for Disease Control and Prevention, National Center for Health Statistics. Hyattsville: Centers for Disease Control and Prevention. Available from: www.cdc.gov/nchs/data/misc/atlasmet.pdf

Putney, N. M., Alley, D. E. and Bengtson, V. L. (2005). Social gerontology as public sociology in action. *The American Sociologist*, 36(3–4), 88–104. Available from: http://burawoy.berkeley.edu/PS/TAS1/Putney.pdf

Relph, E. (1976). *Place and placelessness*. London: Pion, Ltd.

Rollinson, P. (1990a). The story of Edward: The everyday geography of elderly single room occupancy (SRO) hotel tenants. *Journal of Contemporary Ethnography*, 19(2), 188–206. http://dx.doi.org/10.1177/089124190019002002

Rollinson, P. (1990b). The everyday geography of poor elderly hotel tenants in Chicago. *Geografiska Annaler B*, 72(2–3), 47–57. http://dx.doi.org/10.2307/490550

Rosel, N. (2003). Aging-in-place: Knowing where you are. *International Journal of Aging and Human Development*, 57(1), 77–90. http://dx.doi.org/10.2190/AMUD-8XVX-9FPK-MR8G

Rosenberg, M. and Everitt, J. (2001). Planning for aging populations: Inside or outside the walls. *Progress in Planning*, 56(3), 119–168. http://dx.doi.org/10.1016/S0305-9006(01)00014-9

Rosenberg, M. W., Moore, E. G. and Ball, S. B. (1989). Components of change in the spatial distribution of the elderly population in Ontario, 1976–1986. *Canadian Geographer*, 33(3), 218–29. http://dx.doi.org/10.1111/j.1541-0064.1989.tb00905.x

Rowles, G. (1978). *Prisoners of space? Exploring the geographical experiences of older people*. Boulder, CO: Westview Press.

Rowles, G. D. (1981). The surveillance zone as meaningful space for the aged. *The Gerontologist*, 21(3), 304–311. http://dx.doi.org/10.1093/geront/21.3.304

Rowles, G. D. (1983a). Between worlds: A relocation dilemma for the Appalachian elderly. *International Journal of Aging and Human Development*, 17(4), 301–314. http://dx.doi.org/10.2190/X2UF-RQ0V-BLN5-GDCN

Rowles, G. D. (1983b). Place and personal identity in old age: Observations from Appalachia. *Journal of Environmental Psychology*, 3(4), 299–313. http://dx.doi.org/10.1016/S0272-4944(83)80033-4

Rowles, G. D. (1988). What's rural about rural aging? An Appalachian perspective. *Journal of Rural Studies*, 4(2), 115–124. http://dx.doi.org/10.1016/0743-0167(88)90029-0

Rowles, G. D. (2000). Habituation and being in place. *OTJR: Occupation, Participation and Health*, 20(1), 52S–67S. http://dx.doi.org10.1177/15394492000200S105

Rowles, G. D. and Bernard, M. A. (eds). (2013). *Environmental gerontology: Making meaningful places in old age*. New York: Springer Publishing Company.

Rowles, G. D. and Watkins, J. F. (2003). History, habit heart and hearth: On making spaces into places, in K. W. Schaie, H. W-Wahl, H. Mollenkopf and F. Oswald (eds), *Aging independently: Living arrangements and mobility* (pp. 77–96). New York: Springer Publishing Company.

Schwanen, T., Hardill, I. and Lucas, S. (2012). Spatialities of ageing: The co-construction and co-evolution of old age and space. *Geoforum*, 43(6), 1291–1295. http://dx.doi.org/10.1016/j.geoforum.2012.07.002

Schwanen, T. and Páez, A. (2010). The mobility of older people – an introduction. *Journal of Transport Geography*, 18(5), 591–595. http://dx.doi.org/10.1016/j.jtrangeo.2010.06.001

Schwanen, T. and Ziegler, F. (2011). Wellbeing, independence and mobility: An introduction. *Ageing & Society*, 31(5), 719–733. http://dx.doi.org/10.1017/S0144686X10001467

Skinner, M. W., Cloutier, D. and Andrews, G. J. (2015). Geographies of ageing: Progress and possibilities after two decades of change. *Progress in Human Geography*, 39(6), 776–799. http://dx.doi.org/10.1177/0309132514558444

Skinner, M. and Hanlon, N. (eds). (2016). *Ageing resource communities: New frontiers of rural population change, community development and voluntarism*. London and New York: Routledge.

Skinner, M. W., Joseph, A. E., Hanlon, N., Halseth, G. and Ryser, L. (2014). Growing old in resource communities: Exploring the links among voluntarism, aging, and community development. *The Canadian Geographer*, 58(4), 418–428. http://dx.doi.org/10.1111/cag.12087

Skinner, M. W. and Rosenberg, M. W. (2005). Co-opting voluntarism? Exploring the implications of long-term care reform for the nonprofit sector in Ontario. *Environment and Planning C: Politics and Space*, 23(1), 101–121. http://dx.doi.org/10.1068/c0434

Smith, A. E. (2009). *Ageing in urban neighbourhoods: Place attachment and social exclusion*. London: The Policy Press.

Smith, G. C. and Ford, R. G. (1998). Geographical change in residential care provision for the elderly in England, 1988–1993. *Health & Place*, 4(1), 15–31. http://dx.doi.org/10.1016/S1353-8292(97)00029-4

Speare, A. and Meyer, J. (1988). Types of elderly residential mobility and their determinants. *Journal of Gerontology*, 43(3), 574–581. http://dx.doi.org/10.1093/geronj/43.3.S74

Swenson, M. (1998). The meaning of home to five elderly women. *Health Care for Women International*, 19(5), 381–393. http://dx.doi.org/10.1080/073993387246160

Tarrant, A. (2010). Constructing a social geography of grandparenthood: A new focus for intergenerationality. *Area*, 42(2), 190–197. http://dx.doi.org/10.1111/j.1475-4762.2009.00920.x

Tarrant, A. (2013). Grandfathering as spatio-temporal practice: Conceptualizing performances of ageing masculinities in contemporary familial carescapes. *Social & Cultural Geography*, 14(2), 192–210. http://dx.doi.org/10.1080/14649365.2012.740501

Tarrant, A. (2016). The spatial and gendered politics of displaying family: Exploring material cultures in grandfathers' homes. *Gender, Place & Culture*, 23(7), 966–982. http://dx.doi.org/10.1080/0966369X.2015.1073703

Tatnall, A. and Lepa, J. (2003). The Internet, e-commerce and older people: An actor-network approach to researching reasons for adoption and use. *Logistics Information Management*, 16(1), 56–63. http://dx.doi.org/10.1108/09576050310453741

Thrift, N. (2008). *Non-representational theory: Space, politics, affect*. Abington: Routledge.

Townsend, P. (1957). *The family life of older people: An inquiry in East London*. London: Routledge.

Townsend, P. (1962). *The Katherine Buildings collection*. Unpublished document, The British Library, Archive Record. Available from: https://archiveshub.jisc.ac.uk/search/archives/af9a2c6b-d461-3e40-8031-38439122bc1b?component=537fa9e9-32d3-39e5-a896-dd09432db973

Townsend, P. (1981). The structured dependency of the elderly: A creation of social policy in the twentieth century. *Ageing & Society*, 1(1), 5–28. http://dx.doi.org/10.1017/S0144686X1000020

Tulle, E. (2008). Acting your age? Sports science and the ageing body. *Journal of Aging Studies*, 22(4), 340–347. http://dx.doi.org/10.1016/j.jaging.2008.05.005

Vanderbeck, R. M. (2007). Intergenerational geographies: Age relations, segregation and re-engagements. *Geography Compass*, 1(2), 200–221. http://dx.doi.org/10.1111/j.1749-8198.2007.00012.x

Wahidin, A. (2002). Reconfiguring older bodies in the prison time machine. *Journal of Aging and Identity*, 7(3), 177–193. http://dx.doi.org/10.1023/A:1019764506372

Wainwright, S. P. and Turner, B. S. (2004). Epiphanies of embodiment: Injury, identity and the balletic body. *Qualitative Research*, 4(3), 311–337. http://dx.doi.org/10.1177/1468794104047232

Waldbrook, N., Rosenberg, M. W. and Brual, J. (2013). Challenging the myth of apocalyptic aging at the local level of governance in Ontario. *The Canadian Geographer*, 57(4), 413–430. http://dx.doi.org/10.1111/j.1541-0064.2013.12030.x

Walsh, K. and O'Shea, E. (2008). Responding to rural social care needs: Older people empowering themselves, others and their community. *Health & Place*, 14(4), 795–805. http://dx.doi.org/10.1016/j.healthplace.2007.12.006

Ward, R. (1984). *The ageing experience*. New York: Harper & Row.

Warnes, A. M. (1987). The ageing of Britain's population: Geographical dimensions. *Espace, Populations, Sociétiés*, 5(2), 317–327. http://dx.doi.org/10.3406/espos.1987.1205

Warnes, A. M. (1999). UK and western European late-age mortality: Trends in cause-specific death rates, 1960–1990. *Health & Place*, 5(1), 111–118. http://dx.doi.org/10.1016/S1353-8292(98)00044-6

Warnes, A. M. and Law, C. M. (1984). The elderly population of Great Britain: Locational trends and policy implications. *Transactions of the Institute of British Geographers*, 9(1), 37–59. http://dx.doi.org/10.2307/621866

Watkins, J. F. (1990). Appalachian elderly migration: Patterns and implications. *Research on Aging*, 12(4), 409–429. http://dx.doi.org/10.1177/0164027590124003

Wiles, J. L. (2003a). Informal caregivers' experiences of formal support in a changing context. *Health and Social Care in the Community*, 11(3), 189–207. http://dx.doi.org/10.1046/j.1365-2524.2003.00419.x

Wiles, J. L. (2003b). Daily geographies of caregivers: Mobility, routine, scale. *Social Science & Medicine*, 57(7), 1307–1325. http://dx.doi.org/10.1016/S0277-9536(02)00508-7

Wiles, J. L. (2005). Conceptualising place in the care of older people: The contributions of geographical gerontology. *International Journal of Older People Nursing*, 14(s2), 121–129.

Wiles, J. L., Wild, K., Kerse, N. and Allen, R. (2012). Resilience from the point of view of older people: 'There's still life beyond a funny knee'. *Social Science & Medicine*, 74(3), 416–424. http://dx.doi.org/10.10196/j.sociscimed.2011.11.005

Wiles, J. L., Allen, R., Palmer, A. J., Hayman, K. J., Keeling, S. and Kerse, N. (2009). Older people and their social spaces: A study of well-being and attachment to place in Aotearoa New Zealand. *Social Science & Medicine*, 68(4), 664–671.

Yiannakoulias, N., Rowe, B. H., Svenson, L. W., Schopflocher, D. P., Kelly, K. and Voaklander, D. C. (2003). Zones of prevention: The geography of fall injuries in the elderly. *Social Science & Medicine*, 57(11), 2065–2073.

Zhou, Y. R. (2012). Space, time and self: Rethinking aging in the contexts of immigration and transnationalism. *Journal of Aging Studies*, 26(3), 232–242. http://dx.doi.org/10.1016/j.jaging.2012.002.002

Part II

Geographical perspectives on ageing

3 Health geographies of ageing

Janine L. Wiles

Introduction

Ageing shapes, and is shaped by, geographies of health. Health geographies of ageing have moved from a biomedical emphasis on health as the absence of disease to embrace more holistic socio-ecological understandings of ageing and health in social, physical and symbolic contexts. Health geographers examine the relationship between people and their physical, social and symbolic environments. They focus on the distribution of and engagement with health, illness and a wide variety of resources for health, from the most intimate level of the body through to regional and global scales.

Just as health geography is a substantive sub-discipline of social geography, health geographies of ageing are a substantive sub-theme of the geography of ageing; geography is a crucial element of the ageing process. In this chapter, I explore health geographies of ageing as a substantive research tradition, from early concerns about the health trends of older populations to contemporary concerns for how space and place impact older people's health and health care and how older people co-experience their health and place. I address three geographical areas of spatial, relational and critical health geographies of ageing. Throughout, three key themes that define health geographies of ageing are incorporated. These themes include inequities in health geographies of ageing, the health aspects of ageing in place and attachment to place and landscapes of care, including the provision of health care and other forms of support and service for ageing adults.

Spatial science geographies of ageing and health

Most spatial science approaches to ageing and health geography focus on measuring and understanding the implications of the changing distribution of ageing populations at global, regional and urban levels (Cook and Halsall, 2011; Davies and James, 2011; Moore and Rosenberg, 2001). Researchers show how trends in disease, technology and social conditions shape population ageing at these different spatial scales (Davies and James, 2011; McCracken and Phillips, 2005). They investigate patterns and especially inequalities in distributions of older people and in health outcomes and quality of life in old age (Warnes, 1999). Such information is useful in being able to indicate likely demand for a range of services and supports associated with old age and to plan for future needs (Joseph and Cloutier, 1990; Rosenberg, 2014).

A problem for this research is the availability of high-quality data, especially in big national-level data sets. Data collection tends to focus on morbidity and mortality patterns of younger people, and there is variation between regions. As well as often being further through the ageing transition and thus having proportionally older populations, wealthier countries such as New Zealand, Australia and those in North America and Europe tend to have better data at a much greater level of detail. Researchers in these areas are therefore able to report more about differences and inequalities between older social groups (e.g., by social class, by ethnicity) in different spaces (such as rural–urban, between cities or across regions; Huisman, Kunst and Mackenbach, 2003; Macintyre, 1993). What data are available from less-wealthy nations point to gross global inequalities and inequities in life expectancy and causes of mortality, with people in poorer countries still tending to die younger from injury and infectious diseases (Lloyd-Sherlock, 2000). There are also differences in morbidity, health and quality of life in old age as rates of cardiovascular and similar chronic diseases increase in less-wealthy countries, pointing to a double burden (Mackay and Mensah, 2004).

More detailed understanding of inequalities and patterns, such as comparative cross-national work showing how socio-economic inequalities shape mortality in old age just as they do in younger age groups (Huisman *et al.*, 2003), highlights the need to improve the socio-economic well-being of older people as a group in order to improve the health of older populations. At the same time, descriptive tools such as disease atlases, on the rare occasions where information related to older people is included, show the importance of recognising the diversity among older people (Feldman, 2007; Kerr *et al.*, 2014). There are differences in morbidity and mortality between 'younger-old' and 'older-old' (80-years-plus) people (Grundy, 1997; Wilkinson and Sainsbury, 1998) and between different ethnic or socio-economic status groups (Grundy and Sloggett, 2003). Both multilevel modelling and detailed qualitative research enables researchers to address more complex questions such as how to understand the relationship between individual and contextual place effects (such as neighbourhood socio-economic status or rural and urban contexts) on health outcomes. An important question is how the accumulated experiences of a lifetime affect health and mortality in old age (Herron and Skinner, 2013; Joseph and Martin-Maithews, 1993; Macintyre, Ellaway and Cummins, 2002).

Geographies of movement and proximity

Another key theme in geographical ageing and health research has been to understand the location and movement of older people, especially as these relate to relevant services and supports for older people (Longino, Perzynski and Stoller, 2002; Walters, 2002; see also Hanlon, Chapter 18 and Gatrell, Chapter 19). Researchers have identified classic patterns of mobility and migration related to retirement age (Bures, 1997; Glaser and Grundy, 1998). Many 'young-old' move to leisure destinations in early old age, while many 'older-old' return to their original areas, probably to be closer to family and friends in later old age as they think about needs for support (Murphy and Zehner, 1988; Stoller and Longino, 2001). These patterns are changing as new opportunities and constraints for older age employment, support and leisure arise (Al-Hamad, Flowerdew and Hayes, 1997; Gustafson, 2001; McHugh and Mings, 1996). However, for health geographers, these moves need to be understood as part of the complexities of people's lives (Rowles, 1986; Rowles and Watkins, 1993) and are relevant to exchanges of care and support and impacts on health (G. Smith, 1998a). Moves motivated by 'negative' life events or circumstances, for example, can have depressive effects or be particularly stressful (Bradley and Van Willigen, 2010). Decisions made by both older people and their families about geographic proximity among kin often have implications for the nature of their interactions (Smith, 1998b). Greater proximity between an older person and family members is typically associated with more provision of care, both in terms of hours and type of support provided. However, this is influenced by characteristics such as health status, gender, marital and parental status, and employment of both the older person and potential supporters (Hallman and Joseph, 1997, 1999; Joseph and Hallman, 1996, 1998).

Increasingly, technology mediates geographic proximity (Milligan, Roberts and Mort, 2011). The ability to interact with greater ease and less cost via video and audio connections can enable active and supportive caring relationships across long distances on a daily basis. It is possible to live on the other side of the world and provide direct emotional and social support or to liaise with paid support agencies (Atkinson, Lawson and Wiles, 2011; Milligan and Wiles, 2010). Developing technologies allow remote surveillance and monitoring, which, in turn, potentially permit people to stay 'at home' rather than move to residential care (Schillmeier and Domènech, 2010). Such innovations have critical implications for both the health and the geographies of older people (Milligan *et al.*, 2011). This technology might enable older people who are experiencing frailty to retain a greater level of autonomy and independence. However, many ethical and social issues have been raised in this ever-more salient debate (Milligan, Mort and Roberts, 2010). Who should have the right to access such private information (e.g., should adult children automatically be able to see their parents' data?)? Is there potential for even greater social and spatial isolation and loneliness for older people who are monitored remotely rather than in person? Might these technologies ultimately reconfigure the experience of home and the way care occurs?

At the more individual, local level, there has been considerable assessment of interactions between ageing people and their living environments. For example, the environmental gerontologist Lawton argued that person–environment 'fit' could be understood as a relationship between personal capacities or 'competence' and environmental 'press' or stresses (Lawton, 1980). As an individual's personal capacity declines, their physical environment becomes more important in enabling or disabling them; those with greater levels of 'competence' are better able to utilise resources in the environment to meet their personal needs (Lawton, 1982). Researchers have drawn on this theory to examine housing environments of community-dwelling older people, for example, by developing tools for assessing both discrete, objective aspects of physical environments and of personal characteristics and comparing and contrasting these across national boundaries (Iwarsson, 2005). These instruments can be used to identify risks or opportunities in housing and the immediate environment. Such research has practical implications in terms of improving built design to create more enabling and functional living environments and improving the health outcomes of older people, both at the collective policy level and at the immediate level of assessing and improving individual homes.

Relational geographies of ageing and health

More relational approaches to ageing and health geographies seek to understand older people's interactions with social and physical environments, asserting that the indivisible relationship between people and their environments means we cannot measure these separately (Kontos, 1998; Rowles, 1978, 1986; Rubinstein, 1990; Rubinstein and Parmalee, 1992; Sugihara and Evans, 2000). They often include the views and voices of older people. In particular, geographical researchers emphasise that every aspect of every older person must always be understood as embedded or situated in socially constructed, dynamic places (Cutchin, 2001, 2003, 2005). Perhaps the most important themes in this field of research are sense of place and attachment to place: the idea that older people with good connections to place are more likely to feel secure and in control and have a positive sense of self, all of which enhances well-being and the process of ageing (Rowles, 1986; Rowles and Ohta, 1983; Rowles and Ravdal, 2002; Rubinstein, 1989). Many geographic researchers have demonstrated how attachment to place seems to grow stronger as people age, with a growing investment of meaning in both objects and places over time (Christoforetti, Gennai and Rodeschini, 2011; Gilleard, Hyde and Higgs, 2007; Wiles *et al.*, 2009; see also Rowles, Chapter 16). Older people seem to be much more likely than younger people to say they like their home or their neighbourhood, for example. The reasons this pattern is observed so consistently range from pride and personal investment over time to the immediate environment becoming more important as mobility declines, although recent work suggests the nature and size of older people's social worlds are far more diverse than initially thought (Wiles, Allen *et al.*, 2009). Familiarity, emotional connections, networks of friends and family, and a personal sense of contextualised identity or history linked to a place are all likely to contribute to what Rowles refers to as 'being in place' or 'insideness' (Rowles, 1993). There may also be selective recall of positive experiences over a longer time (Christoforetti *et al.*, 2011; Golant, 1984). While attachment to place or investment in place is generally understood as having positive effects on health and well-being (Morita *et al.*, 2010), there is also evidence that negative associations with place can have a damaging effect on health and well-being (Golant, 2008).

These geographic themes have gained practical relevance in recent decades with the growing political emphasis on 'ageing in place' as a response to the supposed 'crisis' of ageing populations in wealthier and now less-wealthy nations (Lehning, 2012; World Health Organisation, 2007). There is a stated preference on the part of many governments for people to age at home or in 'the community' as an alternative to moving to expensive institutionalised care, and to reduce moves between institutional settings with environments that offer different, consecutively higher levels of care and support. Older people themselves do tend to report they prefer to age in place or to 'stay put', and policy tends to place emphasis on concepts like autonomy and independence as positive ideals.

Critical health geographies of ageing

Health geographers have argued for much more critical understanding of what it means to 'age in place', both from the perspectives of older people and urban, regional and global governance (Boyle, Wiles and Kearns, 2015; Cutchin, 2001). This includes a need to understand the implications of living in disadvantaged or difficult places (Scharf, Phillipson and Smith, 2005; Smith *et al.*, 2004). Moreover, an overemphasis on ageing in place as a policy mechanism may prevent people from seeking better alternatives (Golant, 2015; see also Golant, Chapter 15).

Research with older people suggests many of them do not know what 'ageing in place' means, contra to the ubiquity of the term among policy makers and service providers, and older people have much more complex ideas about what 'staying put' or ageing in the community means (Wiles *et al.*, 2011). Studies with older people suggest that a wide range of factors including the quality and adaptability of their housing and the availability of social and physical resources in their neighbourhoods are important, but so, too, are their emotional and social and symbolic perceptions of these things (Peace, Holland and Kellaher, 2006). Thus, once a house or neighbourhood is no longer 'comfortable' for residents (for example, in their ability to carry out daily functions or to manage stairs or heat large empty rooms or engage with the social and physical resources available in a neighbourhood such as shops or transport), many older people report they are quite likely to try to adapt their housing or move to more accessible or functionally appropriate housing (Peace, Holland and Kellaher, 2011). Others may draw on diverse accommodative and adaptive coping mechanisms and repertoires to achieve a sense of residential normalcy (Golant, 2015; Mansvelt, 2012). Likewise, research shows that many older people's reluctance to move to institutionalised living environments stems from negative perceptions of those environments, and quite often people's actual experience of them is much more positive and complex (Wiles and Rosenberg, 2003).

Of course, access to any living environment is influenced by the resources people have available to them. Typically, we might think of this at an individual level, such as whether they own or rent their homes or have strong social networks. However, a critical geographical perspective on ageing and health geography suggests that we must also think more broadly about whether older people have access to appropriate resources for adapting their current living circumstances, including financial resources to renovate a home or emotional and practical resources to deal with planning consents and managing contractors (Means, 2007; Wiles and Rosenberg, 2009).

For example, at a neighbourhood level this will be influenced by the amount of support available – such as adequate and good quality home-based care when needed – as well as the resources available, such as accessible, adequate and affordable public transport for older people who are no longer driving, or appropriate leisure spaces and opportunities (Pain, Mowl and Talbot, 2000). At even broader scales the extent to which health services and home-based support are collectively provided (e.g., publicly funded) will shape the degree to which older people must rely on other sources of support such as family caregiving and the nature of intergenerational exchanges of wealth (Wild, Wiles and Allen, 2011; Wiles *et al.*, 2012).

In short, for governments to encourage healthy ageing in place requires much more than simplistic rhetoric around independence and autonomy. Such rhetoric is often framed in highly individualistic, middle-class and middle-aged terms and fails to recognise that in fact all of us are interdependent and that concepts like 'autonomy' can mean quite different things in varying social and cultural contexts. One potentially positive move in this direction is the global Age-friendly Cities initiative (WHO, 2007). This initiative draws on discussions with older people in many different cities around the world to develop frameworks supporting policy makers in creating and sustaining age-friendly cities; it includes thinking about issues such as street design and use, housing and transport policy and innovations around social networking.

In general, these critical geographical approaches to understanding ageing and health challenge the stigmatisation and homogenisation of old age, ageing and ageing bodies (Grenier, 2005; Wiles, 2011). This challenge applies both to society, in general, and to research that treats older people as passive and dependent, frail or asexual, research which talks about older people but not with them or which casts old age as a medical problem to be 'treated'. Critical geographical approaches recognise the heterogeneity of older age and pay more attention to ageing bodies and how they both relate to and shape a wide variety of spaces and social contexts, ranging from homes, neighbourhoods and communities to gardens and sheds, leisure spaces, workspaces, educational spaces, and political spaces. For example, there is growing recognition that care and support is not unidirectional but is, rather, a two-way exchange and that older people make significant contributions to both their families and the places in which they live (Milligan and Wiles, 2010; Stephens, Breheny and Mansvelt, 2015; Wiles and Jayasinha, 2013).

Looking forward

In all aspects of geographical research on ageing and health there is a growing and exciting emphasis on a strengths-based approach to understanding what helps older people achieve well-being, together with necessary attention to understanding and addressing inequalities and inequities among older people and between older people and other groups. Increasingly, attention is being paid to different aspects of ageing, including cultural and class differences and the experiences of those in advanced age as opposed to younger cohorts. A wide range of approaches and methods are deployed in this highly relevant and necessary area of geographical gerontology research. These range from sophisticated geographical information systems (GIS) and spatial analytic techniques in a post-positivist theoretical framework to emancipatory methodological frameworks that include participatory and transformative research strategies working with older people, as well as established and proven techniques such as good quality surveys and in-depth interviews. At global and local levels, with rapidly growing proportions and numbers of older people and particularly those in advanced age, there is growing urgency for research that helps us to mitigate associated problems and, more important, to understand and maximise the many advantages of an ageing population.

References

Al-Hamad, A., Flowerdew, R. and Hayes, L. (1997). Migration of elderly people to join existing households: Some evidence from the 1991 household sample of anonymised records. *Environment and Planning A, 29*(7), 1243–1255. http://dx.doi.org/10.1068/a291243

Atkinson, S., Lawson, V. and Wiles, J. (2011). Care of the body: Spaces of practice. *Social and Cultural Geography, 12*(6), 563–572. http://dx.doi.org/10.1080/14649365.2011.601238

Boyle, A., Wiles, J. L. and Kearns, R. A. (2015). Rethinking ageing in place: The 'people' and 'place' nexus. *Progress in Geography (China), 34*(12), 1495–1511. http://dx.doi.org/10.18306/dlkxjz.2015.12.002

Bradley, D. E. and Van Willigen, M. (2010). Migration and psychological well-being among older adults: A growth curve analysis based on panel data from the health and retirement study, 1996–2006. *Journal of Aging and Health, 22*(7), 882–913. http://dx.doi.org/10.1177/0898264310368430

Bures, R. (1997). Migration and the life course: Is there a retirement transition? *International Journal of Population Geography, 3*(2), 109–119. http://dx.doi.org/10.1002/psp.341

Christoforetti, A., Gennai, F. and Rodeschini, G. (2011). Home sweet home: The emotional construction of places. *Journal of Aging Studies*, 25(3), 225–232. http://dx.doi.org/10.1016/j.jaging.2011.03.006

Cook, I. G. and Halsall, J. (2011). *Aging in comparative perspective: Processes and policies*. New York: Springer.

Cutchin, M. (2001). Deweyan integration: Moving beyond place attachment in elderly migration theory. *The International Journal of Aging and Human Development*, 52(1), 29–44. http://dx.doi.org/10.2190/AF2D-A0T4-Q14C-1RTW

Cutchin, M. (2003). The process of mediated aging-in-place: A theoretically and empirically based model. *Social Science & Medicine*, 57(6), 1077–1090. http://dx.doi.org/10.1016/S0277-9536(02)00486-0

Cutchin, M. (2005). Spaces for inquiry into the role of place for older people's care. *Journal of Clinical Nursing*, 14(S2), 121–129. http://dx.doi.org/10.1111/j.1365-2702.2005.01280.x

Davies, A. and James, A. (2011). *Geographies of aging: Social processes and the spatial unevenness of population aging*. Surrey: Ashgate.

Feldman, H. (2007). *Atlas of Alzheimer's disease*. Boca Raton, FL: Taylor and Francis.

Gilleard, C., Hyde, M. and Higgs, P. (2007). The impact of age, place, aging in place, and attachment to place on the well-being of the over 50s in England. *Research on Aging*, 29(6), 590–605. http://dx.doi.org/10.1177/0164027507305730

Glaser, K. and Grundy, E. (1998). Migration and household change in the population aged 65 and over, 1971–1991. *International Journal of Population Geography*, 4(4), 323–339. http://dx.doi.org/10.1002/(SICI)1099-1220(199812)4:4<323::AID-IJPG111>3.0.CO;2-M

Golant, S. M. (1984). *A place to grow old: The meaning of environment in old age*. New York: Columbia University Press.

Golant, S. M. (2008). Commentary: Irrational exuberance for the aging in place of vulnerable low-income older homeowners. *Journal of Aging and Social Policy*, 20(4), 379–397. http://dx.doi.org/10.1080/08959420802131437

Golant, S. M. (2015). *Aging in the right place*. Baltimore, MD: Health Professions Press.

Grenier, A. M. (2005). The contextual and social locations of older women's experiences of disability and decline. *Journal of Aging Studies*, 19(2), 131–146. http://dx.doi.org/10.1016/j.jaging.2004.07.003

Grundy, E. (1997). Demography and gerontology: Mortality trends among the oldest old. *Ageing & Society*, 17(6), 713–725. http://dx.doi.org/10.1017/S0144686X97006715

Grundy, E. and Sloggett, A. (2003). Health inequalities in the older population: The role of personal capital, social resources and socio-economic circumstances. *Social Science & Medicine*, 56(5), 935–947. http://dx.doi.org/10.1016/S0277-9536(02)00093-X

Gustafson, P. (2001). Retirement migration and transnational lifestyles. *Ageing & Society*, 21(4), 371–394. http://dx.doi.org/10.1017/S0144686X01008327

Hallman, B. C. and Joseph, A. (1997). Exploring distance and caregiver gender effects in eldercare: Towards a geography of family caregiving. *Great Lakes Geographer*, 4(2), 15–29.

Hallman, B. C. and Joseph, A. (1999). Getting there: Mapping the gendered geography of caregiving to elderly relatives. *Canadian Journal on Aging*, 18(4), 397–414. http://dx.doi.org/10.1017/S0714980800010011

Herron, R. V. and Skinner, M. W. (2013). Using care ethics to enhance qualitative research on rural aging and care. *Qualitative Health Research*, 23(12), 1697–1707. http://dx.doi.org/10.1177/1049732313509893

Huisman, M., Kunst, A. E. and Mackenbach, J. P. (2003). Socioeconomic inequalities in morbidity among the elderly: A European overview. *Social Science & Medicine*, 57(5), 861–873. http://dx.doi.org/10.1016/S0277-9536(02)00454-9

Iwarsson, S. (2005). A long-term perspective on person-environment fit and ADL dependence among older Swedish adults. *The Gerontologist*, 45(3), 327–326. http://dx.doi.org/10.1093/geront/45.3.327

Joseph, A. and Cloutier, D. (1990). A framework for modeling the consumption of health services by the rural elderly. *Social Science & Medicine*, 30(1), 45–52. http://dx.doi.org/10.1016/0277-9536(90)90328-P

Joseph, A. and Hallman, B. C. (1996). Caught in the triangle: The influence of home, work, and elder location on work-family balance. *Canadian Journal on Aging*, 15(3), 393–412. http://dx.doi.org/10.1017/S0714980800005845

Joseph, A. and Hallman, B. C. (1998). Over the hill and far away: Distance as a barrier to the provision of assistance to elderly relatives. *Social Science & Medicine*, 46(6), 631–639. http://dx.doi.org/10.1016/S0277-9536(97)00181-0

Joseph, A. and Martin-Maithews, A. (1993). Growing old in aging communities. *Journal of Canadian Studies*, 28(1), 14–29. http://dx.doi.org/10.3138/jcs.28.1.14

Kerr, A., Exeter, D. J., Hanham, G., Grey, C., Zhao, J., Riddell, T., Lee, M., Jackson, R. and Wells, S. (2014). Effect of age, gender, ethnicity, socioeconomic status and region on dispensing of CVD secondary prevention medication in New Zealand: The Atlas of Health Care Variation CVD cohort (VIEW-1). *New Zealand Medical Journal*, 127(1400), 39–69.

Kontos, P. C. (1998). Resisting institutionalization: Constructing old age and negotiating home. *Journal of Aging Studies*, 12(2), 167–184. http://dx.doi.org/10.1016/S0890-4065(98)90013-5

Lawton, M. (1980). *Environment and aging*. Belmont: Brooks-Cole.

Lawton, M. (1982). Competence, environmental press, and the adaptation of older people, in M. Lawton, P. Windley and T. Byerts (eds), *Aging and the environment: Theoretical approaches* (pp. 33–59). New York: Springer.

Lehning, A. J. (2012). City governments and aging in place: Community design, transportation and housing innovation adoption. *The Gerontologist*, 52(3), 345–356. http://dx.doi.org/10.1093/geront/gnr089

Lloyd-Sherlock, P. (2000). Population aging in developed and developing regions: Implications for health policy. *Social Science & Medicine*, 51(6), 887–895. http://dx.doi.org/10.1016/S0277-9536(00)00068-X

Longino, C. F., Perzynski, A. and Stoller, E. P. (2002). Pandora's briefcase: Unpacking the retirement migration decision. *Research on Aging*, 24(1), 29–49. http://dx.doi.org/10.1177/016402750302400100

Macintyre, S. (1993). Gender differences in longevity and health in Eastern and Western Europe, in S. Platt, H. Thomas, S. Scott and G. Williams (eds), *Locating health: Sociological and historical explorations* (pp. 57–73). Aldershot: Avebury.

Macintyre, S., Ellaway, A. and Cummins, S. (2002). Place effects on health: How can we conceptualise, operationalise, and measure them? *Social Science & Medicine*, 55(1), 125–139. http://dx.doi.org/10.1016/S0277-9536(01)000214-3

Mackay, J. and Mensah, G. (2004). *The atlas of heart disease and stroke*. Geneva: World Health Organisation and US Centers for Disease Control and Prevention. Available from: www.who.int/cardiovascular_diseases/resources/atlas/en/

Mansvelt, J. (2012). Consumption, aging and identity: New Zealander's narratives of gifting, ridding and passing on. *New Zealand Geographer*, 68(3), 187–200. http://dx.doi.org/10.1111/j.1745-7939.2012.01233.x

McCracken, K. and Phillips, D. R. (2005). International demographic transitions, in G. Andrews and D. R. Phillips (eds), *Aging and place: Perspectives, policy, practice* (pp. 36–60). London and New York: Routledge.

McHugh, K. E. and Mings, R. C. (1996). The circle of migration: Attachment to place in aging. *Annals, Association of American Geographers*, 86(3), 530–550. http://dx.doi.org710.1111/j.1467-8306.1996.tb01765.x

Means, R. (2007). Safe as houses? Aging in place and vulnerable older people in the UK. *Social Policy and Administration*, 41(1), 65–85. http://dx.doi.org/10.1111/j.1467-9515.2007.00539.x

Milligan, C., Mort, M. and Roberts, C. (2010). Cracks in the door? Technology and the shifting topology of care, in M. Schillmeier and M. Domènech (eds), *New technologies and emerging spaces of care* (pp. 19–38). London: Routledge.

Milligan, C., Roberts, C. and Mort, M. (2011). Telecare and older people: Who cares where? *Social Science & Medicine*, 72(3), 347–354. http://dx.doi.org/10.1016/j.socscimed.2010.08.014

Milligan, C. and Wiles, J. L. (2010). Landscapes of care. *Progress in Human Geography*, 34(6), 736–754. http://dx.doi.org/10.1177/0309132510364556

Moore, E. and Rosenberg, M. W. (2001). Canada's elderly population: The challenges of diversity. *The Canadian Geographer*, 45(1), 145–150. http://dx.doi.org/10.1111/j.1541-0064.2001.tb01179.x

Morita, A., Takano, T., Nakamura, K., Kizuki, M. and Seino, K. (2010). Contribution of interaction with family, friends and neighbours, and sense of neighbourhood attachment to survival in senior citizens: 5-year follow-up study. *Social Science & Medicine*, 70(4), 543–549. http://dx.doi.org/10.1016/j.socscimed.2009.10.057

Murphy, P. A. and Zehner, R. B. (1988). Satisfaction with sunbelt migration. *Australian Geographical Studies*, 26(2), 320–334. http://dx.doi.org/10.1111/j.1467-8470.1988.tb00582.x

Pain, R., Mowl, G. and Talbot, C. (2000). Difference and the negotiation of 'old age'. *Environment and Planning D: Society and Space*, 18(3), 377–393. http://dx.doi.org/10.1068/d31

Peace, S. M., Holland, C. and Kellaher, L. (2006). *Environment and identity in later life*. New York: Open University Press.

Peace, S. M., Holland, C. and Kellaher, L. (2011). 'Option recognition' in later life: Variations in aging in place. *Ageing & Society*, 31(5), 734–757. http://dx.doi.org/10.1017/S0144686X10001157

Rosenberg, M. W. (2014). Health geography I: Social justice, idealist theory, health and health care. *Progress in Human Geography*, 38(3), 466–475. http://dx.doi.org/10.1177/0309132513498339

Rowles, G. D. (1978). *Prisoners of space? Exploring the geographical experiences of older people*. Boulder, CO: Westview Press.

Rowles, G. D. (1986). The geography of aging and the aged: Toward an integrated perspective. *Progress in Human Geography*, 10(4), 511–539. http://dx.doi.org/10.1177/030913258601000403

Rowles, G. D. (1993). Evolving images of place in aging and 'aging in place'. *Generations*, 17(2), 65–70.

Rowles, G. D. and Ohta, R. J. (eds). (1983). *Aging and Milieu: Environmental perspectives on growing old*. New York: Academic Press.

Rowles, G. D. and Ravdal, H. (2002). Aging, place, and meaning in the face of changing circumstances, in R. Weiss and S. Bass (eds), *Challenges of the third age: Meaning and purpose in later life* (pp. 81–114). New York: Oxford University Press.

Rowles, G. D. and Watkins, J. F. (1993). Elderly migration and development in small communities. *Growth & Change*, 24(4), 509–538. http://dx.doi.org/10.1111/j.1468-2257.1993.tb00136.x

Rubinstein, R. L. (1989). The home environments of older people: A description of the psycho-social processes linking person to place. *Journal of Gerontology, Social Sciences*, 44(2), S45–S53. http://dx.doi.org/10.1093/geronj/44.2.S45

Rubinstein, R. L. (1990). Personal identity and environmental meaning in later life. *Journal of Aging Studies*, 4(2), 131–147. http://dx.doi.org/10.1016/0890-4065(90)90011-V

Rubinstein, R. L. and Parmalee, P. A. (1992). Attachment to place and representation of life course by the elderly, in I. Altman and S. Low (eds), *Human behaviour and environment vol. 12: Place attachment* (pp. 139–163). New York: Plenum Press.

Scharf, T., Phillipson, C. and Smith, A. (2005). Social exclusion of older people in deprived urban communities of England. *European Journal of Aging*, 2(2), 76–87. http://dx.doi.org/10.1007/x10433-005-0025-6

Schillmeier, M. and Domènech, M. (2010). *New technologies and emerging spaces of care*. London: Routledge.

Smith, A., Sim, J., Scharf, T. and Phillipson, C. (2004). Determinants of quality of life amongst older people in deprived neighbourhoods. *Ageing & Society*, 24(5), 793–814. http://dx.doi.org/10.1017/S0144686X04002569

Smith, G. (1998a). Geographic separation and patterns of social interaction between residents of senior citizen apartment buildings and their adult children. *The Canadian Geographer*, 42(2), 145–158. http://dx.doi.org/10.1111/j.1541-0064.1998.tb01561.x

Smith, G. (1998b). Residential separation and patterns of interaction between elderly parents and their adult children. *Progress in Human Geography*, 22(3), 368–384. http://dx.doi.org/10.1191/030913298673626843

Stephens, C., Breheny, M. and Mansvelt, J. (2015). Volunteering as reciprocity: Beneficial and harmful effects of social policies to encourage contribution in older age. *Journal of Aging Studies*, 33, 22–27. http://dx.doi.org/10.1016/j.jaging.2015.02.003

Stoller, E. P. and Longino, C. F. (2001). "Going home" or "leaving home"? The impact of person and place ties on anticipated Counterstream Migration. *The Gerontologist*, 41(1), 96–102. http://dx.doi.org/10.1093/geront/41.1.96

Sugihara, S. and Evans, G. W. (2000). Place attachment and social support at continuing care retirement communities. *Environment and Behavior*, 32(3), 400–409. http://dx.doi.org/10.1177/00139160021972586

Walters, W. H. (2002). Later-life migration in the United States: A review of recent research. *Journal of Planning Literature*, 17(1), 37–66. http://dx.doi.org/10.1177/088541220201700103

Warnes, A. M. (1999). UK and western European late-age mortality: Trends in cause-specific death rates, 1960–1990. *Health & Place*, 5(1), 111–118. http://dx.doi.org/10.1016/S1353-8292(98)00044-6

Wild, K., Wiles, J. L. and Allen, R.E.S. (2011). Resilience: Thoughts on the value of the concept for critical gerontology. *Ageing & Society*, 33(1), 137–158. http://dx.doi.org/10.1017/S0144686X11001073

Wiles, J. L. (2011). Reflections on being a recipient of care: Vexing the concept of vulnerability. *Social and Cultural Geography*, 12(6), 573–588.

Wiles, J. L., Allen, R.E.S., Palmer, A. J., Hayman, K. J., Keeling, S. and Kerse, N. (2009). Older people and their social spaces: A study of well-being and attachment to place in Aotearoa New Zealand. *Social Science & Medicine*, 68(4), 664–671. http://dx.doi.org/10.1016/j.socscimed.2008.11.030

Wiles, J. L. and Jayasinha, R. (2013). Care for place: The contributions older people make to their communities. *Journal of Aging Studies*, 27(2), 93–101. http://dx.doi.org/10.1016/j.jaging.2012.12.001

Wiles, J. L., Leibing, A., Guberman, N., Reeve, J. and Allen, R.E.S. (2011). The meaning of 'aging in place' to older people. *The Gerontologist*, 52(3), 357–366. http://dx.doi.org/10.1093/geront/gnr098

Wiles, J. L. and Rosenberg, M. W. (2003). Paradoxes and contradictions in Canada's home care provision: Informal privatisation and private informalisation. *International Journal of Canadian Studies: Special Issue, Health and Well-Being in Canada*, 28(Fall/Autumn), 63–89.

Wiles, J. L. and Rosenberg, M. W. (2009). The role of scale in conceptualizing primary health care practice: Considering social and institutional structures and systems, in V. Crooks and G. Andrews (eds), *Geographies of primary health care services* (pp. 75–92). London: Ashgate.

Wiles, J. L., Wild, K. L., Kerse, N. M. and Allen, R.E.S. (2012). Resilience from the point of view of older people: 'There's still life beyond a funny knee' *Social Science & Medicine*, 74(3), 416–424. http://dx.doi.org/10.1016/j.socscimed.2011.11.005

Wilkinson, T. J. and Sainsbury, R. (1998). The association between mortality, morbidity and age in New Zealand's oldest old. *International Journal of Aging and Human Development*, 46(4), 333–343. http://dx.doi.org/10.2190/9TE4-JCB5-4C8T-PFK9

World Health Organisation. (2007). *Global age-friendly cities project*. Geneva: WHO Press. Available from: www.who.int/ageing/projects/age_friendly_cities/en/

4 Social and cultural geographies of ageing

Christine Milligan and Anna Tarrant

Introduction

In this chapter, we discuss some of the conceptual, theoretical and empirical contributions that geographers have made to the social and cultural geographies of ageing. We do not attempt to be all-encompassing; indeed, a number of reviews in recent years have already done an excellent job of synthesising much of this work (e.g., Andrews *et al.*, 2007, 2009; Cutchin, 2009; Schwanen, Hardill and Lucas, 2012; Skinner, Cloutier and Andrews, 2015). Rather, our aim is to consider the roots of contemporary work on the social and cultural geographies of ageing before focusing in on what we view as some particularly strong themes within the field, as well as some novel and emerging areas of work. We begin our chapter by providing a brief historical account of some of the key geographical contributions to the social and cultural dimensions of ageing that have provided a foundation for the contemporary theoretical and empirical contributions we discuss. Our chapter is then framed around four distinctive themes that range from the intimate and micro-level of the body to the community and urban form. In doing so, we focus first on the inter-relationships among embodiment, ageism and place. We then continue with a discussion of how geographers have contributed to issues of gender and ageing. Our third theme focuses on geographies of intergenerationality, ageing and the life course. Finally, we discuss the specific contributions that geographical gerontology has made to understanding the connections between ageing and the urban form (see also Buffel and Phillipson, Chapter 10). Here we pay particular attention to relationalities among the home, gardens and community in later life.

Situating the social and cultural geographies of ageing

The reviews of geographical gerontology referred to earlier have done important work in scoping the field, in part, because they place current developments within a temporal framework and, in part, because they illustrate how early engagements with social and cultural theory in relation to ageing have been theoretically anchored within social geography (Andrews *et al.*, 2013).

In thinking about these origins, we are drawn to the early work of Golant (1984), Harper and Laws (1995) and Rowles (1978). These are perhaps some of the seminal texts that laid the foundation for contemporary concerns with the sociocultural and spatial dimensions of ageing. Not only did they set a marker for the application of a more theoretically informed lens to the social and cultural processes that influence the construction and experience of ageing, but they also were particularly critical of the unreflective influence of empiricism and positivistic approaches that dominated the emerging discipline at that time. Rowles's (1978) text, in particular, was important for the nuanced ethnographic approach it took to understanding the ever-declining spatial experiences of older people as they age. In a later co-edited book, Rowles and Ohta (1983) drew attention to both the relationalities among time, space and social activity and the importance of understanding the meaning of home, community and neighbourhood – important markers for contemporary work on ageing in place (see Golant, Chapter 15). Rowles's (1986) review not only critiqued the overly descriptive work in this field but also argued for a more relational engagement between ageing and place and for geographical work to recognise and embrace older people's heterogeneity and sociocultural diversity. Harper and Laws (1995) also argued for a more theoretically and methodologically nuanced approach – one informed by the cultural turn in human geography (Andrews *et al.*, 2007). They argued that the adoption of critical social and cultural theory could add important new insights to the nature, social processes and underlying causes at play in age segregation, social inequalities and marginalisation. Feminist and postmodernist theory, they added, could play an important role in illuminating conceptualisations of both ageing and place.

Influenced by the arguments in these reviews, geographers have since attempted to move beyond static and rigid understandings of old age. Hence, in the twenty years since these papers were published, we have seen the emergence of an expanding body of literature that recognises ageing and old age as "culturally variable and underpinned by a range of social and economic processes, lived experiences and spatial practices" (Hopkins and Pain, 2007: 287).

Contemporary developments

In the wake of these early works, geographers have explored the relationship between age, space and place from a range of philosophical and methodological frameworks (see Skinner *et al.*, 2015). Indeed, it is no longer uncommon for ageing to be described as embodied, emplaced and relational. This theoretical diversity has supported sustained challenges to more static views of ageing populations and has led to more nuanced perspectives about old age.

Schwanen *et al.* (2012) suggested five interrelated reasons why a sustained geographical engagement with sociocultural approaches to ageing and old age is important. First, they point to the growth of the ageing population in both advanced and emerging economies and its influence on intergenerational relations. Debate is relatively polarised in this arena, with ageing populations seen either as a threat or opportunity for younger generations and, more broadly, for the socio-economic structure of society (e.g., Vanderbeck, 2007; Tarrant, 2013; Wiles and Jayasinha, 2013). Second, they point to the emergence of new, spatially differentiated configurations of ageing and age as life course identities increasingly blur and become more difficult to categorise or tease apart (e.g., Hopkins and Pain, 2007; Katz and Monk, 2014). Third, neoliberal values – independence, mobility and experimentation – with their associated discourses such as 'active', 'productive' or 'successful ageing' are influencing perceptions and expectations of later life, infusing everyday spaces and how people identify themselves as old (Hardill, 2009). These ideals compete with pervasive normative assumptions, typical of industrial societies, where old age becomes synonymous with dependency, decline and passivity. The fourth and related point is that stereotypes of old age are maintained through processes of ageism, which are widespread and endemic within everyday practices and understandings of old age. Finally, these stereotypes point to significant changes in the "spatial manifestations of the interplay between the socio-structural processes and structures that shape the experiences and practices of care" for older people (Milligan and Wiles, 2010: 739). Despite emerging neoliberal values of independence and ageing well, the increased likelihood of ill health (particularly among the 'old-old') positions older people as the main receivers of support and care and casts them as passive recipients. Tying into debates around threats versus opportunities, these perceptions have led to a 'tide of alarmism' (Bloom, Canning and Fink, 2010) that characterises fears about the potential for an ageing 'time bomb'. More nuanced empirical research, however, highlights a middle ground between these overly negative and optimistic stereotypes of older people by highlighting the varied and active ways in which older people 'care for place', contribute to their communities, through volunteering, activism, advocacy and nurturing (Wiles and Jayasinha, 2013) and provide support to younger generations, for example as grandparents (Tarrant, 2013).

Geographies of embodiment and ageing

As noted earlier, issues of embodiment and the ageing body are pivotal to the social construction of later life (see also Herron, Chapter 14). This was acknowledged in Mowl, Pain and Talbot's (2000) early work, which made an explicit connection between cultural constructions of space and the ageing body. Their qualitative study of older people in North-East England, explored the centrality of the body to older people's definitions and identifications of old age, as well as the way in which notions about gendered and ageing bodies affected the use and meaning of space, in this case the home. Others have also sought to consider embodiment in later life through the lens of bodywork and the home. England and Dyck's (2011) study of home health care in Canada, for example, revealed the social relationships that exist between frail older people and paid and family carers within the home. By unpacking the practices of intimate bodywork, they illustrated how caregivers, care recipients and the home become constituted and reconstituted through the diverse dynamics of care work. This approach, they argued, allows for a more nuanced understanding of the impact of bodywork than more conventional definitions of care work allow.

More recently, attention has shifted to how ageing bodies are experienced in urban environments and in contexts of body modification. Antoninetti and Garrett (2012), for example, indicate that the fixity of otherwise familiar urban spaces and places can be problematic to body habitus, as residents' lives change (by choice, necessity or ill health) as a result of ageing. Others have also sought to consider how social capital, conceived as a set of relations, practices and subjectivities that influence, and are influenced by, the contexts and space in which they operate might be acquired by older adults in the neighbourhood (Lager, Van Hoven and Huigen, 2015). This can be limiting to those with reduced or limited economic, social or cultural body habitus. Morton's (2014) research further notes, that despite attention to the body, a geographical consideration of the emotional and embodied contexts of ageing is absent. As a counterbalance, she examines the multiple ways in which the ageing body might be experienced and modified though anti-ageing management and cosmetic technology and "the spatialities and corporeal scales at which these take place" (Morton, 2014: 1041). Critiquing the way in which existing work has tended to focus on cosmetic surgery practices in a unified manner, she argues for greater consideration of the embodied, emotional and spatial context of anti-ageing practices – something that is particularly pertinent as anti-ageing practices become increasingly normalised and accessible.

A number of interdisciplinary studies have also suggested how the ageing body might be studied empirically by geographers in relation to the bodyspace and embodiment. Twigg's (2015) research on clothing in later life, for example, highlights the centrality of the ageing body and clothing to social identity – and more recently, its continuing importance for individuals with dementia, their carers and care workers (Buse and Twigg, 2015). Similarly, Ward and Holland's (2011) research drew attention to the social symbolism of greying hair, hairstyling and its implications for magnifying processes of ageism in contemporary culture.

Geographical approaches to gender, sexuality and ageing

It is more than twenty years since Harper and Laws (1995) first called for a more nuanced geography on ageing and place informed by feminist theory. They identified four potential avenues for analysis that could take the field in new directions, including a focus on the position of women within an ageing society, the feminisation of ageing societies, the experiences of elderly women and the epistemological lessons of feminism. Despite some progress in this field, this work is still both limited and somewhat diffuse. Perhaps mirroring similar criticisms elsewhere (Krekula, 2007), gendered experiences of older women have either specifically been left out of theoretical discussions or those discussions have failed to address the implications of age, gender and other structural (in)equalities as intertwining structural systems.

While this is certainly a valid methodological criticism, there have nevertheless been some notable contributions that have sought to reinvigorate this field of inquiry (see for example Joseph and Joseph, Chapter 21). Caregiving, the body and emotion are key themes linking much of this research. Joseph and Hallman's (1998) research is an early example of work that sought to advance a gendered geography of family eldercare. They noted that while caregiving is strongly influenced by distance and proximity between the caregiver and the older care receiver, it almost always becomes another aspect of the unpaid family labour of women. Wasielewski's (2015) research with expat women in Oaxaca, Mexico, notes that while women of the 'boomer' generation are more likely to be bound by caregiving responsibilities in later life than men, they attempt to elude traditional gendered and generational expectations through an emotional geography of ageing. For these women, being an expatriate in Oaxaca makes transgressions of age, gender and nation possible, providing them with opportunities to redefine their self-perceptions and emotions. Morton's (2014) study (referred to earlier) also draws on feminist geographies of the body, emotions and corporeal ideology to examine how such anti-ageing practices bring about alternative embodied identities and spatialities, particularly for women. Indeed, as Twigg (2004) notes, practices constructing the body as a space to be fashioned, worked upon and controlled are a product of both a culture that ages bodies through its valorisation of youth and of a consumer culture that is particularly harsh on women.

Despite Harper and Laws's (1995) agenda for greater focus on feminist-informed ageing research, developments across a number of interdisciplinary fields have also highlighted that old age is gendered for men too (Arber, Davidson and Ginn, 2003). Concerns about older men's experiences of ageing are increasingly reflected in a growth of empirical interest in the geographies of older men and conceptual advances in relation to ageing masculinities (Milligan *et al.*, 2013). Tarrant, for example, has explored the relationships between space, time and gender, by drawing on an empirical study about grandfathers (e.g., Tarrant, 2010, 2013, 2016). Her work examines these relationships through consideration of older men's carescapes (Bowlby *et al.*, 2010) where, paradoxically, the contradictions of men's identities are worked out within constructs and processes of sexism and ageism. Some spaces, she notes, open opportunities for more caring and nurturing expressions of masculinity, while others, such as the school playground, are arenas in which their ageing bodies and assumed heterosexuality are viewed with caution. Her recent work explores the home-making practices of grandfathers and is linked to an emerging literature that explores the relationships among men, masculinities and home – a theme also explored by Milligan and Morbey (2016) in relation to older male caregivers. The theme of home and masculinity is also evident in a small but emerging literature around older gay men. Pilkey (2014) for example explored the importance of older gay men's material possessions at home in their construction of subjectivity, while Waitt and Gorman-Murray (2007) considered the homemaking processes of older gay men in small town Australia. These studies not only highlight challenges to heteronormative understandings of home, but also the paradox of the home for older gay men, particularly those who do not live their lives through their sexuality.

Beyond the home, research has also begun to explore masculinity and ageing at the meso and macro scales. Work by Milligan *et al.* (2013) for example has focused on the Men in Sheds initiative to illustrate how gender-specific spaces can play an important role in the promotion and maintenance of health and well-being among older men that are not being met through other community-based interventions. Christou (2016), in her work on second-generation Greek-American return migrants, moves beyond community social and gendered spaces to examine the intersectionalities of masculinities, mobilities and identities as part of an analytic agenda. Such an agenda, she argues, can help develop socio-spatial understandings of the processes of migration, belonging and the fluctuation of agency in ageing lives.

Relational geographies of ageing, intergenerationality and the life course

Much of the early work that began to consolidate a more coherent geographical understanding of old age as influenced by cultural and social dynamics has been linked to the relational turn in social geography (Andrews *et al.*, 2013). In 2007, Hopkins and Pain sought to advance a more relational approach to the geographies of age more generally, arguing that rather than focusing on specific age groups alone, it is also important to examine the relations within and among different generations (Hopkins and Pain, 2007). Reflecting the patchiness of research on older people at the time, their work predominantly reviewed existing research about younger people yet proposed an important framework for thinking about age relationally, a framework consisting of intergenerationality, intersectionality and life course (Hopkins and Pain, 2007). Together, these concepts illustrate how social and cultural understandings of ageing also play out in space and place. Attention to the life course, for example, has highlighted the unique and varied set of social geographies experienced by different age groups (Hopkins and Pain, 2007). In combination with intergenerationality and intersectionality, it is also possible to tease out some of the potential complexities that must be managed and negotiated in transitions to old age. Our discussion of older men and ageing masculinities is just one example of where this framework has effectively highlighted gaps in empirical research and elevated analysis of diversity in later life.

Recognition that research on ageing was much more limited than that of childhood resulted in some geographers shifting their empirical focus to exploring transitions both into, and within, old age and the implications for later-life geographies. In 2010, for example, Tarrant reflected that later-life transitions, such as those into grandparenthood, are under-researched, yet they hold potential for considerable conceptual development in ageing research (Tarrant, 2010). Riley's (2012) research on farming couples' adjustments to retirement is situated within a similar framework, highlighting how gendered roles and differentiated work–life histories intricately shape specific geographies of retirement for men and women. These approaches not only draw attention to how ageing is worked out through time and space, but how these processes intersect and are worked out in the meso- and macro-scale contexts of relations across generations.

These developments point to a notable shift in recognising both ageing *and* place as co-evolving, dynamic and relational. These new ways of theorising offer exciting possibilities for rethinking how ageing and later life are emplaced. They also have significance for encouraging a more complex understanding of the spatialities of ageing and how both old age and spaces are co-evolving social categories that are made and remade over time. For Schwanen *et al.* (2012) space and ageing can be conceptualised as 'entwined becomings'; such thinking does important work in moving us away from conceptions of space as a featureless, neutral surface upon which life unfolds, toward something that is relational, dynamic and co-evolving. It further forces us to think about the embodiment of ageing in more nuanced ways. According to Andrews, Evans and Wiles (2013) relational thinking also requires further consideration of how ageing is relationally performed in space and time. One way of advancing knowledge in the field, they suggest, is through non-representational theory, which allows us to consider "how affective spaces and places constitute, and impact on, older people's lives" (Andrews *et al.*, 2013: 1352).

Ageing and environments

The 'relational turn' draws attention to the ways in which older people's social and cultural geographies are inextricably embedded in particular places (Skinner *et al.*, 2015). Here, geographers have drawn attention to the importance of particular scales and environments ranging from the home and garden (Milligan, 2009; Milligan and Bingley, 2015), to the community (Wiles *et al.*, 2011; Means and Evan, 2012), to the rural (Meijering *et al.*, 2017; Skinner and Hanlon, 2016) and to the urban form (Lager, Van Hoven and Huigen, 2013). Such works variously draw attention not just to the importance of specific environments for supporting positive ageing but also to the ways in which changes in older people's relationships with these environments (because of factors ranging from declining health and mobility and demographic change to gentrification of the urban form, etc.) can impact an older person's sense of place and belonging. These issues are further explored elsewhere in this book (see Diaz Moore, Chapter 7 and Rowles, Chapter 16).

The home has been a particular focus of contemporary work within geographies of ageing. Driven in large part by policies designed around ageing in place, the concern has been to explore the (sometimes changing) meaning of home for older people (Wiles, 2005). As a site removed from public scrutiny, the home represents a place where older people can exert greater control and independence. As a site of security, familiarity and nurture the home can provide an important buttress to older people's sense of self and identity. However, with increasing age and frailty, the requirements for care and support, their associated technologies and requirements for access to even the most personal areas of the home, can impact the nature and meaning of home. This work highlights the juxtaposition of private and public space and the shifting relationships of power, independence, and autonomy that accrue for older people as their physical and cognitive abilities shift and change (Milligan, 2009).

As an extension of the home, commentators have also drawn attention to the ways in which the meaning attached to the domestic garden, and how gardening activity is performed within that space, can play an important social and cultural role in the lives of older people (Bhatti, 2006). Drawing on Stenner, Church and Bhatti's (2012) modes of engagement, Milligan and Bingley (2015) examined the role of gardens as sites of meaning and identity as well as people's (sometimes gendered) relational engagement with gardens and gardening activity in later life. Their work pointed to the paradox of the domestic garden as a site of affect and creativity, leisure and work and pleasure and heartache for older people.

Beyond the home, community and neighbourhood environments play an important role in how successfully people are able to age in place. Housing, neighbourhood satisfaction and conditions, as well as an older person's ability to negotiate that neighbourhood, are important (Windle, Burholt and Edwards, 2006; Wiles *et al.*, 2011; see also Golant, Chapter 15). So, too, are issues of transportation, recreational opportunities and the emplacement of amenities that can facilitate physical activity, social interaction and cultural engagement in older people (Wahl and Weisman, 2003). Yet, irrespective of the physical structure of the environment, subjective feelings about a neighbourhood can still be a significant source of satisfaction, regardless of any objective measures of suitability or safety (La Gory, Ward and Sherman, 1985). But behind the seemingly simple concept of ageing in place lies a complex process, where older people can find themselves continually reintegrating with places, renegotiating meanings and identity as they come face-to-face with dynamic landscapes manifest through social, political, cultural, and personal change (Andrews *et al.*, 2007). Lager, Van Hoven and Huigen (2015) noted how some of these processes are played out through the complex mechanisms that hinder or foster the creation of older people's social capital in the context of neighbourhood. In doing so, they revealed how different age groups create different neighbourhood rhythms and speeds, raising questions about how these might be synchronised to create spaces that could facilitate meaningful encounters among these different groups.

Beyond the urban – ageing and migration

Dominant ideas about ageing in place has meant that the relationship between ageing, migration and transnationality has received relatively limited attention (Horn and Schweppe, 2016; Walsh and Näre, 2016). Recently, however, the emergence of a small body of work has begun to draw out the relationship between ageing and migration processes – predominantly as a counterbalance to the framing of migration as a process experienced by young people (Lulle and King, 2016). Relatively early work by Warnes *et al.* (2004) and others demonstrated that older people are a diverse group who migrate for a variety of reasons, an issue that is increasingly important in a transnational and globalised context. Indeed, Lulle and King (2016) illustrate how subjective constructions of ageing become transformed when people move and (re)construct their lives in one or more places. While currently lacking a common framework, other research in this vein has examined retirement migration (including second-home owners, international tourists and 'snowbirds'), ageing migration in wealthier countries, transnational families and the implications of migration for family ties and intergenerational relations and the challenges faced by health and social services in a context of transnational ageing (Horn and Schweppe, 2016).

Future directions in social and cultural geographies of ageing

In this chapter we have explored some of the conceptual, theoretical and empirical advances in the social and cultural geographies of ageing. The empirical focus and theoretical developments outlined earlier, have transformed social and cultural understandings about the relationships between ageing and place. This focus has highlighted that over the life course, our everyday interactions across a range of spaces and places, influence both self-identity and how we behave in particular places (see Peace, Holland and Kellaher, 2005). This theoretical knowledge has informed a distinct body of research ranging from the micro-scale spaces of the body, the meso-scale space of the home and community, through to the macro-scale spaces of the urban and beyond.

Exploration of these themes indicates several areas for further development in order to move the field forward in new directions. In terms of substantive themes, there are opportunities to explore the implications of global migration processes on experiences of micro and everyday spaces in later life, following the work of Walsh and Näre (2016) and Lulle and King (2016). As research about ageing men expands, there might also be an opportunity to focus once again on the specifically gendered experiences of middle-aged and elderly women within a contemporary framework that is attentive to the affective and relational dimensions of ageing. There is also great potential for methodological innovation in the field. Rhythm analysis such as that proposed by Lager, Van Hoven and Huigen (2016) represents just one possibility. Additionally, the increasing availability of qualitative longitudinal data in archives (e.g., the ESRC Timescapes Archive and others) represent untapped resources that geographers and other social scientists interested in issues of space and place might exploit to explore the longitudinal and historical reach of existing data sets.

References

Andrews, G. J., Cutchin, M., McCracken, K., Phillips, D. R. and Wiles, J. (2007). Geographical gerontology: The constitution of a discipline. *Social Science & Medicine*, 65(1), 151–168. http://dx.doi.org/10.1016/j.socscimed.2007.02.047

Andrews, G. J., Evans, J. and Wiles, J. (2013). Re-spacing and re-placing gerontology: Relationality and affect. *Ageing & Society*, 33(8), 1339–1373. http://dx.doi.org/10.1017/SO144686X12000621

Andrews, G. J., Milligan, C., Phillips, D. R. and Skinner, M. W. (2009). Geographical gerontology: Mapping a disciplinary intersection. *Geography Compass*, 3(5), 1641–1659. http://dx.doi.org/10.1111/j.1749-8198.2009.00270.x

Antoninetti, M. and Garrett, M. (2012). Body capital and the geography of aging. *Area*, 44(3), 364–370. http://dx.doi.org/10.1111/j.1475-4762.2012.01089.x

Arber, S., Davidson, K. and Ginn, J. (eds). (2003). *Gender and ageing: Changing roles and relationships*. Berkshire: Open University Press.

Bhatti, M. (2006). "When I'm in the garden, I can create my own paradise": Homes and gardens in later life. *The Sociological Review*, 54(2), 318–341. http://dx.doi.org/10.1111/j.1467-954X.2006.00616.x

Bloom, D., Canning, D. and Fink, G. (2010). The greying of the global population and its macroeconomic consequences. *21st Century*, 5(3), 233–242. http://dx.doi.org/10.1080/17450144.2010.480826

Bowlby, S. R., McKie, L., Gregory, S. and MacPherson, I. (2010). *Interdependency and care over the lifecourse*. Oxford: Routledge.

Buse, C. and Twigg, J. (2015). Clothing, embodied identity and dementia: Maintaining the self through dress. *Age, Culture, Humanities*, 1(2), 71–96. Available from: http://ageceulture humanities.org/WP/wp-content/uploads/2015/02/Issue2_MM_BuseTwigg.pdf

Christou, A. (2016). Ageing masculinities and the nation: Disrupting boundaries of sexualities, mobilities and identities. *Gender, Place and Culture*, 23(6), 801–816. http://dx.doi.org/10.1080/0966369X.2015.1058760

Cutchin, M. P. (2009). Geographical gerontology: New contributions and spaces for development. *The Gerontologist*, 49(3), 440–444. http://dx.doi.org/10.1093/geront/gnp095

England, K. and Dyck, I. (2011). Managing the body work of home care. *Sociology of Health and Illness*, 33(2), 206–219. http://dx.doi.org/10.1111/j.1467-9566.2010.01331.x

Golant, S. M. (1984). The geographical literature on aging and old age: An introduction. *Urban Geography*, 5(3), 262–272. http://dx.doi.org/10.2747/0272-3638.5.3.262

Hardill, I. (2009). Introduction: Geographies of aging. *The Professional Geographer*, 61(1), 1–3. http://dx.doi.org/10.1080/003301200802577509

Harper, S. and Laws, G. (1995). Rethinking the geography of ageing. *Progress in Human Geography*, 19(2), 199–221. Available from: http://dx.doi.org/10.1177/030913259501900203

Hopkins, P. and Pain, R. (2007). Geographies of age: Thinking relationally. *Area*, 39(3), 287–294. http://dx.doi.org/10.1111/j.1475-4762.2007.00750.x

Horn, V. and Schweppe, C. (2016). *Transnational aging: Current insights and future challenges*. New York: Routledge.

Joseph, A. E. and Hallman, B. C. (1998). Over the hill and far away: Distance as a barrier to the provision of assistance to elderly relatives. *Social Science & Medicine*, 46(6), 631–639. http://dx.doi.org/10.1016/S0277-9536(97)00181-0

Katz, C. and Monk, J. (2014). Making connections: Space, place and the life course, in C. Katz and J. Monk (eds), *Full circles: Geographies of women over the life course* (pp. 264–278). Abingdon: Routledge.

Krekula, C. (2007). The intersection of age and gender: Reworking gender theory and social gerontology. *Current Sociology*, 55(2), 155–171. http://dx.doi.org/10.1177/0011392107073299

La Gory, M., Ward, R. and Sherman, S. (1985). The ecology of aging: Neighborhood satisfaction in an older population. *Sociological Quarterly*, 26(3), 405–418. http://dx.doi.org/10.1111/j.1533-8525.1985.tb00235.x

Lager, D., Van Hoven, B. and Huigen, P. (2013). Dealing with change in old age: Negotiating working-class belonging in a neighbourhood in the process of urban renewal in the Netherlands. *Geoforum*, 50, 54–61. http://dx.doi.org/10.1016/j.geoforum.2013.07.012

Lager, D., Van Hoven, B. and Huigen, P. (2015). Understanding older adults' social capital in place: Obstacles to and opportunities for social contacts in the neighbourhood. *Geoforum*, 59, 87–97. http://dx.doi.org/10.1016/j.geoforum.2014.12.009

Lager, D., Van Hoven, B. and Huigen, P. (2016). Rhythms, ageing and neighbourhoods. *Environment and Planning A*, 48(8), 1565–1580. http://dx.doi.org/10.1177/0308518X16643962

Lulle, A. and King, R. (2016). Ageing well: The time-space of possibility for older female Latvian migrants in the UK. *Social & Cultural Geography*, 17(3), 444–462. http://dx.doi.org/10.1080/14649365.2015.1089584

Means, R. and Evan, S. (2012). Communities of place and communities of interest? An exploration of their changing role in later life. *Ageing & Society*, 32(8), 1300–1318. http://dx.doi.org/10.1017/SO144686X11000961

Meijering, L., Lettinga, A. T., Nanninga, C. S. and Milligan, C. (2017). Interpreting therapeutic landscape experiences through rural stroke survivors' biographies of disruption and flow. *Journal of Rural Studies*, 51, 275–283. Online first: June 2016. http://dx.doi.org/10.1016/j.jrurstud.2016.06.004

Milligan, C. (2009). *There's no place like home: Place and care in an ageing society*. Surrey: Ashgate e-Book.

Milligan, C. and Bingley, A. (2015). Gardens and gardening in later life, in J. Twigg and W. Martin (eds), *Routledge handbook of cultural gerontology* (pp. 321–329). London: Routledge.

Milligan, C. and Morbey, H. (2016). Care, coping and identity: Older men's experiences of spousal care-giving. *Journal of Aging Studies*, 38, 105–114. http://dx.doi.org/10.1016/j.aging.2016.05.002

Milligan, C., Payne, S., Bingley, A. and Cockshott, Z. (2013). Place and well-being: Shedding light on activity interventions for older men. *Ageing & Society*, 35(1), 124–149. http://dx.doi.org/10.1017/S0144686X13000494

Milligan, C. and Wiles, J. L. (2010). Landscapes of care. *Progress in Human Geography*, 34(6), 736–754. http://dx.doi.org/10.1177/0309132510364556

Morton, K. (2014). Emerging geographies of disciplining the ageing body: Practising cosmetics technologies in the aesthetic clinic. *Gender, Place & Culture*, 22(7), 1041–1057. http://dx.doi.org/10.1080/0966369X.2014.939151

Mowl, G., Pain, R. and Talbot, C. (2000). The ageing body and the homespace. *Area*, 32(2), 189–197. http://dx.doi.org/10.1111/j.1475-4762.2000.tb00129.x

Peace, S., Holland, C. and Kellaher, L. (2005). Making space for identity, in G. Andrews and D. Phillips (eds), *Ageing and place: Perspectives, policy, practice*. Routledge Studies in Human Geography, v9 (pp. 188–204). Oxford: Routledge.

Pilkey, B. (2014). Queering heteronormativity at home: Older gay Londoners and the negotiation of domestic materiality. *Gender, Place & Culture*, 21(9), 1142–1157. http://dx.doi.org/10.1080/0966369X.2013.832659

Riley, M. (2012). Moving on? Exploring the geographies of retirement adjustment amongst farming couples. *Social & Cultural Geography*, 13(7), 759–781. http://dx.doi.org/10.1080/14649365.2012.725136

Rowles, G. D. (1978). *Prisoners of space? Exploring the geographic experience of older people*. Boulder, CO: Westview.

Rowles, G. D. (1986). The geography of ageing and the aged: Towards an integrated perspective. *Progress in Human Geography*, 10(4), 511–539. http://dx.doi.org/10.1177/0733464815614917

Rowles, G. D. and Ohta, R. J. (eds). (1983). *Aging and milieu: Environmental perspectives on growing old*. New York: Academic Press.

Schwanen, T., Hardill, I. and Lucas, S. (2012). Spatialities of ageing: The co-construction and co-evolution of old age and space. *Geoforum*, 43(6), 1291–1295. http://dx.doi.org/10.1016/j.geoforum.2012.07.002

Skinner, M., Cloutier, D. and Andrews, G. (2015). Geographies of ageing: Progress and possibilities after two decades of change. *Progress in Human Geography*, 39(6), 776–799. http://dx.doi.org/10.1177/0309132514558444

Skinner, M. and Hanlon, N. (2016). *Ageing resource communities: New frontiers of rural population change, community development, and voluntarism*. London: Routledge.

Stenner, P., Church, A. and Bhatti, G. (2012). Human-landscape relations and the occupation of space: Experiencing and expressing domestic gardens. *Environment and Planning A*, 44(7), 1712–1727. http://dx.doi.org/10.1068/a44378

Tarrant, A. (2010). Constructing a social geography of grandparenthood: A new focus for intergenerationality. *Area*, 42(2), 190–197. http://dx.doi.org/10.1111/j.1475-4762.2009.00920.x

Tarrant, A. (2013). Grandfathering as spatio-temporal practice: Conceptualizing performances of ageing masculinities in contemporary familial carescapes. *Social & Cultural Geography*, 14(2), 192–210. http://dx.doi.org/10.1080/14649365.2012.740501

Tarrant, A. (2016). The spatial and gendered politics of displaying family: Exploring material cultures in grandfathers' homes. *Gender, Place and Culture*, 23(7), 969–982. http://dx.doi.org/10.1080/096639X.2015.1073703

Twigg, J. (2004). The body, gender, and age: Feminist insights in social gerontology. *Journal of Aging Studies*, 18(1), 59–73. http://dx.doi.org/10.1016/j.jaging.2003.09.001

Twigg, J. (2015). Dress and age: The intersection of life and work. *International Journal of Ageing and Later Life*, 10, 55–67. http://dx.doi.org/10.3384/ijal.1652-8670.15-55

Vanderbeck, R. (2007). Intergenerational geographies: Age relations, segregation, and re-engagements. *Geography Compass*, 1(2), 200–221. http://dx.doi.org/10.1111/j.1749-8198.2007.00012.x

Wahl, H. W. and Weisman, J. (2003). Environmental gerontology at the beginning of the new millennium: Reflections on its historical, empirical, and theoretical development. *The Gerontologist*, 43(5), 616–627. http://dx.doi.org10.1093/geront/43.5.616

Waitt, G. and Gorman-Murray, A. (2007). Homemaking and mature-age gay men 'down under': Paradox, intimacy, subjectivities, spatialities, and scale. *Gender, Place and Culture*, 14(5), 569–584.

Walsh, K. and Näre, L. (2016). *Transnational migration and home in later life*. London: Routledge.

Ward, R. and Holland, C. (2011). 'If I look old, I will be treated old': Hair and later-life image dilemmas. *Ageing & Society*, 31(2), 288–307. http://dx.doi.org/10.1017/SO144686X10000863

Warnes, A. M., Friedrich, K., Kellaher, L. and Torres, S. (2004). The diversity and welfare of older migrants in Europe. *Ageing & Society*, 24(3), 307–326. http://dx.doi.org/10.1017/SO144686X04002296

Wasielewski, P. L. (2015). The magic of Oaxaca: An emotional geography of age, gender and nation. *Symbolic Interaction*, 38(4), 557–574. http://dx.doi.org/10.1002/symb.186

Wiles, J. L. (2005). Conceptualizing place in the care of older people: The contribution of geographical gerontology. *Journal of Clinical Nursing*, 14(8B), 100–108. http://dx.doi.org/10.1111/j.1365-2702.2005.01281.x

Wiles, J. L. and Jayasinha, R. (2013). Care for place: The contributions older people make to their communities. *Journal of Aging Studies*, 27(2), 93–101. http://dx.doi.org/10.1016/j.jaging.2012.12.001.

Wiles, J. L., Leibing, A., Guberman, N., Reeve, J. and Allen, R.E.S. (2011). The meaning of 'Aging in Place' to older people. *The Gerontologist*, 52(3), 357–366. http://dx.doi.org/10.1093/geront/gnr098

Windle, G. S., Burholt, V. and Edwards, R. T. (2006). Housing related difficulties, housing tenure and variations in health status: Evidence from older people in Wales. *Health & Place*, 12(3), 267–278. http://dx.doi.org/10.1016/j.healthplace.2004.08.010

5 Population geographies of older people

Mark W. Rosenberg and Kathi Wilson

Introduction

From a non-geographic perspective, the familiar foci of research on older people (e.g., the impact of the older population on health care systems) often seem the same regardless of where the research is done. The recurring theme of this chapter, however, is to argue that the geography of where older people live matters analytically because depending on the geographic scale (global to the local), the issues and responses of older people, their families, friends, volunteers, service providers and decision-makers play themselves out very differently but in ways that are connected to other geographic, demographic, economic and political scales, trends and policies.

Throughout the chapter, we follow the United Nations' definition of the older population as 60 and older except where we cite data or a quote from authors who use other statistical definitions as noted. We also acknowledge that any age (e.g., 60 or 65) is an arbitrary and artefactual way of dividing the older population from other population groups across the life course. On the other hand, it is also important to acknowledge that for all the limitations inherent in defining the older population simply through whether an individual is above a certain age, it is the way governments and institutions at all geographic scales ultimately use age as a key criterion of eligibility for free or discounted services (e.g., for prescription drugs or long-term care) and the distribution of benefits (e.g., eligibility for a public pension). For example, one automatically becomes eligible for Old Age Security at the age of 65 in Canada (Government of Canada, 2016).

Linking population geography and geographies of the older population

At its core, population geography provides a spatial interpretation of basic demographic concepts such as fertility, mortality and migration. As Newbold (2002: 10) points out, however, population geography has "grown to draw upon a multiplicity of methods and theoretical approaches . . . recognizing the importance of place and drawing upon the diverse insights provided by geography and related social science disciplines".

In his book, Newbold (2002) did not focus on the older population *per se*. Instead, he provocatively examined "the (un)certainty of death and disease", drawing together global and national data on variations in mortality rates, life expectancy, leading causes of death and theories such as the epidemiological transition model, all staples of population geography. He then recast the ideas of population geography to raise a series of thought-provoking questions about how the ageing of populations in the twenty-first century will challenge governments at various geographic scales to address their needs for health care, housing and services.

In essence, there are no better examples of the overlap between population geography and the geographies of ageing than the books by Rogers (1992) and Moore, Rosenberg and McGuinness (1997). The former is an edited collection divided into four parts. In part 1, the chapters focus on why older people move. Is it lifestyle (e.g., moving to areas with high levels of amenities and warm climates), the anticipation of health issues or because of health issues that older people move? In part 2, the importance of ageing in place, living near to children and the challenges of living alone are major themes. Part 3 charts the migration trends of the older population through the 1980s and into the 1990s. In part 4, case studies from Japan, the United Kingdom and the United States are used to highlight how population composition (i.e., the characteristics of the older population) and context (the characteristics of the places) play key roles in understanding the geographies of ageing. In the Foreword to *Growing Old in Canada* (Moore *et al.*, 1997), Dr. Monica Boyd, one of Canada's most noted demographers, wrote that this

> study includes a spatial analysis of where the elderly live focusing on major migration trends . . . elderly people's income, health status, living arrangements and other characteristics . . . accompanied by analysis focusing on the capacities of communities to provide support and services the elderly need.
>
> (p. xvii)

In the remainder of this chapter, we follow in this tradition of the overlap between population geography and geographical gerontology to examine the population geographies of the older population. The remainder of the chapter is divided into six parts: global trends; comparing regions and countries; national geographies; ageing in place, migration and immigration; other geographies of the older population; and other older populations, other geographies.

Global trends

If we think about the defining demographic and geographic trends of the twentieth century, it can easily be argued that the defining demographic was the "population bomb" (Ehrlich, 1968) that followed World War II. Literally, thousands of scholarly and popular works were devoted to documenting the explosion in the growth of the global population, what its underlying causes were, what was required to address it and what would be the implications socially, economically, politically and environmentally should it not be defused. It would, however, be much more difficult to argue that there was one over-arching geographic trend that similarly captured the thinking of scholars and policy-makers alike at the global scale.

In contrast, there are demographic and geographic trends that are already shaping the twenty-first century in profound ways that are likely to become increasingly important in how we think about changes from the global to the local as the century progresses (see the United Nations, 2015). As global fertility rates continue to decline, ageing of the global population will increase rapidly (United Nations, Department of Economic and Social Affairs, Population Division [UNDESAPD], 2015a). According to UNDESPAD, the number of people aged 60 and over will almost equal the number of children under the age 15 by 2050 for the first time in history. By 2100, depending on the scenario chosen, assuming either a global population of 10 billion or 12.5 billion, approximately 3.2 billion people will be over the age of 60, comprising, respectively, between 32 per cent or 26 per cent of the global population (UNDESAPD, 2015a). Moreover, by 2100, almost 1 billion people on Earth will be aged 80 and over (UNDESAPD, 2015a). In other words, the over-arching demographic trend of the twenty-first century will be the ageing of the global population.

In parallel with this remarkable demographic trend – the ageing of the global population – the twemty-first century will also be characterised geographically as the century of urbanisation. What went largely unnoticed is that in 2007, for the first time in history, more than half the population of the world was classified as urban (UNDESAPD, 2015b). Comparing the 1950 global population to the projected 2050 global population, the world will go from one-third urban and two-thirds rural to two-thirds urban and one-third rural (UNDESAPD, 2015b). Together, what these two trends mean is that globally more thinking will be required to take into account a growing older population and their needs living in mainly urban settings, especially if one focuses on the absolute numbers. Paradoxically, the relative proportions of older people living in rural areas are likely to be even greater than in urban areas, but the absolute numbers will be smaller. For the research community, policy-makers and those responsible for the delivery of services the older rural population will pose its own unique set of challenges (more about this later).

Thus, it is surprising that at a global policy level, it is the needs and concerns about younger age groups (especially children) that continue to dominate the global policy discourse (United Nations, 2015). For example, children – and especially children under the age of 5 – are mentioned often throughout the 17 Sustainable Development Goals (SDGs). The older population is only specifically noted twice throughout the SDGs: under Goal Target 2.2, linked to hunger, and Goal Target 11.7 linked to Sustainable Cities and Communities. Under the remaining goals, specific goal targets relating to the older population cannot be found. There are no targets for improving the lives of older people in areas such as social security and retirement systems, health, literacy, elder abuse or women as caregivers for their spouses, their parents and in-laws.

Comparing regions and countries

Global trends, however, mask remarkable regional and national differences when we compare the developed to the developing world, regional groups of countries or country by country. While the regions and countries of the developed world already have (Japan) or will soon have (Europe, Japan, Canada, the United States, Australia and New Zealand – all by 2030) a quarter or more of their populations aged 60 and over, the countries of Latin America, the Caribbean and Asia will see their populations 60 and older grow to 25 per cent or more by 2050 (UNDESAPD, 2015a). Only in the less-developed countries of Africa will the percentage of the population aged 60 and older remain below double digits by 2050.

In the countries of the developed world, policy-makers have become obsessed with the 'old component' of the dependency ratio or what the UNDESPAD calls the Potential Support Ratio (PSR – the number of people aged 20 to 64 divided by the number of people 65 and over). Japan and seven other countries in Europe already have PSRs under 4 as do countries such as Canada and the United States (UNDESAPD, 2015a). While the majority view is that such low PSRs will place undue strains on public social security and pension programmes, as well as health and social care systems, a few voices have argued against such alarmist views (e.g., Gee and Guttman, 2000). In the BRIC countries (Brazil, Russia, India and China), not only are their populations ageing rapidly but also traditional systems of supporting the older population are declining as young people move to the cities and are no longer geographically located nearby to provide care and/or support for older persons (Yang *et al.*, 2011). In sub-Saharan Africa, where most, if not all, of the countries where the older population makes up less than 10 per cent of the total population are located, the challenges are different (see Rishworth and Elliot, Chapter 9). Infectious diseases (especially HIV/AIDS) and war have hollowed out the working age cohorts of the population (20 to 64). This, coupled with grinding poverty in many of the countries, has resulted in many older people becoming the *de facto* carers while governments have few or no resources to support them or the children for whom they are caring. While in recent years, more attention is beginning to be paid to the challenges faced by older people in sub-Saharan Africa, there is much that yet needs to be learned and done (Aboderin, 2016; Apt, 2011; Maharaj, 2013).

National geographies

The higher-order trends and policy issues already alluded to earlier play themselves out ever more poignantly at the national level and within countries. There are three geographies of a country at play when thinking about the older population. First, there is no country on Earth where the older population is geographically evenly distributed throughout – either in absolute or relative terms. Second, how we think about the geographies of the older population is often determined by the administrative divisions of the country (i.e., how a country is divided up for collecting information about the people who live there). The third geography is how a country defines its political boundaries and apportions political responsibilities to those divisions (Jones *et al.*, 2015).

Canada is a classic example of all three of these geographies. If you simply look at the geographic distribution of the older population in *absolute* terms, the vast majority live in urban Canada (see the *Atlas of the Aging Population of Canada*, www.queensu.ca/geographyandplanning/gaplab/atlas-aging) with the largest concentrations found in the largest cities. If you look at the geographic distribution in *relative* terms, rural parts of Canada have much higher *percentages* of older people compared to urban parts of Canada.

The differences between urban and rural Canada are to some extent magnified and amplified by the administrative units used to analyse the geographic distribution of the older population. In Canada, Statistics Canada, the federal government statistical agency, divides the country into various hierarchical units for analysis. For example, below the level of the province or territory, one administrative division is census divisions (used in the *Atlas of the Aging Population of Canada*) subdivided into census subdivisions. A second way that Statistics Canada divides up the country is into census metropolitan categories: census metropolitan areas (CMAs), census agglomerations (CAs), outside CMAs and CAs and census subdivisions (CSDs). The largest cities in Canada (CMAs) are further divided into smaller geographic units, census tracts. In contrast to the administrative units used in the *Atlas of the Aging Population of Canada*, using CMAs, CAs and CSDs provides a much more complex and nuanced perspective on the geographic distribution of the older population of Canada. While the older population dominated the census tracts at the centres of CMAs in the middle of the twentieth century, today, it is the census tracts of the older suburbs that have the highest concentrations of older people compared to the census tracts of the central city and the new suburbs. So, not only do the population geographies of the older population change as geographic scale changes but also the geographies of the older population change over time (see Hodge, 2008; Moore *et al.*, 1997; Northcott and Petruik, 2013).

As mentioned earlier, a third geography is how countries divide themselves up politically and where the political responsibility for the older population rests within those divisions. Again, using Canada as an example, the country is divided into provinces and territories within a federal political system. Constitutionally, the federal government is responsible for funding and administering a universal social transfer programme called the Old Age Security programme and the Canada Pension Plan (CPP) to all older people except in the province of Quebec which has its own plan: Quebec Pension Plan (QPP; see Banting, 1987). CPP/QPP is an employment-based pension plan that workers and employers pay into during their working years and then receive as monthly payments after the age of 65. The federal government also transfers funds directly to the provinces and territories in support of their social and health care programmes but provides little in the way of direct services to most of the older population (Forest, Marchildon and McIntosh, 2004; Marchildon, McIntosh and Forest, 2004; McIntosh, Marchildon and Forest, 2004). The exceptions are for veterans and the Aboriginal population where, for constitutional reasons, the federal government is responsible for providing services directly. At the provincial and the territorial level, governments have as one of their main responsibilities providing health and social services to their respective populations; increasingly, this means providing services for the older population. Provincial and territorial governments also indirectly fund services for the older population through transfers to local governments. Finally, local governments are responsible for some direct services (e.g., publicly funded homes for the aged) and indirectly through programmes offered at local publicly owned facilities (e.g., seniors' recreational programmes at locally funded public recreation centres) and public transit (e.g., seniors' discounted bus fares, adapted buses and specially adapted transit vans; see Kitchen, 2015).

All three geographies make the analysis of the older population complex. Lack of agreement among researchers about the appropriate scale(s) for analysis, definitional debates (e.g., what constitutes urban and rural) and debates about the artefactual or socially constructed nature of the geographic distribution of the older population further contribute to this complexity. For older people, their family members and friends, how a country is divided politically into multiple administrative layers makes understanding *where* to go and what level of government is responsible for transfers of funds and the provision of services a complex and often exasperating process. In response, governments have often tried to *regionalise* and *centralise* services to provide a local response and a single entry point for accessing services for the older population. For example, in the province of Ontario, the provincial government divided the province into 14 regions (called Local Health Integration Networks; LHINs) with each region having community care access centres (CCACs) to organise the delivery of home care services mainly, but not exclusively, for older people in need of medical and non-medical home care services.

Ageing in place, migration and immigration

What factors underpin the geographic distribution of the older population? First and foremost is ageing in place, the very fact that at some point in most people's adult lives they make a decision to remain where they are throughout their older years. While it is taken for granted that most older people prefer to age in place and governments strongly favour ageing in place, there remain many unresolved issues related to ageing in place. In particular, the ability of governments to provide the various services needed when older people can no longer maintain themselves (e.g., they need home care services) or their homes (e.g., they need someone to do the heavy household chores) is arguably the issue that has received the most attention. A second set of issues is what is to be done when it is no longer safe for older people to remain in their homes. A third set of issues arises when older people can no longer afford to remain in their homes but cannot afford to move out of them even if they wished to do so. This has recently led some researchers to argue that the focus should be on *ageing in the right place* (e.g., Golant, 2015).

Second are the various types of migration of the older population. A relatively small proportion of the older population moves around the time of retirement or when they exit the labour force. Often, these 'movers' are the 'healthy and wealthy' older population who move to places known for their mild winter climates and high levels of amenities. A second motivation for older movers is to be near to other family members (usually their adult children) either in anticipation of a time when they will need informal support or after a serious health event which necessitates the need for that support. Another motivation for the preceding types of moves, often occurring in conjunction with these moves, is the process of 'downsizing' from a house to some form of congregate living or retirement community

While the preceding moves are often analysed as permanent, another type of mover among the older population are the 'seasonal' movers, who spend part of the year in the place where they live after retirement or their exit from the labour force and part of the year (usually associated with winter) in another part of the country or another country known for its warm winters. In North America, we refer to this groups as 'snowbirds'. Similar movers can be found in Europe who leave Northern Europe for locations in Southern Europe, North Africa, the Caribbean or South America. As the wealth of East Asian retirees has grown, there are also growing numbers who fit the descriptions given earlier, but there has been little research to document their numbers and their similarities and differences in contrast to North American and European older movers.

A final type of mover among the older population is those who return to their countries of origin to live out their lives. Often these movers take advantage of the wealth they have accumulated, public transfers and public and private pension plans to which they have access in the countries where they have spent their working lives. They choose their countries of origin for retirement because of differences in the cost of living and to reunite with family members who never emigrated with them. How important this type of migration is to both the countries of return and to the countries of exit goes well beyond accounting for the transfer of wealth and deserves more attention among researchers.

The geographic distribution is not only determined by ageing in place and the older population who are movers but is also determined by the changing behaviour of younger age cohorts. As alluded to above, all things being equal, the older population will inevitably grow in size relative to other age cohorts because of declining fertility rates. What has hastened the growth of the older population virtually everywhere, however, is the effects of rural-to-urban migration of younger working age populations which leaves rural areas with the paradoxical challenge of supporting an older population who might be few in absolute numbers but very high in relative percentage terms. Even in countries such as China that try to control the movement of people through its residential registration system (*Hukou*), the flow of young, working-age people to the cities continues unabated (Yu, Rosenberg and Cheng, 2015).

The other aspect of international migration of younger working-age cohorts that needs to be taken into account is how it affects the older population in both the countries from which emigrants originate and those countries receiving immigrants. Often younger working-age emigrants from developing countries leave behind parents who will no longer have them as informal caregivers in their later years. What is also troubling is that younger working-age emigrants tend to be among the best educated and are often health care professionals, thus robbing developing countries of much need leadership and health care workers (Organisation for Economic Co-operation and Development [OECD], 2010). For the receiving countries, particularly in the developed world, international migration mediates the PSR, especially in those cities where immigrants concentrate. Immigrant health care workers are often specifically recruited to provide care for older people either in private homes or in public and private residential care facilities (see Kelly *et al.*, 2012).

Other geographies of the older population

While the idea of age-restricted housing, especially in the form of institutional settings and nursing homes, has a history that dates back well before the 1950s, the idea of the purpose-built retirement community took off with the Sun City developments in Florida and Arizona in the early 1960s (Laws, 1993). Today, residential communities for the older population can be found throughout the developed countries of the world, and they are increasing in the fast-developing countries of Asia and South America (see www.Internationalliving.com).

The essential features of purpose-built residential communities for the older population remain largely the same today, with their emphasis on standardised architecture taking into account the needs of an older and perhaps less mobile population, wider streets to accommodate various forms of transportation (e.g., walking, cycling, golf carts and automobiles), recreational facilities (e.g., golf courses, swimming pools and clubhouses) and sometimes various levels of health care services. Over time, what has changed, however, is a greater emphasis on security, especially in the form of gated communities, and who lives inside the walls of gated communities (see Rosenberg and Everitt, 2001).

Paradoxically, the emphasis on living in gated communities has generated a growing debate about the value of age-segregated communities that exclude children and sometimes even working age adults below an age that someone has decided defines who is an older person (see McHugh, Gober and Borough, 2002). It is a complex, multidimensional debate that extends beyond gated communities – that are often their own jurisdictions – to publicly subsidised social housing that is the responsibility of local governments. Even in publicly subsidised social housing, those who favour age-segregated communities argue for them as a way of protecting older people from those who might take advantage of them and as a way to provide targeted services to older people. Those who argue for age-*integrated* communities argue that age-segregated communities are socially unhealthy and may even have negative health outcomes for older persons isolated from other age groups (see www.encyclopedia.com/education/encyclopedias-almanacs-transcripts-and-maps/housing-age-segregated).

Another new geography of the older population now receiving increasing attention is the reinvention of communities in rural areas as retirement communities or simply as communities increasingly dominated by an older population. One particular example of this phenomenon is known as ageing resource communities (Skinner and Hanlon, 2016). Skinner and Hanlon (2016) characterize these communities as rural communities that lose their major employer – often a mine or lumber mill – resulting in much of the working age population departing, leaving a mainly older population with the challenges of how to maintain services and even grow in new directions (Gjertsen, Ryser and Halseth, 2016; Lovell, Gray and Boucher, 2016; Markey, Halseth and Ryser, 2016).

Purpose-built residential communities for the older population, gated communities and ageing resource communities are, in many ways, reflections of a broader movement towards creating 'age-friendly communities' everywhere. In 2006, the World Health Organization (WHO) and a small set of countries and cities worked together to define the dimensions of age-friendly communities (WHO, 2007). These dimensions include community support and health services, outdoor spaces and buildings, transportation, housing, social participation, respect and social inclusion, civic participation and employment and communication and information. Even with the efforts of the WHO and individual countries to promote the idea of age-friendly communities, Eisenberg (2015) notes that according to the WHO there are only 258 cities in 30 countries around the world that have been designated age-friendly, leading Eisenberg to the title of his article: "Why Are There so Few Age-Friendly Cities?"

Other older populations – other geographies

As the result of state or colonial action or through treaties, the geographies of Indigenous Populations are often distinct from the rest of the population. What is common to many of the Indigenous Populations around the world is that they have lost much of their traditional lands, and in some cases (e.g., in Canada and the United States), they were forced onto reserves often in rural and remote places (Gracey and King, 2009; King, Smith and Gracey, 2009). What is also common to Indigenous populations is that life expectancy is lower, fertility rates are higher and on metrics such as health status, education and income, Indigenous Populations generally have poorer outcomes than the rest of the population (Gracey and King, 2009; King *et al.*, 2009).

Taking the Indigenous Peoples of Canada as an example. The government of Canada recognizes three distinct groups: the First Nations, the Inuit and the Métis. Their geographies are also distinct. For example, the First Nations live on-reserve, off-reserve and some move back and forth between reserves and urban places (Wilson and Rosenberg, 2002). Health Canada (2014) goes even further in classifying reserves as non-isolated, semi-isolated, isolated and remote-isolated.

The complex geographies of Indigenous Peoples and what those geographies mean in the everyday lives of older Indigenous Peoples remains one of the least-researched topics on older people in both the literature on the geographies of ageing and the gerontology literature. Among the few examples to be found are the papers by Wilson and Rosenberg (2002), Wilson *et al.* (2010), Wilson, Rosenberg and Abonyi, (2011) and Rosenberg and Wilson (2014) who try to capture quantitatively how the complex geographies of the Indigenous Peoples of Canada affect the general health, the mental health, access to health care and traditional health practices of older Indigenous Peoples in Canada.

Conclusion

The ageing of the global population will be a transcendent phenomenon of the twenty-first century. It has social, economic and political implications at every geographic scale both for research and the development of public policy. Population ageing will challenge governments at all geographic scales to find ways to treat their respective older populations with dignity and respect. The public, private and non-profit sectors will be challenged to find new ways to deliver health and social services, transportation, housing and recreation services. Incorporating the concepts and methods of population geography into the geographies of ageing provides a platform on which to build our understanding of other geographical gerontologies.

What is missing from this chapter but taken up in other chapters of this book is that these challenges will be made even more complex by the growing diversity of the older population. Not only will gender and socio-economic differences need to be taken into account, but race, ethnicity, disability, morbidity and mental health will also add to the geographic complexities of creating age-friendly geographies at all scales of analysis and public policy.

References

Aboderin, I. (2016). Coming into its own? Developments and challenges for research on aging in Africa. *Journals of Gerontology: Social Sciences*, 72(4), 643–645. Online first. http://dx.doi.org/10.1093/geronb/gbw017

Apt, N. A. (2011). Aging in Africa: Past experiences and strategic directions. *Ageing International*, 37(1), 93–103. http://dx.doi.org/10.1007/x12126-011-9138-8

Banting, K. (1987). *The Welfare state and Canadian federalism* (2nd edn). Kingston: McGill-Queen's University Press.

Ehrlich, P. R. (1968). *The population bomb*. New York: Ballantine Books.

Eisenberg, R. (2015). Why are there so few age-friendly cities? *Forbes*. Available from: www.forbes.com/sites/nextavenue/2015/08/12/why-are-there-so-few-age-friendly-cities/#547d2aca91fd

Forest, P-G., Marchildon, G. P. and McIntosh, T. (eds). (2004). *Changing health care in Canada. The Romanow Papers, Vol. 2.* Toronto: University of Toronto Press.

Gee, E. M. and Guttman, G. M. (eds). (2000). *The overselling of population aging: Apocalyptic demography, intergenerational challenges, and social policy.* Don Mills: Oxford University Press.

Gjertsen, T-A., Ryser, L. and Halseth, G. (2016). Gamvik, 'a good place to grow old': The role of voluntary organizations in an ageing resource-dependent municipality in northern Norway, in M. Skinner and N. Hanlon (eds), *Ageing resource communities: New frontiers of rural population change, community development and voluntarism* (pp. 106–118). London: Routledge.

Golant, S. M. (2015). *Aging in the right place.* Baltimore, MD: Health Professions Press.

Government of Canada. (2016). *Old age security.* Available from: www.canada.ca/en/services/benefits/publicpensions/cpp/old-age-security.html

Gracey, M. and King, M. (2009). Indigenous health part 1: Determinants and disease patterns. *The Lancet,* 374(9683), 65–75. http://dx.doi.org/10.1016/S0140-6736(09)60914-4

Health Canada. (2014). *A statistical profile on the health of first nations in Canada: Determinants of health, 2006 to 2010.* Ottawa: Health Canada. Available from: http://health.chiefs-of-ontario.org/sites/default/files/attachments/Determinants%20of%20Health%202006-2010-EN-FINAL.pdf

Hodge, G. (2008). *The geography of aging.* Montreal: McGill-Queen's University Press.

Jones, M., Jones, R., Woods, M., Whitehead, M., Dixon, D. and Hannah, M. (2015). *An introduction to political geography: Space, place and politics* (2nd edn). London: Routledge.

Kelly, P., Garcia, M., Esguerra, E. and the Community Alliance for Social Justice (2012). Filipino immigrants in the Toronto labour market: Towards a qualitative understanding of deprofessionalization, in R. S. Coloma, B. McElhinny, E. Tungohan, J. P. Catungal and L Davidson (eds), *Filipinos in Canada: Disturbing invisibility* (pp. 68–88). Toronto: University of Toronto Press.

King, M., Smith, A. and Gracey, M. (2009). Indigenous health part 2: The underlying causes of the health gap. *The Lancet,* 374(9683), 76–85. http://dx.doi.org/10.1016/S0140-6736(09)60827-8

Kitchen, H. (2015). *No seniors' specials: Financing municipal services in aging municipalities.* IRPP Study, No. 51, 32 pp., Montreal. Available from: http://irpp.org/wp-content/uploads/2015/02/study-no51.pdf

Laws, G. (1993). "The land of old age": Society's changing attitudes toward urban built environments for elderly people. *Annals of the Association of American Geographers,* 83(4), 672–693. http://dx.doi.org/10.1111/j.1467-8306.1993.tb01960.x

Lovell, S., Gray, A. and Boucher, S. (2016). Experiences of economic change in small town New Zealand: Implications for voluntarism and community capacity, in M. Skinner and N. Hanlon (eds), *Ageing resource communities: New frontiers of rural population change, community development and voluntarism* (pp. 119–130). London: Routledge.

Maharaj, P. (ed.). (2013). *Aging and health in Africa.* New York: Springer.

Marchildon, G. P., McIntosh, T. and Forest, P-G. (eds). (2004). *The fiscal sustainability of health care in Canada: The Romanow papers,* vol. 1. Toronto: University of Toronto Press.

Markey, S., Halseth, G. and Ryser, L. (2016). Planning for all ages and stages of life in resource hinterlands: Place-based development in northern British Columbia, in M. Skinner and N. Hanlon (eds), *Ageing resource communities: New frontiers of rural population change, community development and voluntarism* (pp. 131–145). London: Routledge.

McHugh, K., Gober, P. and Borough, D. (2002). The Sun City wars: Chapter 3. *Urban Geography,* 23(7), 627–648. http://dx.doi.org/10.2747/0272-3638.23.7.627

McIntosh, T., Marchildon, G. P. and Forest, P-G. (eds). (2004). *The governance of health care in Canada: The Romanow Papers,* vol. 3. Toronto: University of Toronto Press.

Moore, E. G., Rosenberg, M. W. and McGuinness, D. (1997). *Growing old in Canada.* Toronto: ITP Nelson.

Newbold, K. B. (2002). *Six billion plus: Population issues in the twenty-first century.* New York: Rowman & Littlefield.

Northcott, H. C. and Petruik, C. R. (2013). Trends in the residential mobility of seniors in Canada, 1961–2006. *The Canadian Geographer,* 57(1), 43–55. http://dx.doi.org/10.1111/j.1541-0064.2012.00449.x

Organisation for Economic Co-operation and Development (OECD). (2010). *International migration of health workers.* Paris: OECD Observer, Policy Brief. Available from: www.who.int/hrh/resources/oecd-who_policy_brief_en.pdf

Rogers, A. (ed.). (1992). *Elderly migration and population redistribution: A comparative study.* London: Bellhaven Press.

Rosenberg, M. W. and Everitt, J. (2001). Planning for aging populations: Inside or outside the walls. *Progress in Planning,* 56(3), 119–168. http://dx.doi.org/10.1016/S0305-9006(01)00014-9

Rosenberg, M. W. and Wilson, K. (2014). Older adults: Shedding light on an invisible population, in P. Menzies and L. Lavallée (eds), *Journey to healing: Aboriginal people with addiction and mental health issues* (pp. 173–184). Toronto: Centre for Addiction and Mental Health (CAMH).

Skinner, M. and Hanlon, N. (eds). (2016). *Ageing resource communities: New frontiers of rural population change, community development and voluntarism.* London: Routledge.

UNDESAPD. (2015b). *World urbanization prospects: The 2014 revision.* Publication No. ST/ESA/SER.A/366, Author, New York. Available from: https://esa.un.org/unpd/wup/publications/files/wup2014-report.pdf

United Nations. (2015). *Transforming our world: The 2030 agenda for sustainable development.* New York: Author. Available from: www.un.org/ga/search/view_doc.asp?symbol=A/RES/70/1&Lang=E

United Nations, Department of Economic and Social Affairs, Population Division (UNDESAPD). (2015a). *World population prospects: The 2015 revision, key findings and advance tables.* Working Paper No. ESA/P/WP 241, Author, New York. Available from: https://esa.un.org/unpd/wpp/publications/files/key_findings_wpp_2015.pdf

Wilson, K. and Rosenberg, M. W. (2002). Exploring the determinants of health for first nations peoples in Canada: Can existing frameworks accommodate traditional activities? *Social Science & Medicine,* 55(11), 2017–2031. http://dx.doi.org/10.1016/S0277-9536(01)00342-2

Wilson, K., Rosenberg, M. W. and Abonyi, S. (2011). Aboriginal peoples, health and healing approaches: The effects of age and place on health. *Social Science & Medicine,* 72(3), 355–364. http://dx.doi.org/10.1016/j.socscimed.2010.09.022

Wilson, K., Rosenberg, M. W., Abonyi, S. and Lovelace, R. (2010). Aging and health: An examination of differences between older Aboriginal and non-Aboriginal people. *Canadian Journal on Aging,* 29(3), 369–382. http://dx.doi.org/10.1017/S0714980810000309

World Health Organization. (2007). *Global age-friendly cities: A guide.* Geneva: Author. Available from: www.who.int/ageing/publications/Global_age_friendly_cities_Guide_English.pdf?ua=1

Yang, C., Rosenberg, M. W., Wang, W., Yang, L. and Li, H. (2011). Population aging and residential care resources in Beijing, China: Analysing the spatial distribution of the elderly population and residential care facilities. *Asian Journal of Gerontology & Geriatrics,* 6, 14–21.

Yu, J., Rosenberg, M. W. and Cheng, Y. (2015). Aging at home and the intent to relocate in Beijing. *Progress in Geography,* 34(12), 1577–1585. http://dx.doi.org/10.18306/dlkxjz.2015.12.006

6 Planning and design of ageing communities

Judith E. Phillips

Introduction

How spaces and places are designed for, as well as interpreted, used and experienced by older people has an impact on their activity, quality of life, health and identity (Michael, Green and Farquhar, 2006; Day, 2008; Sugiyama and Ward Thompson, 2007). The impact the environment has in terms of where we live, what lifestyle choice we have and the community we inhabit is critical to ageing well (Annear *et al.*, 2014). Yet to date, planners and designers have done little to explicitly enhance and support the well-being of older people. Age is generally subsumed under an inclusive design label, covering older people's issues, which are mainly expressed as accessibility and mobility (Hockey, Phillips and Walford, 2013). There is, however, increasing recognition of the need to design and plan for ageing populations and communities across the globe in terms of designing suitable, sustainable environments providing the opportunity to age well in place, to retain independence and to be mobile and socially connected.

Planning and design incorporate and require a broad, holistic approach – not just the built environment but also the natural environment and health, urban design, transportation, housing and land-use patterns, as well as social care services. Therefore, considering what is an 'optimal' or 'ideal' environment for 'ageing well' is complex and multifaceted. Good planning and design must ensure key elements are in place including the participation of older people themselves in the shaping of place. Consequently, the importance of community-level planning is crucial in this respect. Given that older people spend the majority of time in their local communities it is imperative that the community level be a focus of concern for geographical gerontology. Different communities will pose different kinds of issues to the planner. For example, deprived communities will challenge planners to use limited resources effectively; unequal and discriminatory communities will challenge planners to redistribute resources and empower citizens. Yet at the community level there is potential for the ownership of issues and participation of older people in addressing these challenges themselves.

Internationally, the World Health Organization's (WHO) Global Age-friendly Cities initiative sets out the blueprint for the design of age-friendly communities and provides a top-down framework for planners and designers (WHO, 2007). This initiative focuses on tangible aspects of the physical fabric of the urban environment (outdoor spaces and the built environment) as well as community support and health services, civic participation and employment, communication and information, respect, and social inclusion (WHO, 2007). More recently the age friendly cities initiative has expanded to include communities rather than focusing on cities alone, despite reservations and shortcomings. The criticism focuses first on the use of universal standards despite the diversity of cities and populations and second on the almost clinical problem–solution type of approach, which ignores context or difference (Phillipson, 2011; Handler, 2015).

Macro-environmental problems

Top-down approaches raise further problematic environmental aspects, particularly at a macro level.

Lag between policy at national and global levels and practice at local level in planning and design

The Age-friendly Cities initiative acts as a framework for understanding change that needs to occur in the environment. However, the translation from global to local requires an interdisciplinary approach, and not all stakeholders are on the same journey. Developing an age-friendly landscape amid competing agendas is often seen as too complex an issue. The implications of ageing and urbanisation, in particular, for services, infrastructure, technology and planning, are vast (Global City Indicators Facility: GCIF, 2013). A lack of funding, political commitment and coordination and fragmentation among different levels of government do not easily translate into actions at a local level. The concept of 'Lifetime homes and Lifetime neighbourhoods' (Communities and Local Government, 2008) in the United Kingdom – a move toward sustainable development – is a similar example which has not been wholeheartedly adopted universally and is seen as a longer-term investment requiring political will and commitment (HAPPI 3, 2016). Additionally, planners have recently recognised the need to incorporate older people in planning at a local level. Without local population assessment and design guidance, planners otherwise risk missing the needs of an ageing population. Where assessment does exist, it is outcomes or 'tick-box' focused. Moreover, profitability and land use often take priority in the minds of developers and builders (unpublished report, Welsh Government, 2017). Designers and planners are not necessarily concerned with the integration of health, housing and social care, and they are not planning with a blank canvas but are often retrofitting and regenerating areas that have a long history of not suiting the physical needs of older people despite long and deep emotional attachments to the place that older people living there might have (Buffel, Phillipson and Scharp, 2012).

Rather than focusing on prescribed 'outcome' models of age-friendliness, Phillipson (2015) argues that future developments would benefit from evaluation of specific interventions. In addition, he suggests that radical intervention is required to construct age-friendly communities – developing new forms of citizenship which recognise and support changing needs across the life course. Moreover, he argues for an understanding of the dilemmas of the urban environment in terms of changing needs of the body (supporting autonomy, safety and inclusion), developing an age-friendly approach within the context of lifetime communities (accessibility of the built environment) and ensuring the engagement of older people in the planning and regeneration of their community.

'Mainstreaming' ageing issues within urban development represents a major challenge for social and public policy (Buffel et al., 2012)

Questions are raised (Golant, 2014) around whether the age-friendly community initiative has its own identity or is like other community-based initiatives (e.g., 'walkable communities'; see Golant, Chapter 15). A further problem arises in the question, 'Who to plan for?' That is, what kind of older people are age-friendly communities trying to reach, and how do we know what works? Should the initiative be looking at frail older people as well or just concentrating on preventative approaches for fit and active older people? The concept of age-friendly communities is framed around active or successful ageing (Lui *et al.*, 2009; Plouffe and Kalache, 2010), and while it is refreshing to see ageing and later life framed in more positive terms, it has meant that it can further stigmatize and exclude those who are not active and those who may be dependent. The needs of older people with dementia and the need to create vibrant spaces for them have been ignored in design and planning (Mitchell and Burton, 2006). Certain groups may also become linked with unlivable communities and cities that are stigmatized and go into decline. There is a complex balance of regeneration and gentrification alongside shifting inequality. There is a need to renew areas to avoid the creation of 'livable bubbles' or pockets that are isolated with few interconnections.

Planning and designing across generations or specifically for older people

Given the emerging trends in many Western countries including solo living, the need for dementia-supportive environments and the use of assistive technologies, the questions are whether we should provide a mixture of housing which includes integration (into mainstream housing) and segregation (such as gated retirement communities) and whether societies should follow a mainstreaming approach and engineer balanced communities.

There has also been a gap between the development of livability and age-friendly cities with the two initiatives developing in parallel without reference to each other. Several models exist to convey the livable city with the Phillips Centre for Health and Wellbeing (www.newcommunityparadigms.pbworks. com), stressing the importance of resilience (defined as incorporating flexibility and balance), inclusion (defined as equal opportunities to participate in the city), diversity and authenticity (such as identity/branding) in the construction of the city. The livability model of cities, however, addresses business and developers as well as investors and tourists with little focus on older inhabitants. Successful cities and urban areas are judged on the basis of their charisma and the extent to which they provide employment, not necessarily on inclusivity of an older population. Similarly, the debate has largely ignored ageing communities in suburban, rural and other locations where differential and unequal experiences of ageing have a significant impact on quality of life and where residents face unique challenges including depopulation, remoteness and reduced access to transportation links (Keating, Eales and Phillips, 2013).

Theoretical problems

Environmental gerontology has traditionally expressed the individual's relationship with their environment through the Person-Environment fit model (Lawton and Nahemow, 1973). Increasingly, it is recognised that older people have agency (both individually and collectively) to shape spaces and make places to suit their needs as they age over the life course. Environmental gerontology – and the attendant P-E (Person–Environment) Fit model – has focused on individual problems and technological solutions for the individual, following a medical model (such as aids and adaptations). Geographical gerontology has focused on the relational aspects (social and spatial) with greater emphasis on the collective as well as the diversity of *how* and *why* older people use space and place (often missing from the dominant P-E fit model).

Designers and planners have drawn on such deficit and individualistic models in their practice, seeing ageing as a problem to be solved by changes to the environment (and person). Seeing older people beyond the stereotypical care/medical model to older people as full citizens engaging in their environment (Andrews *et al.*, 2007) is critical in designing for the older population of the twenty-first century. Phillipson (2015) argues that we need to incorporate debates about social justice and rights to the city when developing new theoretical approaches and public policy responses. In addition, understanding peoples' relationship(s) with their community using ethnographic approaches is crucial. In his view, we have yet to develop a coherent theoretical model.

Micro-environmental problems

Designing urban environments has increasingly become a focus of research particularly across Europe through research programmes such as the United Kingdom's New Dynamics of Ageing (www.newdynamics.group.shef.ac.uk/showcase.html), Life Long Health and Well-Being (www.mrc.ac.uk/research/

initiatives/lifelong-health-wellbeing/), and projects such as ENABLE-AGE (www.psychologie.uni-heidelberg.de/ae/apa/research/enable.html), OPUS, I'DGO (www.idgo.ac.uk/), and Mobility, Mood and Place (https://sites.eca. ed.ac.uk/mmp/). Many of the projects under these programmes addressed the indoor and outdoor environments, public spaces, and green areas for older people through an interdisciplinary lens, including geriatric medicine, ergonomists, architects, planners and engineers, gerontologists, practitioners and policy makers. What many of the studies focusing on the environment illustrate is that how spaces are designed – public, private and liminal – is critical if the environment is to act as a facilitator and not a barrier to active ageing.

Drawing on two studies (OPUS and Go-FAR), a number of barriers, which posed safety concerns for older people, were identified in relation to the physical environment. These concerns included walking barriers such poor signage, badly maintained streets, clutter and trip hazards, confusing 'shared' spaces and poor paving, and they included 'sensory overload' factors such noise, traffic busyness, smell and complexity of the environment (Hammond and Musselwhite, 2013; Ormerod *et al.*, 2015; Phillips *et al.*, 2013). Driving conditions (poorly designed cars and signage, levels of pollution), as well as infrastructure (lack of toilets, poor lighting, dark alleyways) and public transport (insufficient routes, lack of punctuality, careless drivers), deterred older people in their use of the local environment. Additionally, negative attitudes towards older people were a barrier to going outdoors.

Measures to overcome environmental barriers, support older people and create safe and inclusive spaces that can make a difference to older peoples' lives can be achieved by small changes to the environment. Research from the previously mentioned studies showed that improvements in access along a journey including good pavements, many stopping points enabling social interaction, routes with interesting features, place distinctiveness, green spaces (Holland *et al.*, 2007), legible and easy to understand crossings (Dommes and Cavallo, 2011), places which have meaning and self-explanatory cues for drivers (Box, Gandolfi and Mitchell, 2010), bus stops with real-time information and clean, safe toilet facilities are important.

Such changes to the environment have demonstrated that they can lead to an increase in activity levels (Ståhl *et al.*, 2008). Adaptations and adjustments focusing on feedback from a range of older people, using smart technology (e.g., to enhance transport information systems and better navigation) and drawing on a range of research into age-friendly urban environments can lead to accessible and inclusive environments that benefit older people and stimulate economic development and regeneration (such as demonstrated in Manchester, United Kingdom).

Key designs of ageing communities emerging in urban spaces

At a European level, innovation in design of ageing communities has revolved around housing and urban design, ambient assisted living, urban environments, dementia-supportive environments, physical activity, age-friendly businesses and services, tourism, transport and the voices of older people (EU Age Innovation Platform, www.age-platform.eu). In 2015, 10 European cities were spotlighted for how well they were responding to the needs of an ageing population (ARUP, 2015). Key developments of note included innovations in the built environment, mobility and the digital environment. Other developments are taking place, for example, the shaping of dementia friendly communities (Brussels; Dobner, 2014), smart growth planning (Barcelona) and improved governance for collaborative action (Manchester). The success of planning and designing for an ageing population, however, will be its sustainability in meeting the changing needs of older people. The future city may look very different from today, with increasingly older populations (Buffel *et al.*, 2012) and new forms of vulnerability in the city such as heat waves, floods and the threat of terrorism.

With increased urbanisation and ageing populations, designers and planners must think and act differently in order to accommodate the diversity of needs and demands of older populations in urban contexts. The traditional concept of the city with concentric circles and radial roads and networks is no longer a predominant model. Boundaries are increasingly circumvented and discontinuous; more and more, spaces are segregated by vested interests and wealth, income and occupation (Royal Institute of British Architects - RIBA, 2013). The future city as a "kaleidoscopic mosaic of fragments and corridors" (Scholl, Dubbeling and Perić: UPAT, 2015: 134) with neighbourhoods separated by transport corridors and other infrastructure with little opportunity for citizen participation is a realistic future. Scholl *et al.* (2015) suggest that in such a context, new styles of management and design are required which are likely to be intense and require a new style of planning and planning tools. There is optimism however that the possibilities of new technologies and the mix of information-rich environments and age-friendly digital solutions will open the way for smart environments to meet a diverse range of needs of older people.

There is increasing recognition that planning for later life is not about the siting of residential homes or retirement communities that promotes segregated living for those over 75. It is also not about interpreting 'Ageing in Place' as staying put across the life course. Instead, planning is now focusing on 'active ageing'. Such planning is for people who are seen as consumers and in employment or volunteer work, whether that be for grandchildren, spouses or in the formal economy. Planning also is for sustainability, in terms of age (and energy), as well as the recognition of different and varied lifestyles in older age are key to future-proofing.

In the United Kingdom, the Royal Institute of British Architects (RIBA) have addressed the issues of a changing urban demographic landscape in a paper, *Silver Linings* (2013), which scopes out how ageing populations could shape cities in the United Kingdom by 2030, and addresses issues planners and architects should be considering. Anticipating the urban future was based on the following scenarios: more transient lifestyles and fewer fixed residences, the rise of consumers and increasing individualism. RIBA's vision incorporates a number of considerations for planning for an ageing population:

- Reinventing the family home (offering independence and co-dependence between third-agers and their families)
- Reclaiming the high street (acting as a catalyst for new public amenity, private enterprise and intergenerational exchange to complement existing retail, i.e., spaces for caring, learning, entrepreneurship and consumption)
- Rejuvenating coastal towns (providing opportunities for work and leisure to older people who still migrate in retirement to seaside towns)
- Reinventing spaces such as 'pop-up' spaces for older people (e.g., university 'pop-ups' sharing skills between generations, public buses taking on a different function as social mobile spaces; Handler, 2015)
- Reenergising the city through city networks of health hubs, connected by routes promoting exercise in public spaces, thereby offering opportunities to stay healthy and socialise while moving around the city.

Relevance to geographical gerontology

The ways in which space, place, time and distance relate to the experiences of older people are at the heart of geographical gerontology. Understanding ageing from a broad spatial and relational perspective offers a wider lens on the environment than the narrower focus of environmental gerontology, which may lack innovation and new theorising (Wahl and Weisman, 2003). However, there is a need to make the geographical/spatial more explicit and create an identity that is *geographical* rather than just environmental gerontology and urban sociology. Geography is an interdisciplinary-based subject, a mosaic with a conceptual *bricolage* and is necessary to reframe designing ageing communities from perspectives of time, space and place. How should we understand ageing communities within a geographical gerontology framework?

One of the understated elements in the geographical *bricolage* has been the emphasis on time. The importance of conceptualising time-space and life course space is crucial to our understanding of the relationship between person and environment yet we under-theorise time and its relationship to space and distance. Interdisciplinarity, new methods of collecting data such as geographical information systems (GIS) and the use of 'big data' provide means to remedy this with more nuanced approaches to time to be factored into design and planning.

Schwanen and Zeigler (2011) also argue for a more relational geographical approach to designing for older people because the built environment plays a role in the construction of social relationships – where people can meet and how they connect. The key question is how does this become routinely approached in the practices of planners? Can the design of buildings facilitate social connectivity and expand the relational qualities of designed spaces and their impact on everyday life?

One way to achieve this is to directly engage the community. A city's own sense of a livable community is key; it needs to be owned as a vision by its citizens, irrespective of age. Well-designed policies and environments can maintain structures and infrastructure that encourage engagement at individual and community level to create viable and involved communities. Additional data collection is required *directly from older adults and stakeholders within these communities* to understand particular issues and ensure inclusion and full access to spaces, structures and services and to tap into the potential that older people represent (Plouffe and Kalache, 2010). Specific processes and approaches need to be adopted to proactively identify and empower those older people who are seldom heard within the local community.

Empowerment is a key feature in a model of participative collaboration, respect and inclusion; it shifts the emphasis from place to people and their perspectives. Addressing inequalities as well as viewing older adults as social actors with 'rights to the city' is a crucial consideration (Handler, 2014).

Conceptual scope

The grain of the environment (i.e., the morphology of the community, the pattern of streets, the layout of building blocks, dimensions and densities) and the deconstruction of space (suburban and rural hinterland) that define the urban form of the built environment all have influence in the lives of older people. Similarly, the focus on different modes of transport, accessibility issues, proximity and social networks are all incorporated under the umbrella of geographical gerontology. It is this combination of physical and social features that distinguishes geographical gerontology and gives it the potential to develop transdisciplinary perspectives, concepts and methods.

Understanding optimal environments for older people requires an inter- and transdisciplinary approach (Phillipson, 2015) and conceptual breadth as it requires tackling issues of inequality, poverty, urban decay and renewal, as well as technological challenges. Yet the narrow focus on concepts such as 'ageing in place' and 'staying put' may be conceptually limiting to creative and innovative design and planning. Such planning must take into account growing inequalities and changing mobility and lifestyles of future cohorts of older people. Policy has embraced such concepts, enshrined them in guidance and yet in reality many people are forced to live in unsuitable environments as a consequence (Smith, 2009). All the preceding demonstrate new ways of thinking about city landscapes and offer creative ways of designing and planning for older people.

Methodological scope

Geographical gerontology opens new ways of capturing the dynamics of peoples' experience and the environment. There have been a number of measures assessing the impact of the environment on the well-being of older people on which to build interventions (e.g., Burton, Mitchell and Stride,

2011). Non-traditional methods such as the use of photovoice, social media and poetry as well as new modes of representation are emerging within the disciplines. Social media and new analytical approaches can help us understand our interaction with place; 'walkability' can be assessed using data from social media, photographs, GIS, glocal positioning system (GPS) and personalised technologies to look at the walking experience. Data from different kinds of venues has been a good predictor of street safety. Google Street View and social media are informing architecture and planning (Davies, 2015). There is also potential for new methodological exploration through qualitatively informed three-dimensional GIS (Kwan, 2004). Although many tools are available to assess livability, they fail to include the voices of older people themselves. Assessment is based on assumptions that older people have little agency, contribute little to the economy or can help solve their own problems.

In contrast to many tools for assessing livability, the UK Urban Ageing Consortium produced a research and evaluation framework for age-friendly cities based on the original WHO domains. Handler (2015), in a guest blog for the International Longevity Centre UK, reframed the landscape of age-friendly by calling on creative urban practitioners (architects, planners and designers) to engage in the urban ageing agenda (at a variety of scales including retrofitting neighbourhood areas) through community-led auditing, "temporary interventions, participative design processes, [and] speculative What ifs." In effect, "creating scenarios as a catalyst for re-imagining spaces," designing for flexibility and "making time and space for . . . mapping hidden experiences within the city, using these to inform designs and interventions which become context-specific" (Handler, 2015: 9).

The combination of the digital, empirical (observing how people experience place) and the use of 'big data' has the potential to provide us with exciting methodological avenues to design and plan a future age-friendly community. Such communities would provide more integrative services for older people incorporating housing, energy efficiency and service infrastructure.

Key gaps and opportunities for advancing the field

Sustainable urban development requires a better understanding of the space and time (spatio-temporal) characteristics of urbanization. Spatio-temporal sustainability (Schwanen and Kwan, 2012) has given us a hook but needs to go further. Different temporal contexts of design and planning offer the opportunities to develop at different levels: in terms of the individual, such as time-space and use (public space used during different times of day; e.g., Holland *et al.*, 2007); life course and intergenerational space, such as 'streets for life' (Mitchell and Burton, 2006); and political time frames given the potential for short termism which can constrain the creativity of planners and designers. The OECD (2015) highlights that policy strategies for actions in ageing societies need to: develop a long-term vision for their societies; develop indicators to measure ageing societies so that cities can understand where they are on the journey;

build consensus across generations; redesign the urban area to increase attractiveness, walkability and well-being; and consequently revitalise the economy. Underpinning this is the importance of governance for collaborative action involving new actors, ensuring policy synergies among levels of government and creating a horizontal governance structure to overcome policy silos. What geographical gerontology can bring is a much broader understanding of how these factors can integrate to produce action.

Conclusion

Geographical gerontology has potential as an area of study to develop new models of planning and design that accommodate the diversity of older people, as well as the diversity of cities, and to reflect the change that can occur in both over time. To progress, there is need to strip down the complexity of the city to manageable necessary design solutions. Collaboration, co-design and participation incorporating the voices of older people are crucial because place is inextricably tied up with human relationships. Yet planning the environment doesn't mean people use it as planned (e.g., social routing – people choose routes based on their social value rather than the most direct) or imbue place with similar meanings. To fully understand the interaction between place and person, there is a need to address the issue from an interdisciplinary perspective.

References

Andrews, G., Cutchin, M., McCracken, K., Phillips, D. and Wiles, J. (2007). Geographical gerontology: The constitution of a discipline. *Social Science & Medicine*, 65(1), 151–168. http://dx.doi.org/10.1016/j.socscimed.2007.02.047

Annear, M., Keeling, S., Wilkinson, T., Cushman, G., Gidlow, B. and Hopkins, H. (2014). Environmental influences on healthy and active ageing: A systematic review. *Ageing & Society*, 34(4), 590–622. http://dx.doi.org/10.101/0144686X1200116X

ARUP. (2015). *Shaping ageing cities: 10 European case studies.* Help Age International, Available from: www.dataplan.info/img_upload/5c84ed46aa0abfec4ac40610dde11285/shaping-ageing-cities_a4_web.pdf [accessed 9 July 2016].

Box, E., Gandolfi, J. and Mitchell, K. (2010). *Maintaining safe mobility for the ageing population: The role of the private car.* London: RAC Foundation. Available from: www.racfoundation. org/assets/rac_foundation/content/downloadables/maintaining%20safe%20mobility%20-%20rac%20foundation%20-%20140410%20-%20report.pdf

Buffel, T., Phillipson, C. and Scharf, T. (2012). Ageing in urban environments: Developing age-friendly cities. *Critical Social Policy*, 32(4), 597–617. http://dx.doi.org/10.1177/02610 18311430457

Burton, E., Mitchell, L. and Stride, C. (2011). Good places for ageing in place: Development of objective built environment measures for investigating links with older people's well-being. *BMC Public Health*, 11, 839–851. http://dx.doi.org/10.1186/1471-2458-11-839

Communities and Local Government. (2008). *Lifetime homes, lifetime neighbourhoods: A national strategy for housing in an ageing society.* London: Department of Health, Department for Works and Pensions. Available from: www.cpa.org.uk/cpa/lifetimehomes. pdf

Davies, J. (2015). The virtual and the physical. *NESTA*, web log post, 22 October 2015. Available from: www.nesta.org.uk/blog/virtual-and-physical [accessed 22 December 2016].

Day, R. (2008). Local environments and older people's health: Dimensions from a comparative qualitative study in Scotland. *Health & Place*, 14(2), 299–312. http://dx.doi.org/10.1016/j.healthplace.2007.07.001

Dobner, S. (2014). Bruges: A dementia-friendly city. *The Protocity*, 3 February 2014. Available from: http://theprotocity.com/bruges-dementia-friendly-city/

Dommes, A. and Cavallo, V. (2011). The role of perceptual, cognitive, and motor abilities in street-crossing decisions of young and older pedestrians. *Ophthalmic and Physiological Optics*, 31(3), 292–301. http://dx.doi.org/10.1111/j.1475-1313.2011.00835.x

Global City Indicators Facility. (2013). *GCIF policy snapshot No 2: Cities and ageing*, September. Available from: http://media.wix.com/ugd/672989_42ce3f3c825e15eb6e498bd601110a4e.pdf.

Golant, S. M. (2014). Age-friendly communities: Are we expecting too much? *IRPP Insight 5*, February, Institute for Research on Public Policy (IRPP), Montreal. Available from: http://irpp.org/wp-content/uploads/assets/research/faces-of-aging/age-friendly/golant-feb-2014.pdf.

Hammond, V. and Musselwhite, C. (2013). The attitudes, perceptions and concerns of pedestrians and vulnerable road users to shared space: A case study from the UK. *Journal of Urban Design*, 18(1), 78–97. http://dx.doi.org/10.1080/13574809.2012.739549

Handler, S. (2014). *An alternative age-friendly handbook*. Manchester: University of Manchester Library/UK Urban Ageing Consortium. Available from: http://hummedia.manchester.ac.uk/institutes/micra/Handbooks/Age-friendly%20Handbook_LARGE%20PRINT%20VERSION.pdf

Handler, S. (2015). The future of design: What makes a designer age-friendly? *International Longevity Centre – UK*, blog post, 15 November. Available from: http://blog.ilcuk.org.uk/2015/11/30/guest-blog-sophie-handler-chair-of-the-riba-working-group-on-research-and-ageing-the-future-of-design-what-makes-a-designer-age-friendly/ [accessed 12 February 2016].

HAPPI 3 (2016). *Making retirement living a positive choice*. All Party Parliamentary Group on Housing and Care for Older People. Retrieved from: www.housinglin.org.uk/HAPPI3 [accessed 9 July 2016].

Hockey, A., Phillips, J. and Walford, N. (2013). Planning for an ageing society: Voices from the planning profession. *Planning Practice and Research*, 28(5), 527–543. http://dx.doi.org/10.1080/02697459.2013.820039

Holland, C., Clark, A., Katz, J. and Peace, S. (2007). *Social interaction in urban public places*. York: Joseph Rowntree Foundation.

Keating, N., Eales, J. and Phillips, S. J. (2013). Age-friendly rural communities: Conceptualizing 'best-fit'. *The Canadian Journal on Aging*, 32(4), 319–332. http://dx.doi.org/10.1017/S0714980813000408

Kwan, M-P. (2004). GIS methods in time-geographic research: Geo computational and geo visualization of human activity patterns. *Geografiska Annaler: Series B, Human Geography*, 86(4), 267–280. http://dx.doi.org/10.1111/j.0435-3684.2004.00167.x

Lawton, M-P. and Nahemow, L. (1973). Ecology and the aging process, in C. Eisdorfer and M. P. Lawton (eds), *The psychology of adult development and aging* (pp. 619–674). Washington, DC: American Psychological Association.

Lui, C., Everingham, J., Warburton, J., Cuthill, M. and Bartlett, H. (2009). What makes a community age-friendly: A review of international literature. *Australasian Journal of Ageing*, 28(3), 116–121. http://dx.doi.org/10.1111/j.1741-6612.2009.00355.x

Michael, Y., Green, M. and Farquhar, S. (2006). Neighbourhood design and active ageing. *Health & Place*, 12(4), 734–740. http://dx.doi.org/10.1016/j.healthplace.2005.08.002

Mitchell, L. and Burton, E. (2006). Neighbourhoods for life: Designing dementia – friendly outdoor environments. *Quality in Ageing and Older Adults*, 7(1), 26–33. http://dx.doi.org/10.1108/14717794200600005

OECD. (2015). *Ageing in cities*. Paris: OECD Publishing. http://dx.doi.org/10.1787/9789264231160-en [Accessed 16 July 2016].

Ormerod, M., Newton, R. A., MacLennan, H. A., Faruk, M., Thies, S.B.A., Kenney, L.P.J., Howard, D. and Nester, C. J. (2015). Older people's experiences of using tactile paving. *Municipal Engineer*, 168(1), 3–10. http://dx.doi.org/10.1680/muen.14.00016

Phillips, J., Walford, N., Hockey, A., Lewis, M. and Foreman, N. (2013). Older people and outdoor environments: Pedestrian anxieties and barriers in the use of familiar and unfamiliar spaces. *Geoforum*, 47, 113–124. http://dx.doi.org/10.1016/j.geoforum.2013.04.002

Phillipson, C. (2011). Developing age friendly communities: New approaches to growing old in urban communities, in R. Settersten and J. Angel (eds), *Handbook of sociology of aging* (pp. 279–293). New York, Springer.

Phillipson, C. (2015). Developing age-friendly urban communities: Critical issues for public policy. *Public Policy and Aging Report*, 25(1), 4–8. http://dx.doi.org/10.1093/ppar/pru052

Plouffe, L. and Kalache, A. (2010). Towards global age-friendly cities: Determining urban features that promote active aging. *Journal of Urban Health*, 87(5), 733–739. http://dx.doi.org/10.1007/s11524-010-9466-0

Royal Institute of British Architects (RIBA). (2013). *Silver linings: The active third age and the city*. London: RIBA. Available from: www.architecture.com/files/ribaholdings/policyand internationalrelations/buildingfutures/projects/silverlinings.pdf

Scholl, B., Dubbeling, M. and Perić, A. (2015). *Urban Planning Advisory Team (UPAT) ten years of UPATS: Reflections and results*. Zurich: Hochschulverlag. Available from: http://isocarp.org/app/uploads/2014/05/upat_10sept15_final.pdf

Schwanen, T. and Kwan, M. (2012). Critical space-time geographies. *Environment and Planning A*, 44(9), 2043–2048. http://dx.doi.org/10.1068/a45437

Schwanen, T. and Zeigler, H. (2011). Wellbeing, independence and mobility: An introduction. *Ageing & Society*, 31(5), 719–733. http://dx.doi.org/10.1017/S0144686X10001467

Smith, A. (2009). *Ageing in urban neighbourhoods*. Bristol: Policy Press.

Ståhl, A., Carlsson, G., Hovbrandt, P. and Iwarsson, S. (2008). Let's go for a walk: Identification and prioritisation of accessibility and safety measures involving elderly people in a residential area. *European Journal of Ageing*, 5(3), 265–273. http://dx.doi.org/10.1007/s10433-008-0091-7

Sugiyama, T. and Ward Thompson, C. (2007). Outdoor Environments, activity and the well-being of older people: Conceptualising environmental support. *Environment and Planning A*, 39(8), 1943–1960. http://dx.doi.org/10.1068/a38226

Wahl, H. W. and Weisman, G. (2003). Environmental gerontology at the beginning of the new millennium: Reflections on its historical, empirical and theoretical development. *The Gerontologist*, 43(5), 616–627. http://dx.doi.org/10.1093/geront/43.5.616

Welsh Government (2017). *Our Housing Agenda: meeting the aspirations of older people*. Report by the Expert Group on Housing and an Ageing Population in Wales to Welsh Government. Available from: http://gov.wales/topics/housing-and-regeneration/housing-supply/expert-group-on-housing-an-ageing-population/?lang=en [accessed 21 September 2017].

World Health Organisation. (2007). *Global age-friendly cities: A guide*. Geneva: WHO. Available from: www.who.int/ageing/publications/Global_age_friendly_cities_Guide_English.pdf?ua=1

7 Environment and ageing

Keith Diaz Moore

Introduction

The concept of place is central to geography, with some referring to the domain as the study of places (e.g., Cresswell, 2013). Within gerontology, research on place has occurred over the past four decades from numerous disciplinary perspectives, including sociology (Gubrium, 1978), architecture (Howell, 1983), anthropology (Rubinstein, 1989) and geography (Rowles, 1978). Kendig observes, "The increasing emphasis on 'place' – within and across anthropological, psychological, and sociological theory – arguably reflects underlying strength rather than fragmentation of the environmental field" (Kendig, 2003: 612). This chapter begins with this shared belief – that place as a construct has great strength and the potential to unify disparate inquiries in regard to the person-place relationship of older adults. Starting with the roots of environmental gerontology – another field within gerontology that explores the person-place relationship, albeit from a different tradition than geographical gerontology (see also Andrews, Evans and Wiles, 2013), it suggests that a natural bridge between environmental and geographical gerontology is the powerful construct of place. Tracing the evolution of conceptual development in environmental gerontology, this chapter focuses on four critical propositions that underlie the concept of place from an environmental gerontology perspective.

Environmental gerontology

Many authors (Lawton, 1983; Peace *et al.*, 2007; Wahl, 2001) cite the psychologist Robert Kleemeier's 1959 chapter titled "Behavior and the organization of the bodily and the external environment" as the beginning of the rigorous investigation of ageing–environment issues. Bootstrapping his chapter together, Kleemeier began with the changes in the human body associated with ageing. In so doing, he identified concerns for obesity and inactivity, as well as cosmetic and other interventions we may now refer to as 'anti-ageing' techniques. Moving out from the body, Kleemeier focused on sensory loss and speculated how this might change engagement with the proximate environment. On this trajectory, it 'followed naturally' to consider the living space of the elderly person, which he suggested should be considered at four different scales: the house, the objects within the house, the community and the working place. The need to address the geography of ageing at scales ranging from the body to that of community or settlement foreshadows the observation made by Harper and Laws (1995) and reinforced by Skinner, Cloutier and Andrews regarding the possibilities for geographical gerontology.

Yet the progeny of Kleemeier is environmental gerontology which has its primary roots, not surprisingly, in environmental psychology. Leading names in the field such as M. Powell Lawton, Eva Kahana, Hans-Werner Wahl and Gerald Weisman bespeak this psychological influence. Wahl and Weisman define environmental gerontology as concerned with "the description, explanation, and modification or optimization of the relation between the elderly person and his or her environment" (Wahl and Weisman, 2003: 616). Thus, environmental gerontology is centered on the pursuit of scientific knowledge and directed toward improving environments for ageing. This desire to both understand as well as inform action-taking is central to the theoretical formulations found within the field.

Individual competence and environmental press: the ecological model of ageing

The theoretical cornerstone of environmental gerontology is Lawton and Nahemow's (1973) Ecological Model of Ageing (EMA) which posits that an individual's observable behavior and affect are "a function of the competence of the individual and the environmental press of the situation" (Lawton, 1982: 26). Hence the theory is often referred to as the 'competence-press' model (cf. Diaz Moore, 2005; Kendig, 2003). The theory suggests that individuals with greater competency are able to adapt successfully to a wider range of environmental press (or 'stress'), as compared to individuals with lower or diminished competency. One takeaway from the theory is that the less competent the individual, the greater the impact of environmental factors on that individual. This came to be known as the environmental docility hypothesis (Lawton, 1980). Empirical testing of the competence-press model has yielded mixed results (Lieberman and Tobin, 1983; Nahemow, 2000), though considerable empirical support for the environmental docility hypothesis has been found (cf. Morgan *et al.*, 1984; Pruchno, Burant and Peters, 1997). Given the model's breadth and construct imprecision, a lack of empirical support is perhaps predictable. Undoubtedly, though, the greatest utility of the EMA is conceptual, as it highlighted the recognition of the following:

1 Older adults are heterogeneous (here, in terms of competence);
2 Environments have differential impacts upon individuals and that different environments may have varying impacts; and
3 That those impacts may be positive or negative in nature.

Person–environment congruence or 'fit'

Building upon the recognition of the heterogeneity of both older adults as well as of environments, there have been efforts to theorize person–environment congruence or 'fit.' Perhaps the best known is that of Eva Kahana (1982) which suggests that behavior (in terms of well-being and 'adequate functioning') is influenced not only by individual and environmental characteristics but also by the degree of congruence, or 'goodness of fit', between the preferences of the individual and perceived attributes of the environment (e.g., safety, stimulation, interaction, degree of homogeneity). Importantly, a lack of congruence will eventually contribute to poor well-being unless 'adaptive strategies' are undertaken to increase the person–environment fit. Thus, the focus becomes the adaptation or negotiation between a thinking and perceiving person and an environment possessing perceivable attributes. Two dimensions of person–environment processes receiving the most recent attention are agency and belonging (Geboy, Diaz Moore and Smith, 2012; Wahl, 2016; Wahl, Iwarsson and Oswald, 2012).

Environmental attributes

The definition of the most salient environmental attributes for older adults has been a focus within environmental gerontology for certainly the past 25 years. Perhaps the leading environmental gerontologist in this area is Gerald Weisman, who first articulated a set of nine therapeutic goals for environments serving people experiencing dementia in the seminal book *Holding on to Home* (Cohen and Weisman, 1991). Throughout the 1990s, Weisman's work may be characterized as focused on enhancing clarity regarding the concept of environmental attributes. Drawing on his own and the work of other environmental gerontologists (Calkins, 1988; Lawton, Fulcomer and Kleban, 1984; Lyman, 1993; Windley and Scheidt, 1980; Zeisel, Hyde and Levkoff, 1994) with respect to qualities or characteristics of the environment, Weisman suggests that environments for ageing be conceptualized in terms of particular 'attributes of place experience,' for example, safety and security, awareness and orientation, support of functional abilities, regulation and quality of stimulation, personal control, privacy and social interaction. Such attributes are at the core of his "Integrative Model of Place" which suggests that such attributes emerge from transactions amongst four components of place: individual, organizational, social and physical contexts that influence experience (Calkins and Weisman, 1999; Weisman, 1997; Weisman, Chaudhury and Moore, 2000). Calkins and Weisman (1999) go so far as to offer an equation of place experience: $P = f(I, E_{sop}, (I \times E_{sop})$, where P = place experience, I = individual, E = environment, s = social environment, o = organizational environment, and p = physical environment.

The cultural turn and the concept of place

Simultaneously, environmental gerontology was increasingly influenced by theoretical critique largely arising from geography and anthropology. As an example, Rubinstein and de Medeiros critique person–environment (P-E) fit models thusly: "P-E fit can be altered by the elder's consciousness of the life world, by how the older person experiences the self, [and] by how the person individually interprets cultural meaning, and the importance of place in later life" (Rubinstein and de Medeiros, 2004: 64). For Rubinstein and de Medeiros, the environmental gerontology models rooted in the field's environmental psychology foundation overlook the important mediating role culture plays in human experience. They highlight two roles of culture in the P-E transaction: (1) as an originating frame of reference colored by assumptions about space, language, narration, and expectations of self and others and (2) as a mediating, or interpretive, lens for ongoing transactions that is shaped by an individual's past experiences, social status, and the like (Rubinstein and de Medeiros, 2004). Such a perspective suggests that environmental experience is socially constructed, informed by various cultural lenses. A core construct for researchers (e.g., Jaber Gubrium, 1975; Graham Rowles., 1983; Robert Rubinstein, 1989) having this perspective is the concept of place. Gubrium, in perhaps the first definition of place given in gerontology, states, "By place, I mean geographic locations ... that are taken for granted to have certain meanings on particular occasions when specific people are gathered there" (Gubrium, 1978: 28).

Wahl and Lang (2004) synthesized Weisman's systemic approach and this constructed view of the ageing-environment experience by positing their "Socio-Physical Place Over Time" (SPOT) theory. Therein, Wahl and Lang explicitly suggest that place should be viewed as encompassing three premises: (1) "behavior is embedded" in places, which "combine both a physical-spatial as well as social-cultural dimension"; (2) places are "socially-constructed ... socially shaped" physical environments; and (3) "places are dynamic and show both change and stability over time, as people age" (Wahl and Lang, 2004: 17–18). More recently, the Ecological Framework of Place (EFP) offers a definition of place that meets the three criteria of Wahl and Lang, suggesting that place is "a milieu involving people ('place participants'), the physical setting, and the programme of the place, all catalyzed by situated human activity and fully acknowledging that all four may change over time" (Diaz Moore, 2014: 184). In so doing, the EFP firmly asserts that place is a socio-physical construct, or in the felicitous shorthand of Canter, "places are shared aspects of experience" (Canter, 1986: 218). It also suggests that the $I \times E_{sop}$ interaction proposed by Calkins and Weisman may be understood as the 'program of a place' – "the inherent yet largely implicit socially shared understandings that enable effective co-action and forward the underlying, socio-temporally negotiated purpose of the place" (Diaz Moore, 2014: 186). Finally, the framework contends that not only is behavior embedded in places but also that places are actually catalysed through human co-activity. In so doing, from a geographical perspective, the EFP is perhaps aligned with that which views place as an experiential phenomenon dependent upon human beings acting as agents (e.g., Agnew; 2011; Sack, 1997; Tuan, 1977).

Propositions of place from an environmental gerontology perspective

What follows are four intrinsic propositions suggested by the emergent use of place as a concept in environmental gerontology: places are purposeful, places are catalysed by co-activity, places are bound by the physical setting, and place and self are reciprocal.

Proposition 1: places are purposeful

In an important response to a critique of the environmental docility hypothesis, Lawton (1989) explored the role proactivity has in place experience. In so doing, he highlighted that while competence is necessary for efficacy, the perception of efficacy is based on the comparison between the desired and the resulting outcome. Similarly, Kahana's congruence model recognizes that individuals may vary in terms of their desired environmental attributes (e.g., homogeneity vs. heterogeneity). Both environmental psychologists argue that people have needs and desires and that these drive the purpose with which a person transacts with a given environment.

However, physical settings, through the provision of resources, are also designed to facilitate co-action and thereby the underlying purpose(s) of the place. As Andrews and colleagues discuss it,

> because of people and technologies *in situ*, places possess basic agency – hospitals provide medicine, theatres entertain, neighborhoods host a wide range of human activity, and so on. Such agency can be helpful to older people, or act as a barrier or challenge.
>
> (Andrews *et al.*, 2013: 1344)

The 'programme', our social-shared expectations, also attempts to circumscribe and facilitate certain co-action as is discussed further later. It is therefore possible that these three purposes (those associated with people, programme and physical setting) may not facilitate 'fit' or congruence in either a functional or a perceived sense. An individual may seek to be engaged in an activity not facilitated by the properties of the physical setting and frowned on by the programme of the place (protests exemplify this premise). The physical setting may have been designed for one type of activity at one time but is now hosting a different activity that it may well constrain. This gives rise to the discussion of 'goodness of fit' in the discipline of architecture (Alexander, 1964). Finally, the socially shared understanding of the place may be in conflict with both what a certain group of people may want to do and with what the physical setting may enable.

Proposition 2: places are catalysed by co-activity

The geographer Robert Sack asserts, "Places constrain and enable our actions, and our actions construct and maintain places" (Sack, 1997: 13). In subsequent work, Sack (2001) states that in order to further projects (individual or collective), we create places – physical settings where human activities are, in theory, delimited by rules. This concept of rules as the cognitive schema by which we understand expected/anticipated activity, is also asserted by the environmental psychologist David Canter (1986) in suggesting that places are structured by socially-shared and understood place rules and roles. The EFP uses Barker's (1968) term of *programme* as a shorthand for this concept of shared place rules and roles and has been used in empirical research in environmental gerontology (Diaz Moore, 2005; Diaz Moore et al., 2006; Geboy, 2005).

An interesting advancement in this area is the suggestion by de Medeiros, Rubinstein and Doyle (2013) of the relevance of positioning theory (e.g., Harré and Moghaddam, 2003). This theory suggests that the notions of rules and roles are too static, and that positioning, which describes "the way rights and duties are taken up and laid down, ascribed and appropriated, refused and defended in the fine grain of the encounters of daily lives" (Harré, 2008: 29), captures more closely the dynamic of negotiating one's place. De Medieros and colleagues make two critical assertions in developing what they term 'environmental positioning':

1 "that the environment itself – the objects, the structures, even the location – all contribute to the positioning of the actors and observers (de Medeiros *et al.*, 2013: 88)" – and
2 that such positionings are related to the concept of self as a "culturally constituted individual" and through reference suggest that they perhaps constitute the "interpretive structures" (ibid.) of self, which Herzog and Markus define as "a multi-faceted, dynamic system of interpretive structures that mediates behavior" (Herzog and Markus, 1999: 228).

There may well be fertile ground in furthering a geographical gerontology through linkage to positioning theory, which, over the past decade, has found increasing relevance in fields such as public relations (James, 2014), organizational studies (Zelle, 2009) and gerontology itself (Allen and Wiles, 2013). There are also two additional propositions regarding place that are supported by this discussion: (1) that a place is bounded by the physical setting, and (2) that place and self are conceptually linked.

Proposition 3: physical setting as place boundary

While place is constituted of people, programme and the physical setting, the physical setting is the initial dimension by which we define place. This is made clear by Gubrium's initial definition of place shared earlier. Rapoport offers this useful observation as to why this is: "the situation, the rules, and the ongoing and appropriate behavior are communicated by cues in the setting" (Rapoport, 1990: 12). As such, settings are highly culture-specific, drawing on cultural assumptions as an originating frame of reference as suggested by Rubinstein and de Medeiros (2004). Underlying this entire model of place is the assumption of purposefulness that drives certain activities for which we, as geographical creatures, manipulate physical settings to facilitate (and perhaps which we design to constrain other activities we seek to thwart [e.g., crime]). It is important to note that, as Rapoport suggests, "activity systems take place in systems of settings" (Rapoport, 1990: 12). This proposition is consistent with the so-called relational turn, where places are understood as networked and as organizing social relations (Andrews *et al.*, 2013; Darling, 2009).

It is perhaps useful to note at this juncture that physical settings may be thought of as networked through movement (e.g., systems of activities running through systems of settings), but one should not overlook that places are also nested in terms of scale. Based on the work of Alexander, Ishikawa and Silverstein (1977), Habraken (1998) and Norberg-Schulz (1985), the physical setting may be conceptualized at the proximate (e.g., products, furniture), building/site, neighbourhood/community and settlement scales (Diaz Moore, 2014). If one views these scales as different levels of analysis, it is important to realize that the relationship between scales often proves critical to any specific place experience. For instance, a home may be designed in an exemplary fashion for accessibility, but it may become cluttered by possessions and may exist in a neighbourhood with no sidewalks and in a small town that is a food or service desert. Similarly, often of great impact are the differential rates of change between environmental scales, for example, we are likely to change the furniture in a home more often than we are apt to redesign a neighbourhood for instance.

Proposition 4: the reciprocity between place and self

A core concept in environmental psychology is 'place-identity,' which Proshansky defines as "those dimensions of self that define the individual's personal identity in relation to the physical environment by means of a complex pattern of conscious and unconscious ideas, beliefs, preferences, feelings, values, goals and behavioral tendencies and skills relevant to this environment" (Proshansky, 1978: 155). This reciprocity between place and self is consistent with the assumption that humans are social beings and define self in relation to the socio-physical world (Goffman, 1978; Mead, 1934). As we take action, we assess how that action is received and have it inform how we may proceed immediately. Yet equally important, we consider it within a long view sense of self. This reflexivity Mead (1934) views as constitutive of self, involving both the in-the-moment 'I' and the more stable, socially developed, longer-term 'Me.'

In the language of this chapter, place has aspects of people, programme and the physical setting, and that self is socially defined in relation to others (people); cultural assumptions, norms and values (program); and the physical setting in which co-action takes place. Our 'self' is informed by those places we have experienced previously and is influenced by the place in which we are participating, while simultaneously playing an essential role in that place being reconstituted through (as a result of) our participation.

Conclusion

Environmental gerontology as a field has evolved from its environmental psychology roots with its emphasis on processes, such as competence, stress, person-environment fit and the like to "viewing 'place'" as the key integrative construct in conceptualizing both the environments occupied by older persons and older persons' interactions with these environments" (Wahl and Weisman,

2003: 625). In order to serve this function, it is critical for the concept of place to be viewed as socially-constructed, as opposed to individually subjective. To quote the environmental psychologist Canter (1986: 218) once again, "places are shared aspects of experience" (Canter, 1986: 218). As noted, the critique leading toward this evolution has in no small measure come from the developments found within geographical gerontology wherein Wiles (2005) suggests place has these six characteristics:

1　Places are processes;
2　Places are subject to ongoing negotiation;
3　The many different experiences and contested interpretations of places (some of these may compete or conflict);
4　Power relations are expressed through, and shape, places;
5　Places are interrelated – to other places, at different scales, at different times; and
6　Places are simultaneously material/physical and symbolic.

The Ecological Framework of Place is but one heuristic that attempts to synthesize and accommodate developments within environmental gerontology but also responds well to the characteristics regarding the concept of place found in geographical gerontology. Place is a complex construct involving the dynamic interplay of structural and procedural aspects that result in differential outcomes for individuals, social groups and cultures. However one speaks of the domain, geographical or environmental gerontology, while originating from different historical and disciplinary roots, it may be that the concept of place and how it is central to the ageing experience is arguably the heart of the matter.

References

Agnew, J. (2011). Space and place, in J. Agnew and D. Livingstone (eds), *The SAGE handbook of geographical knowledge* (pp. 316–330). London: Sage.

Alexander, C. (1964). *Notes on the synthesis of form* (vol. 5). Cambridge, MA: Harvard University Press.

Alexander, C., Ishikawa, S. and Silverstein, M. (1977). *A pattern language: Towns, buildings, construction* (Vol. 2). New York: Oxford University Press.

Allen, R. E. and Wiles, J. L. (2013). The utility of positioning theory to the study of ageing: Examples from research with childless older people. *Journal of Aging Studies*, 27(2), 175–187. http://dx.doi.org/10.1016/j.jaging.2013.02.001

Andrews, G. J., Evans, J. and Wiles, J. (2013). Re-spacing and re-placing gerontology: Relationality and affect. *Aging & Society*, 33(8), 1339–1373. http://dx.doi.org/10.1017/S0144686X12000621

Barker, R. G. (1968). *Ecological psychology: Concepts and methods for studying the environment of human behavior*. Stanford, CA: Stanford University Press.

Calkins, M. P. (1988). *Design for dementia*. Baltimore, MD: National Health Pub.

Calkins, M. and Weisman, G. D. (1999). Models for environmental assessment, in B. Schwarz and R. Brent (eds), *Aging, autonomy and architecture: Advances in assisted living* (pp. 130–142). Baltimore, MD: Johns Hopkins University Press.

Canter, D. (1986). Putting situations in their place: Foundations for a bridge between social and environmental psychology, in A. Furnham (ed.), *Social behaviour in context* (pp. 208–239). London: Allyn and Bacon.

Cohen, U. and Weisman, G. D. (1991). *Holding on to home.* Baltimore, MD: Johns Hopkins University Press.

Cresswell, T. (2013). *Place: A short introduction.* New York: Wiley-Blackwell.

Darling, J. (2009). Thinking beyond place: The responsibilities of a relational spatial politics. *Geography Compass*, 3(5), 1938–1954. http://dx.doi.org/10.1111/j.1749-8198.2009262.x

De Medeiros, K., Rubinstein, R. and Doyle, P. (2013). A place of one's own: Reinterpreting the meaning of home among childless older women, in G. Rowles and M. Bernard (eds), *Environmental gerontology: Making meaningful places in old age* (pp. 79–102). New York: Springer Publishing.

Diaz Moore, K. (2005). Using place rules and affect to understand environmental fit: A theoretical exploration. *Environment and Behavior*, 37(3), 330–363. http://dx.doi.org/10.1177/001391650427657

Diaz Moore, K. (2014). An ecological framework of place: Situating environmental gerontology within a life course perspective. *The International Journal of Aging and Human Development*, 79(3), 183–209. http://dx.doi.org/10.2190/AG.79.3.a

Diaz Moore, K., Geboy, L. and Weisman, G. D. (2006). *Designing a better day: Guidelines for adult and dementia day services centers.* Baltimore, MD: Johns Hopkins University Press.

Geboy, L. (2005). *Architecture as a catalyst for organizational change: Facilitating a person-centered approach to care in an adult/dementia day center.* Ph.D. diss., University of Wisconsin, Milwaukee.

Geboy, L., Diaz Moore, K. and Smith, E. K. (2012). Environmental gerontology for the future: Community-based living for the third age. *Journal of Housing for the Elderly*, 26(1–3), 44–61.

Goffman, E. (1978). *The presentation of self in everyday life.* Harmondsworth: Penguin Books.

Gubrium, J. (1975). *Living and dying at Murray Manor.* New York: St. Martin's Press.

Gubrium, J. (1978). Notes on the social organization of senility. *Urban Life*, 7(1), 23–44. http://dx.doi.org/10.1177/089124167800700102

Habraken, J. (1998). *The structure of the ordinary: Form and control in the built environment.* Cambridge, MA: MIT Press.

Harper, S. and Laws, G. (1995). Rethinking the geography of ageing. *Progress in Human Geography*, 19(2), 199–221. http://dx.doi.org/10.1177/030913259501900203

Harré, R. (2008). Positioning theory. *Self-Care, Dependent Care & Nursing*, 16(1), 28–32.

Harré, R. and Moghaddam, F. M. (eds). (2003). *The self and others: Positioning individuals and groups in personal, political, and cultural contexts.* Westport, CT: Greenwood Publishing Group.

Herzog, A. R. and Markus, H. R. (1999). The self-concept in life span and aging research, in V. L. Bengtson and K. W. Schaie (eds), *Handbook of theories of aging* (pp. 227–252). New York: Springer Publishing.

Howell, S. C. (1983). The meaning of place in old age, in G. D. Rowles and R. J. Ohta (eds), *Aging and milieu: Environmental perspectives on growing old* (pp. 97–107). New York: Academic Press.

James, M. (2014). *Positioning theory and strategic communications: A new approach to public relations research and practice.* London: Routledge.

Kahana, E. (1982). A congruence model of person-environment interaction, in M. P. Lawton, P. Windley and T. Byerts (eds), *Aging and the environment: Theoretical approaches* (pp. 97–121). New York: Springer Publishing.

Kendig, H. (2003). Directions in environmental gerontology: A multidisciplinary field. *The Gerontologist*, 43(5), 611–614. http://dx.doi.org/10.1093/geront/43.5.611

Kleemeier, R. W. (1959). Behavior and the organization of the bodily and external environment, in J. E. Birren (ed.), *Handbook of aging and the individual* (pp. 400–451). Chicago: University of Chicago Press.

Lawton, M. P. (1980). *Environment and aging*. Albany: Center for the Study of Aging.

Lawton, M. P. (1982). Competence, environmental press, and the adaptation of older people, in M. P. Lawton, P. Windley and T. Byerts (eds), *Aging and the environment: Theoretical approaches* (pp. 33–59). New York: Springer Publishing.

Lawton, M. P. (1983). Environment and other determinants of well-being in older people. *The Gerontologist*, 23(4), 349–357. http://dx.doi.org/10.1093/geront/23.4.349

Lawton, M. P. (1989). Behavior-relevant ecological factors, in K. Schaie and K. Schooler (eds), *Social structure and aging: Psychological processes* (pp. 57–78). Hillsdale, NJ: LEA Publishers.

Lawton, M. P., Fulcomer, M. and Kleban, M. H. (1984). Architecture for the mentally impaired elderly. *Environment and Behavior*, 16(6), 730–757.

Lawton, M. P. and Nahemow, L. (1973). Ecology and the aging process, in C. Eisdorfer and M. P. Lawton (eds), *The psychology of adult development and aging* (pp. 619–674). Washington, DC: American Psychological Association.

Lieberman, M. A. and Tobin, S. S. (1983). *The experience of old age: Stress, coping, and survival.* New York: Basic Books.

Lyman, K. A. (1993). *Day in, day out with Alzheimer's: Stress in caregiving relationships.* Philadelphia, PA: Temple University Press.

Mead, G. H. (1934). *Mind, self and society.* Chicago: University of Chicago Press.

Morgan, T., Hansson, R., Indart, M., Austin, D., Crutcher, M., Hampton, P., Oppegard, K. and O'Daffer, V. (1984). Old age and environmental docility: The roles of health, support and personality. *Journal of Gerontology*, 39, 240–242. http://dx.doi.org/10.1093/geronj/39.2.240

Nahemow, L. (2000). The ecological theory of aging: Powell Lawton's legacy, in R. Rubinstein, M. Moss and M. Kleban (eds), *The many dimensions of aging*. New York: Springer Publishing.

Norberg-Schulz, C. (1985). *The concept of dwelling: On the way to figurative dwelling.* New York: Rizzoli International.

Peace, S., Wahl, H-W., Mollenkopf, H. and Oswald, F. (2007). Environment and ageing, in J. Bond, S. Peace., F. Dittmann-Kohli and G. Westerhof (eds), *Ageing in Society* (3rd edn, pp. 209–234). London: Sage Publications.

Proshansky, H. M. (1978). The city and self-identity. *Environment and Behavior*, 10(2), 147–169.

Pruchno, R., Burant, C. and Peters, N. (1997). Typologies of caregiving families: Family congruence and individual well-being. *The Gerontologist*, 37, 157–167. http://dx.doi.org/10.1093/geront/37.2.157

Rapoport, A. (1990). Systems of activities and systems of settings, in S. Kent (ed.), *Domestic architecture and the use of space: An interdisciplinary cross-cultural study* (pp. 9–20). Cambridge: Cambridge University Press.

Rowles, G. D. (1978). *Prisoners of space? Exploring the geographical experience of older people.* Boulder, CO: Westview Press.

Rowles, G. D. (1983). Geographical dimensions of social support in rural Appalachia, in G. D. Rowles and R. Ohta (eds), *Aging and Milieu: Environmental perspectives on growing old*, (pp. 111–130). New York: Academic Press.

Rubinstein, R. (1989). The home environments of older people: A description of the psychosocial processes linking person to place. *Journal of Gerontology*, 44(2), S45–S53. http://dx.doi.org/10.1093/geronj/44.2.545

Rubinstein, R. and de Medeiros, K. (2004). Ecology and the aging self. *Annual Review of Gerontology and Ggeriatrics*, 23, 59–84.

Sack, R. D. (1997). *Homo geographicus: A framework for action, awareness, and moral concern*. Baltimore, MD: Johns Hopkins University Press.

Sack, R. D. (2001). The geographic problematic: Empirical issues. *Norsk Geografisk Tidsskrift (Norwegian Journal of Geography)*, 55(3), 107–116. http://dx.doi.org/10.1080/002919501753129862

Skinner, M. W., Cloutier, D. and Andrews, G. J. (2015). Geographies of ageing: Progress and possibilities after two decades of change. *Progress in Human Geography*, 39(6), 776–799. http://dx.doi.org/10.1177/0309132414558444

Tuan, Y-F. (1977). *Space and place: The perspective of experience*. Minneapolis, MN: University of Minnesota Press.

Wahl, H. W., Iwarsson, S. and Oswald, F. (2012). Aging well and the environment: Toward an integrative model and research agenda for the future. *The Gerontologist*, 52(3), 306–316. http://dx.doi.org/10.1093/geront/gnr154

Wahl, H-W. (2001). Environmental influences on aging and behavior, in J. E. Birren and K. W. Schaie (eds), *Handbook of the psychology of aging* (5th edn., pp. 215–237). San Diego, CA: Academic Press.

Wahl, H-W. (2016). Theories of environmental influences on aging and behavior, in N. Pachana (ed.), *Encyclopedia of geropsychology* (pp. 1–8). Singapore: Springer. http://dx.doi.org/10.1007/978-981-287-080-3_132-1

Wahl, H-W. and Lang, F. (2004). Aging in context across the adult life course: Integrating physical and social environmental research perspectives. *Annual Review of* Gerontology *and Geriatrics*, 23, 1–33.

Wahl, H-W. and Weisman, G. D. (2003). Environmental gerontology at the beginning of the new millennium: Reflections on its historical, empirical and theoretical development. The *Gerontologist*, 43(5), 616–627. http://dx.doi.org/10.1093/geront/43.5.616

Weisman, G. D. (1997). Environments for older persons with cognitive impairments, in G. T. Moore and R. W. Marans (eds), *Toward the integration of theory, methods, research, and utilization* (pp. 315–346). New York: Springer.

Weisman, G. D., Chaudhury, H. and Moore, K. D. (2000). Theory and practice of place: Toward an integrative model, in R. L. Rubinstein, M. Moss and M. H. Kleban (eds), *The many dimensions of aging* (pp. 3–21). New York: Springer.

Wiles, J. (2005). Conceptualizing place in the care of older people: The contributions of geographical gerontology. *Journal of Clinical Nursing*, 14(S2), 100–108. http://dx.doi.org/10.1111/j.1365-2702.2005.01281.x

Windley, P. G. and Scheidt, R. J. (1980). Person – environment dialectics: Implications for competent functioning in old age, in L. W. Poon (ed.), *Aging in the 1980s: Psychological issues* (pp. 407–423). Washington, DC: American Psychological Association.

Zeisel, J., Hyde, J. and Levkoff, S. (1994). Best practices: An Environment-Behavior (EB) model for Alzheimer special care units. *American Journal of Alzheimer's Disease Care and Related Disorders Research*, 9(2), 4–21. http://dx.doi.org/10.1177/153331759400900202

Zelle, G. (2009). *Exploring the application of positioning theory to the analysis of organisational change*. Adelaide: Australian and New Zealand Academy of Management Conference.

Part III

Geographical scales of inquiry

8　Global ageing

David R. Phillips and Zhixin Feng

Introduction

Populations are growing older in nearly all countries globally. Demographic ageing is a consequence of the combined influences of increased longevity, decreasing mortality, migration and, sometimes, sharply declining fertility. The pace of demographic ageing often reflects, to some extent, trends in socio-economic development plus social conditions, especially education and aspirations in different areas, countries and regions. Most populations have been ageing demographically for over a century, especially in the more-developed countries, whilst the process of population ageing began relatively more recently in most less-developed countries (Kinsella and Phillips, 2005). Today, as the *World Report on Ageing and Health* notes, "for the first time in history, most people can expect to live into their 60s and beyond" (World Health Organization (WHO), 2015: 3). This chapter focuses on four aspects of global ageing of importance to geographical gerontology. First, it outlines the patterns and diversity of global ageing. Second, it covers impacts on older people in emergency situations across the world. Third, it lays out global climate change issues and health of older people. Fourth, the chapter discusses effects of selected contemporary socio-technological trends, such as increased mobility and interconnectedness for older people.

Global ageing and global variations

Global ageing patterns and diversity

From a geographical gerontology perspective, it is clear that ageing of populations around the world is an established trend although the patterns of ageing are varied. It is conventional to look at the current and projected proportions (usually in percentages) of persons aged 60+ (for most UN population data) or, increasingly, 65+. These data show distinct global variations and patterns and very interesting probable future patterns. However, absolute numbers of older persons may sometimes be more illuminating

For example, data in Figure 8.1 from 2015 (United Nations DESA Population Division: UNDESA, 2017a) show that only Japan has more than 30 per

Figure 8.1 Percentage of population aged 60+ in 2015

Source: authors' calculations, Probabilistic Population Projections based on data in UNDESA (2017a) *World Population Prospects: The 2017 Revision World Population Prospects: The 2017 Revision* (https://esa.un.org/unpd/wpp/Download/Probabilistic/Population/).

cent of its population aged 60+ years. In comparison, we see almost all African nations, most countries in South and South-East Asia, and Central America have under 10 per cent of population in this age group. Most European countries, including Australasia, the United States and Canada, fall in the 20–24 per cent category, with some European countries in the 25–29 per cent category.

However, as the World Report on Ageing and Health (World Health Organization (WHO), 2015) notes, we must be careful because percentage data can be misleading. The data in Figure 8.1 could suggest population ageing is currently less of an issue in Sub-Saharan Africa and China than in, say, Europe. Yet UNDESA (2017a) regional data show that, by weight of total population, Africa as a continent had 65 million people aged 60+ in 2014, more in this age group than either Northern Europe's 24 million or Western Europe's 50 million. Elsewhere, China alone, with 210 million persons aged 60+ in 2014, already had almost double the 114 million in all countries in Northern, Western and Southern Europe combined. This was also fewer than India, which tends to be considered a 'young' country but which had 117 million older persons. Even if Eastern Europe (which includes the Russian Federation) is considered, the European region as a whole had only 177 million people aged 60+, still fewer than China.

Looking forward to 2050, based on the same UNDESA data set, projections point to an even starker picture in numbers. From a global figure of some 901 million persons aged 60+ in 2015, we see a likely total of approximately 2 billion or slightly more than double the 2015 number in a period of just 35 years (which is only half the average life span!). But more important is the geographical change in distribution of that population. Figure 8.2 (UNDESA, 2017a) indicates that countries everywhere will have higher percentages in the 60+ age group with the main percentage increases likely to be in Latin America, East and South-East Asia and countries in Africa. Again, percentages hide very important numbers of older persons. By 2050, of the 2 billion people age 60+, 1.67 billion will be in what are still termed "less developed" regions and only 0.42 billion will be in "more developed" regions (UNDESA, 2017a). This represents a huge regional shift in global ageing and demands far more refined international analysis of ageing in countries according to differentiation in income categories (high-/middle-/lower-middle-/ low-income countries). Nevertheless, it is starkly clear that the spatial focus of global ageing in coming decades will be moving rapidly from what was sometimes called 'north' to 'south'. The differential capacities of countries, societies and families to cope with the increases in older persons is now a major, and growing, focus.

Demographic drivers of ageing

How have these global changes come about and what lies behind them? Sheer population numbers are at the base, but population dynamics then take control. Many people tend to think that increased longevity, people living longer, is the main reason populations age, and this has a grain of truth. However, "the most prominent factor in population ageing has been fertility decline" which together

Figure 8.2 Percentage of population aged 60+ in 2050

Source: authors' calculations, Probabilistic Population Projections based on data in UNDESA (2017a) *World Population Prospects: The 2017 Revision World Population Prospects: The 2017 Revision* (https://esa.un.org/unpd/wpp/Download/Probabilistic/Population/).

with increased life expectancy are the *demographic drivers* of population ageing (Kinsella and Phillips, 2005: 8). Increased longevity is determined to some extent by socio-economic development, social conditions, education, nutrition and improvements in health care. However, in some cases, of which China is the prominent example, drastic fertility decline is also largely a consequence of population policy, particularly China's one-child policy (Phillips and Feng, 2015). Whilst it is likely that, in the future, declining mortality will play an increasingly important role in ageing, today we are still witnessing principally the results of population policy combined with very rapid economic development, especially since 1980. A further factor influencing population age distributions is differential migration. Often, younger populations move away from their home areas, leaving behind fewer migratory persons, often the older groups. The current different economic opportunities in many parts of the world suggest that migratory flows will become increasingly influential in demographic ageing in some regions and countries. Moreover, some currently older countries (for example, Germany and Japan) are seeking younger immigrants as labour to help counterbalance their ageing populations and workforces.

Globally, total fertility rates (TFRs, numbers of children born per woman) have fallen by half from 5.0 to 2.5 over the half-century from 1965 to 2015 (UNDESA, 2017b). However, over this period, the declines had a greater impact in richer countries, where TFRs were already slowing and had often reached lows well below natural replacement fertility (2.1 children). Moreover, many countries' TFRs have stabilised, over the last decade or so, at between 1.3 and 1.8. Changing immigration and ethnic patterns may slightly increase TFRs in the more-developed regions in the next few decades, but such changes are unlikely to be substantial. The main future fertility reductions are likely to be in the currently less-developed regions (excluding China). Because these countries have the highest population growth rates and often the largest populations, continuing declines in TFRs will considerably increase their proportions of older populations and lead to the large numbers aged 60+ by 2050. As outlined earlier, these effects possibly will be exacerbated by outmigration of younger people from some. A brief consideration of Figure 8.3 (UNDESA, 2017b) will lead to the conclusion that, if TFRs continue to fall in most of sub-Saharan Africa, Western Asia and parts of Latin America, as well as a few large countries in the Asia-Pacific region, such as the Philippines and Indonesia, it will be in these areas where future demographic ageing will be greatest.

Older persons in emergency situations

The preceding data may be fascinating in themselves, but what are the human realities globally in ageing? This chapter takes a necessarily selective view of some important issues for older persons, families and communities. First, disasters and emergencies seem ever more frequent and often well reported in the media. As discussed in the following, emergency situations can arise from natural events (e.g., earthquakes, tsunamis), human-induced circumstances (e.g., armed

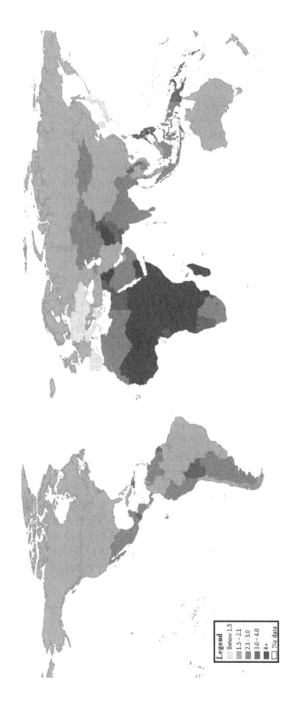

Figure 8.3 Total fertility rates between 2010 and 2015

Source: UNDESA (2017b) *World Population Prospects: The 2017 Revision* (https://esa.un.org/unpd/wpp/Download/Standard/Fertility/).

conflict or war) or combinations (e.g., flooding, fires, famine, drought caused by combined impacts of human activities and – potentially – climate changes). These situations raise serious threats to human security, health and well-being, and apart from direct deaths, crises generally increase the risk of disease and the strain on social services (WHO, 2008). During and after emergency situations, older persons are a particular population group at heightened risk (McCracken and Phillips, 2017). They may have multiple morbidities, disabilities, diminished sensory awareness and social and economic limitations which could impair their ability to respond to and adapt during emergencies (Centers for Disease Control and Prevention, 2012). They also may not be able to escape from emergency situations because of mobility limitations or a lack of physical strength. Many can become targets of reprisals and violence in conflict zones, and some may lack food or transport during natural or induced disasters such as floods, droughts or famines. Significantly, social networks can disintegrate, and families and communities can become displaced, leaving older people isolated and without family or local support (Global Action on Aging, n.d.). After the Nepal earthquake in 2015, for example, many residents in the affected areas were older people and children because working-age populations had often left to find work outside of Nepal. Food and water were also scarce, and rain further hampered the aid effort (Age International, 2015; Howgate, 2015). In these situations, older 'left-behind' people are often rendered extremely vulnerable facing the challenges of surviving and rebuilding.

Older people are also recognised as a vulnerable group in emergency situations both because of their living environments and because they may be less able or well-suited to adapt and tend to be reliant on resources locally available (Robert and Li, 2001). They are sometimes termed 'invisible' to humanitarian agencies because of a lack of age-specific data in emergency situations, a deficiency the WHO (2015) calls to be addressed. With few humanitarian programmes specifically tailored to older people's needs, older people are consulted only minimally in the planning and execution of humanitarian operations. Moreover, their capacities to assist actively in response and recovery is overlooked (HelpAge International: HAI, 2012).

The effects of emergencies on older people vary in severity and duration depending on specific characteristics of both the population and the situation itself (WHO, 2008). Health-related problems may be linked to pre-existing physical illnesses and disabilities (e.g., limited mobility), and/or the emergence of new health problems associated with hazards and environmental barriers, the loss of assistive aids and social support and an impaired access to health services (WHO, 2008). Poorly adapted shelter conditions aggravated health problems of older people after the 1995 Kobe earthquake, the Jamaican hurricanes in 2004–2005; (WHO, 2008) and the 2011 Fukushima tsunami and subsequent nuclear accident. Shelter conditions also hindered access to medications for many people (McCracken and Phillips, 2017). More positively, some studies note the resilience among older persons and that emotional stress effects may be short-term. Older people who have had better access to services and shelter could also experience fewer health problems, as seen in the Sumatra-Andaman earthquake and tsunami in 2004 (WHO, 2008).

The situation for older people who escape from emergency situations and become migrants or refugees in other areas or countries is sometimes no better. Older people may not have access to nor be eligible for public health care; they may have difficulties of knowledge, language and social isolation impeding their access to needed services. They may be less able than younger groups to adapt to a new environment which could create psychological pressure or even result in psychological trauma as well as other physical health problems (Global Action on Aging, n.d.; McCracken and Phillips, 2017). From 2013 onward in Ukraine's conflict, nearly 1.5 million people have fled their homes. Of this number, 60 per cent were older persons (Age International, 2013a). Gillam (2015) reports many older people were reluctant to leave their homes in the war-torn Donbas region. Some were in very poor psychological condition because of shelling and displacement; some had physical injuries and were in very poor health. Many lacked medicines and could not afford to buy more. Others were isolated, without friends or relatives and there were few care workers from aid agencies or charities to assist them. In the unfolding Syrian refugee crisis, many older persons were unable to make or afford the dangerous crossing to Europe; therefore, many fled from Syria to Lebanon (Age International, 2013b). Reports suggest many older Syrian refugees do not receive the help they need, with 61 per cent of older refugees feeling anxious and 10 per cent physically unable to leave their homes; 87 per cent were unable to regularly afford their required medication, and almost all (98.5 per cent) struggled to see a doctor. Psychological pressure was seen among disabled or those unable to perform activities of daily living. Social exclusion has been considerable, with many feeling a burden on their families, feeling guilty that they were unable to help with chores or to provide financially for their family (Age International, 2013b).

In summary, as the WHO (2008, 2015) notes, older persons are often more vulnerable than other ages in emergencies and crises, especially humanitarian crises. Some excellent charities and non-governmental organisations (NGOs) are working globally to redress the relative neglect of older people in such situations; Age International and HelpAge International (HAI) are notable. In a strategy for disaster risk reduction in an ageing world, HAI, like the WHO, calls for gender and age differentiated data. It further suggests stronger codification in international law of human rights-based protection in emergencies and should be included in the UN's *Post-2015 Sendai Framework for Disaster Risk Reduction* (United Nations Office for Disaster Risk Reduction UNISDR, 2015).

Global change: climate change, environment and health

Other global environmental issues are impacting older persons, regardless of whether in very old countries or relatively younger countries. In particular, there is growing interest in potential and actual climate and climate change, as it may affect older people. With regard to vulnerable populations, a 'high confidence' key finding of the 2016 report on the impact of climate change on human health among vulnerable populations in the United States was that impacts vary with age and life stage. "People experience different inherent sensitivities to the impacts of climate change at different ages and life stages. For example, the very young and the very old are particularly sensitive to climate-related health impacts" (Crimmins *et al.*, 2016: 248).

It is already well recognised that there is a strong relationship between climate and health, with effects often being greater or exaggerated among older persons who may have chronic health problems and impaired thermoregulation (Chau, Wong and Woo, 2013; WHO, 2015). Significant change in global climate systems is now widely accepted (McCracken and Phillips, 2016). Specifically, atmospheric and oceanic warming, ice sheets and glaciers shrinking, sea levels rising and ever-more extreme weather events are recognised (Hijioka *et al.*, 2014; Intergovernmental Panel on Climate Change (IPCC), 2014; United Nations Framework Convention on Climate Change (UNFCC, 2015).

Crimmins *et al.* (2016: 257–258) summarise older persons as potentially vulnerable to climate-related health impacts in a number of ways:

Vulnerability to Climate-Related Health Stressors

- Extreme heat events
- Other weather extremes
- Degraded air quality (exacerbates Chronic Obstructive Pulmonary Disease (COPD), lung function, heart attacks, diabetes)
- Vector-borne and waterborne diseases (especially with compromised reduced immunity)

Interactions with Non-Climate Stressors

- Vulnerability related to locations and condition of the built environment
- Vulnerability related to physiological factors
- Vulnerability related to disabilities including dementias

Globally, the nature and consequences of possible changes and local impacts can differ considerably according to location, topography and economic resources to deal with weather consequences. Some areas, for example, will become warmer and sometimes extremely hot, so people will be at risk from the extreme heat and from extended range of some conditions (e.g., malaria and other infectious conditions). Some other places will likely experience more extreme winters as well as summers. Sea-level rise will affect different areas and in some countries (e.g., Bangladesh and the Netherlands) could have very serious consequences for low-lying, densely populated areas. Precipitation will increase in some areas, while other areas will become drier, increasing risks from drought and the consequences of prolonged water shortages. Climate change and its impacts will vary globally, and the implications for geographical gerontology will also vary considerably.

As in emergencies, older persons' risks from climate change have physiological, psychological and social components, and some people may be exposed to multiple vulnerabilities from the overlapping of events. For example, in exceptionally hot or cold periods, there may be storms, floods or droughts, exposing people to the direct health impacts of weather effects and also consequential risk to life and property. Older persons may be least well equipped physically, psychologically and economically to mitigate the effects of change or damage to their home environments, or to escape from danger. The World Health Organization (2008, 2015) has highlighted older persons' immediate and consequent vulnerabilities in emergency situations, many of which may follow from extreme weather events. Examples include local and mass migrations from exceptional flooding in numerous countries and, in 2015–16, in several Horn of Africa areas from extreme drought conditions. One of the strongest recorded super typhoons seen (Haiyan/Yolanda) devastated much of the Central Visayas and destroyed the city of Tacloban in the Philippines in 2013. Of the older people who survived, HelpAge International (2015) reported that 68 per cent said their health had been affected.

Nevertheless, is important to recognise the global and local heterogeneity of older persons and avoid stereotyping all as uniformly vulnerable (Crimmins *et al.*, 2016). As McCracken and Phillips (2016) point out, there are the obvious distinctions by, say, age group ('young-old' versus 'oldest-old', e.g., those aged 60–70 compared with 80+), gender, location (rural/urban; coastal/inland), finances (rich vs. poor), household structure and support (living alone/living with others) and health status (fit vs. frail). But these categories hide a multitude of more subtle differences in terms of risk and ability to respond and cope. Very important, community resources and response readiness may be huge mitigating factors that will improve older persons' chances of surviving and even managing well.

Looking at specific risk factors from weather and climate, physiological vulnerability to heat extremes is important because older people are more sensitive to high or low temperatures and more prone to heat stress (hyperthermia and hypothermia) than younger people. With ageing, the efficiency of the body's self-regulating temperature mechanism declines and, during heat waves, may possibly lead to heat exhaustion or heat stroke. As this is an evolving physiological and psychosocial research area, many statements on impacts of climate change and extreme weather events are rather general or speculative at the moment. However, a notable and researched event was the extreme heat in Europe in August 2003, which caused more than 70,000 excess deaths, mainly among older persons (Robine *et al.*, 2003; WHO, 2013). Cardiovascular and respiratory disease were particularly important as underlying causes of death. In France, deaths of people aged 65+ and living alone were associated with lack of mobility; pre-existing medical conditions were also major risk factors. These were often exacerbated by environmental characteristics and especially living conditions, such as poor thermal insulation in homes, sleeping on the top floor and building temperature (Vandentorren *et al.*, 2006). Such risk factors were confirmed by Poumadère *et al.* (2005), who noted that people aged 75+ made up almost 83 per cent of the excess mortality in that heat wave. Many older people lived alone and in poor economic conditions. They were less able to recognise the risks and, even if they did, were unable to take even simple mitigating actions such as dressing lightly and using cooling devices. Identification of information and advice on specific actions to take in heat waves and heightened official recognition of risk are crucial (Vandentorren *et al.*, 2006). Poumadère *et al.* note that dangerous climate and excess heat events pose an "unambiguous danger" (Poumadère *et al.*, 2005: 1492). Yet how much officials have learned from and can react to such events and assist elderly populations is far from clear. The fear is that if a wealthy country with good administration such as France can be caught unprepared, then poorer countries in which extreme weather events are even more common and resources far fewer will be unable to cope.

Many older people are at extra risk from pre-existing chronic medical conditions and multi-morbidities can increase vulnerability to heat stress (heart disease, hypertension, diabetes, kidney disease). Taking some medications (e.g., beta blockers, diuretics) may affect the body's ability to regulate temperature. Physical mobility, vision or hearing, which often deteriorate with age, can also severely compromise the ability to respond to severe rapid-onset climate dangers such as tropical cyclones and flooding (e.g., by evacuating to safer locations). Very important, older persons with any forms of common mental problems (depression, paranoia) or dementias may be unable to recognise, much less respond to, excess heat or cold. Dementia prevalence approximately doubles every five years after age 65, and major increases are predicted for middle- and lower-income countries, making this a global issue in geographical gerontology.

So, physical and psychological health may be affected directly and indirectly by weather events and climate change and by individual's abilities to cope and adjust. Increasing frequency of intense and longer-lasting extreme weather events will probably have the most direct and immediate effects on morbidity and mortality rates. However, in the longer term, McCracken and Phillips (2016) suggest the greater health burden of climate change may actually come from indirect effects such as changes to food production and nutrition. Indeed, Springmann *et al.* suggest "the health effects of climate change from changes in dietary and weight-related risk factors could be substantial, and exceed other climate-related health impacts that have been estimated" (Springmann *et al.*, 2016: 1938). For example, we may see inundation from rising sea levels, with salinization of coastal land and aquifers; forced mass migration, evident in Somalia and Ethiopia in 2015–16 because of droughts; and movement to slums, refugee camps or to marginal or risky locations without health and welfare services. We know older persons are at greater risk in these circumstances. Zones for malaria, dengue, Zika and other infectious diseases are often expanding or modifying through climate change, meaning some areas may become at greater or lesser risk. Malaria, for example, has been sophisticatedly modelled with respect to potential climate change, and it is clear that socio-economic factors and the global eradication efforts will be very important influences on future risk (Caminade *et al.*, 2014). Air pollution associated with climate change and population concentrations is also increasingly seen a high risk factor, especially for older persons with compromised lung capacity and immune systems (WHO, 2015). Environmental gerontology reviews in Europe and Latin America have highlighted climate change and especially residential environments for older persons as key potential influences on health and well-being (Sánchez-González and Rodríguez-Rodríguez, 2016).

Overall, it is apparent that social and economic resources will strongly affect how well individuals and families can cope with climate change and weather-related events. Those with good economic resources and strong family networks are likely to do relatively well. However, many are poor, have limited education and often live alone or with another older person and so have limited coping capacity. Social isolation and exclusion, which can be a global issue and often exacerbated by economic recession (Phillipson, 2012), therefore become major issues for individuals and for policy-makers (Scharf and Keating, 2012).

All population age groups will be affected by these changes but, as McCracken and Phillips (2016) note, age-based physiological and psychological sensitivities can be exacerbated. A mental health burden can therefore be expected (Page and Howard, 2010; Doherty and Clayton, 2011) with older persons who have pre-existing mental disorders most likely affected. This burden can also occur *within* age groups, for example, older women tend to be more physiologically sensitive than men to high temperatures. Risk of mental and physical disorders can be worsened as older women in some societies may be more socially and economically disadvantaged than their male counterparts. Moreover, older females, especially in poorer countries, tend to have the lowest education levels and fewest economic resources, becoming dependent on the support of their families and local communities. Such gender differences can be important as the 'feminization' of old age – women's greater average longevity – is well recognised (Mujahid, 2006; WHO, 2015), even if the gender gap is generally narrowing.

Older people, technology and interconnectedness

Access to, and use of, technology and related activities is a key topic of concern in global ageing, even if it can only be briefly introduced here. 'Interconnectedness' is frequently mentioned in terms of improving social inclusion or deterring social isolation of older persons. Globally, such interconnectedness is increasingly facilitated by access to information and, especially, access to technology. This can be as basic as access to information on health, welfare and others services, but it often relies on health literacy. Persons must be able to understand information that is available. Indeed, at least four of the eight domains in the WHO's Age-friendly Cities movement focus on 'interconnectedness': social participation, respect and social inclusion, civic participation and employment and communication and information. The sources of information can be oral, written, radio, TV or Internet, among many others. Residents in urban and, crucially, rural areas in both richer and poorer nations have increasingly come to expect interconnectedness.

Older persons in many nations are sometimes stigmatised as being less ready and able than younger groups to cope with information and especially information and communications technology (ICT). However, this is being addressed, and attitudes may be changing. Education in the use of computers and of ICT generally is popular among many older groups and is often central to the theme of lifelong education. Worldwide, ICT is being used to improve access to health care via health information and e-Health, better record keeping, access to information on drugs and treatments and the like (WHO, 2015). Telemedicine, and applications such as remote monitoring of older persons (e.g., for heart problems, diabetes and medical emergencies) is increasingly common and not necessarily confined to richer countries, even though currently advances tend to be far greater in wealthier societies and groups (Marasinghe, Lapitan and Ross, 2015; Sixsmith *et al.*, 2013). Remote diagnosis and advice in health and other topics by linking via the Internet are becoming key topics (McCracken and Phillips, 2017) and are of considerable interest to geographical gerontology.

In many countries, a major development facilitating access to information over the past decade has been advances in mobile phone networks and cheap communication methods (e.g., Skype, Viber, Whatsapp, social media, etc.). Mobile phone networks are now cheaper and more reliable than before and have especially been a boon in poorer and large countries, where they have often eliminated the need for expensive land-line telecommunication systems. Today, remote villages can be connected via mobile and satellite networks and solar sources of power can assist when regular electric supplies are unreliable. Looking forward over the next decade or two, such technological advances and cheap availability of mobile communication networks are likely to have the greatest global impact on older persons' connectedness and access to information and, hence, social participation. Increasingly, older people can network with friends and families even when they are working or living away. In some countries such as China, and many others in Asia and Africa, indirect personal communications are even replacing direct family responsibilities and modifying traditional filial piety (Phillips and Feng, 2015; WHO, 2015). Certainly, many older people can feel more part of their family even if they are a long way apart and can even receive transfers of funds via these means. However, we must be aware of the potential for increased feelings of social isolation and exclusion amongst older people who do not like or cannot use or afford such technology.

A further rapidly developing if not yet widespread major area is assistive technologies in ageing (Sixsmith and Gutman, 2013), which includes the modification of housing to enable them to be incorporated in people's accommodation. Currently, assistive technologies are principally available in richer countries but are also appearing in many poorer countries, where they are sometimes tentatively being produced at low cost (Marasinghe *et al.*, 2015). Such modifications and technologies can sometimes either make ageing in place easier, or sometimes they can be the permissive factor, which can then reduce pressure for institutional care. This is an exciting and rapidly evolving area of the application of technology, which involves research, development, industry and consumers (see, for example www.agein placetech.com). In some countries, notably in East Asia, especially Korea and Japan, futuristic use of robotics for household chores and even for social interactions are being introduced (Sixsmith *et al.*, 2013). Various types of health-related robotic applications with the potential to assist older persons and people with disabilities are also being developed in Europe and the United States (Pearce *et al.*, 2012). In many places, technologies are being applied to home monitoring for safety, fires, falls and the like, as well as devices to improve hearing and walking and alarms, are available to contact remote assistance. It appears that, perhaps contrary to some stereotypes, many older persons are open to owning and using such devices, depending on costs and especially if privacy and security concerns are addressed (WHO, 2015). Smart communication technologies used as assistive ageing devices may not only help the social connectedness of older people but can also help with ageing in the *right* place (Morris *et al.*, 2014; Golant, 2015).

Conclusion

This chapter has introduced global ageing patterns and diversity from the perspective of geographical gerontology. With the main exception of China, current percentages of older populations are generally smaller in lower-income than in higher-income countries. However, this is likely to change quite rapidly over the next few decades, as population ageing will grow much faster in currently 'less-developed' regions than that in richer countries. Indeed, three-quarters of the 2 billion people age 60+ in 2050 will be in today's 'less-developed' regions. Therefore, we must expect distinctive changes emerging in global geographical patterns of ageing over the next few decades.

Whilst looking at selected social occurrences within the realities of ageing globally, we should acknowledge and emphasise the heterogeneity within the 'older cohorts'. Not all older persons are equally at risk but, as a group, older people are generally considered more vulnerable than younger groups in emergency situations, and they may be less able to escape from local dangers and adapt less readily to new environments. Many older persons are resilient but others may experience psychological pressure. In spite of their large numbers, older people can be 'invisible' to humanitarian agencies even to the extent that their needs are not be recognised. Older people are also potentially vulnerable to climate-related health impacts, which are expected to become more marked in the coming years. Last but not least, many contemporary socio-technological trends such as increased interconnectedness may improve access to information and improve social participation for older people. However, these trends have the potential to increase feelings of social isolation and exclusion for those older people who cannot afford them or who are unable to use such technology. All these factors mean that greater attention should be given to older people in such circumstances. They make global ageing and geographical gerontology of major importance now and in the coming decades.

References

Age International. (2013a). *Ukraine conflict.* Available from: www.ageinternational.org.uk/what-we-do/emergencies/previous-emergencies/ukraine/

Age International. (2013b). *Older Syrian refugees 'forgotten' by aid agencies.* Available from: https://blog.ageinternational.org.uk/older-syrian-refugees-forgotten/

Age International. (2015). *Nepal earthquake.* Available from: www.ageinternational.org.uk/what-we-do/emergencies/ongoing-emergencies/Nepal-Earthquake/

Caminade, C., Kovats, S., Rocklov, J., Tompkins, A. M., Morse, A. P., Colón-González, F. J., Stenlund, H., Pim Martens, P. and Lloyd, S. J. (2014). Impact of climate change on global malaria distribution. *PNAS: Proceedings of the National Academy of Sciences,* 111(9), 3286–3291. http://dx.doi.org/10.1073/pnas.1302089111

Centers for Disease Control and Prevention (2012). *Identifying vulnerable older adults and legal options for increasing their protection during all-hazards emergencies: A cross-sector guide for states and communities.* Atlanta: CDC, U.S. Department of Health and Human Services. Available from: https://stacks.cdc.gov/view/cdc/11763

Chau, P. H., Wong, M. and Woo, J. (2013). Living environment, in J. Woo (ed.), *Aging in Hong Kong: A comparative perspective* (pp. 31–67). New York: Springer.

Crimmins, A., Balbus, J., Gamble, J. L., Beard, C. B., Bell, J. E., Dodgen, D., Eisen, R. J., Fann, N., Hawkins, M. D., Herring, S. C., Jantarasami, L., Mills, D. M., Saha, S., Sarofim, M. C., Trtanj, J. and Ziska, L. (eds). (2016). *The impacts of climate change on human health in the United States: A scientific assessment.* Washington, DC: U.S. Global Change Research Program (USGCRP). http://dx.doi.org/10.7930/J0R49NQX

Doherty, T. J. and Clayton, S. (2011). The psychological impacts of global climate change. *American Psychologist,* 66(4), 265–276. DOI: 10.1037/a0023141

Gillam, S. (2015). *Ukraine crisis has huge impact on older people.* 25 February 2015. Available from: www.helpage.org/newsroom/latest-news/ukraine-crisis-has-huge-impact-on-older-people/

Global Action on Aging. (n.d.). *Older persons caught in armed conflict and other emergency situations.* http://globalag.igc.org/armedconflict/

Golant, S. M. (2015). *Aging in the right place.* Baltimore, MD: Health Professions Press.

HelpAge International. (2012). *Older people in emergencies – identifying and reducing risks.* Available from: www.helpage.org/silo/files/older-people-in-emergencies – identifying-and-reducing-risks.pdf

HelpAge International. (2015). *Disaster risk reduction and climate change policy.* Available from: www.helpage.org/what-we-do/climate-change/climate-change-policy/

Hijioka, Y., Lin, E., Pereira, J. J., Corlett, R. T., Cui, X., Insarov, R. D., Lasco, E., Lindgren, E. and Surjan, A. (2014). Asia, in C. B. Field, V. R. Barross, D. J. Dokken, K. J. Mach, M. D. Mastrandrea, T. E. Billir, M. Chatterjee, K. L. Ebi, Y. O. Estrada, R. C. Genova, B. Girma, E. S. Kissel, A. H. Levy, S. McCracken, P. R. Mastrandrea and L. L. White (eds), *IPCC global climate change 2014: Impacts, adaptation and vulnerability: Part B: Regional aspects. Contribution of working group II to the fifth assessment report of the Intergovernmental Panel on Climate Change* (pp. 1327–1370). New York: Cambridge University Press. Available from: www.ipcc.ch/report/ar5/wg2/ (Chapter 24).

Howgate, B. (2015). *Nepal earthquake: Responding to older people's needs.* Available from: www.helpage.org/newsroom/latest-news/nepal-earthquake-older-people-among-most-at-risk-/

Intergovernmental Panel on Climate Change (IPCC). (2014). Global climate change 2014: Impacts, adaptation and vulnerability: Part A – Global and sectoral aspects, in C. B. Field, V. R. Barross, D. J. Dokken, K. J. Mach, M. D. Mastrandrea, T. E. Billir, M. Chatterjee, K. L. Ebi, Y. O. Estrada, R. C. Genova, B. Girma, E. S. Kissel, A. H. Levy, S. McCracken, P. R. Mastrandrea and L. L. White (eds). *IPCC global climate change 2014: Impacts, adaptation and vulnerability: Part A: Global and sectoral aspects. Working Group II contribution to the fifth assessment report of the Intergovernmental Panel on Climate Change.* Cambridge: Cambridge University Press. Available from: www.ipcc.ch/report/ar5/wg2/ (Chapters 1–20).

Kinsella, K. and Phillips, D. R. (2005). Global aging: The challenge of success. *Population Bulletin,* 60(1), 1–44. Washington, DC: Population Reference Bureau. Available from: www.prb.org/pdf05/60.1GlobalAging.pdf

Marasinghe, K. M., Lapitan, J. M. and Ross, A. (2015). Assistive technologies for ageing populations in six low-income and middle-income countries: A systematic review. *BMJ Innovations,* 1, 182–195. http://dx.doi.org/10.1136/bmjinnov-2015-000065

McCracken, K. and Phillips, D. R. (2016). Climate change and the health of older people in Southeast Asia, in R. Akhtar (ed.), *Climate change and human health scenario in South and Southeast Asia* (pp. 29–52). New York: Springer.

McCracken, K. and Phillips, D. R. (2017). *Global health: An introduction to current and future trends* (2nd edn). Abingdon: Routledge.

Morris, M. E., Adair, B., Ozanne, E., Kurowski, W., Miller, K. J., Pearce, A. J., Santamaria, N., Long, M., Ventura, C. and Said, C. M. (2014). Smart technologies to enhance social connectedness in older people who live at home. *Australasian Journal on Ageing,* 33(3), 142–152. http://dx.doi.org/10.1111/ajag.12154

Mujahid, G. (2006). *Population ageing in East and Southeast Asia: Current situation and emerging challenges,* Papers in Population Ageing no. 1, 100pp., UNFPA (United Nations Population Fund), Bangkok. Available from: www.eldis.org/vfile/upload/1/document/0708/DOC22434.pdf

Page, L. A. and Howard, L. M. (2010). The impact of climate change on mental health (but will mental health be discussed at Copenhagen? *Psychological Medicine,* 40(2), 177–180. http://dx.doi.org/10.1017/S0033291709992169

Pearce, A. J., Adair, B., Miller, K., Ozanne, E., Said, C., Santamaria, N. and Morris, M. E. (2012). Robotics to enable older adults to remain living at home, *Journal of Aging Research,* 2012, 12pp. Online first. http://dx.doi.org/10.1155/2012/538169

Phillips. D. R. and Feng, Z. X. (2015). Challenges for the aging family in the People's Republic of China. *Canadian Journal on Aging,* 34(3), 290–304. http://dx.doi.org/10.1017/S0714980815000203

Phillipson, C. (2012). Globalisation, economic recession and social exclusion: Policy challenges and responses. In T. Scharf and N. Keating (eds), *From exclusion to inclusion in old age: A global challenge* (pp. 17–32). Bristol: Policy Press.

Poumadère, M., Mays, C., Le Mer, S. and Blong, R. (2005). The 2003 heat wave in France: Dangerous climate change here and now. *Risk Analysis,* 25(6), 1483–1494. http://dx.doi.org/10.1111/j.1539-6924.2005.00694.x

Robert, S. A. and Li, L. W. (2001). Age variation in the relationship between community socioeconomic status and adult health. *Research on Aging,* 23, 233–258. http://dx.doi.org/10.1177/0164027501232005

Robine, J. M., Cheung, S.L.K., Le Roy, S., Van Oyen, H., Griffiths, C., Michel, J-P. and Herrmann, F. R. (2003). Death toll exceeded 70,000 in Europe during the summer of

2003. *Les Comptes Rendus/Série Biologies*, 331(2), 171–178. http://dx.doi.org/10.1016/j.crvi.2007.12.001

Sánchez-González, D. and Rodríguez-Rodríguez, V. (eds). (2016). *Environmental gerontology in Europe and Latin America: Policies and perspectives on environment and aging*. New York: Springer. Available from: www.researchgate.net/publication/277775613_Environmental_Gerontology_in_Europe_and_Latin_America_Policies_and_perspectives_on_environment_and_aging

Scharf, T. and Keating, N. (2012). Social exclusion in later life: A global challenge, in T. Scharf and N. Keating (eds), *From exclusion to inclusion in old age: A global challenge* (pp. 1–16). Bristol: Policy Press.

Sixsmith, A., Carrillo, M., Phillips, D. R., Lansley, P. and Woolrych, R. (2013). International initiatives in technology and aging, in A. Sixsmith and G. Gutman (eds), *Technologies for active aging* (pp. 201–222). New York: Springer. Available from: http://web.shahed.ac.ir/Parastari/DocLib4/Technologies_for_%20active%20aging.pdf

Sixsmith, A. and Gutman, G. (eds). (2013). *Technologies for active aging*. New York: Springer. Available from: http://web.shahed.ac.ir/Parastari/DocLib4/Technologies_for_%20active%20aging.pdf

Springmann, M., Mason-D'Croz, D., Robinson, S., Garnett, T., Godfray, H.C.J., Gollin, D., Rayner, M., Ballon, P. and Scarborough, P. (2016). Global and regional health effects of future food production under climate change: A modelling study. *Lancet*, 387, 1937–1946. doi: 10.1016/S0140-6736(15)01156-3

United Nations DESA Population Division (UNDESA). (2017a). *Probabilistic population projections based on the world population prospects: The 2017 revision*. New York: United Nations. Available from: https://esa.un.org/unpd/wpp/Download/Probabilistic/Population

United Nations DESA Population Division (UNDESA). (2017b). *World population prospects: The 2017 revision*. New York: United Nations. Available from: https://esa.un.org/unpd/wpp/Download/Standard/Fertility/

United Nations Framework Convention on Climate Change (UNFCCC). (2015). *Adoption of the Paris agreement proposal by the president*. New York: United Nations. Available from: https://unfccc.int/resource/docs/2015/cop21/eng/l09r01.pdf

United Nations Office for Disaster Risk Reduction (UNISDR). (2015). *Towards a post-2015 framework for disaster risk reduction*, 10pp. Available from: www.unisdr.org/we/inform/publications/25129

Vandentorren, S., Bretin, P., Zeghnoun, A., Mandereau-Bruno, L., Crosier, A., Cochet, C., Ribéron, J., Siberan, I., Declercq, B. and Ledrans, M. (2006). August 2003 heat wave in France: Risk factors for death of elderly people living at home. *European Journal of Public Health*, 16(6), 583–591. doi:10.1093/eurpub/ck1063

World Health Organization (WHO). (2008). *Older people in emergencies: An active ageing perspective*. Geneva: WHO. Available from: www.who.int/ageing/publications/Emergencies English13August.pdf

World Health Organization (WHO). (2013). *Climate change and health*. Fact Sheet no. 266, WHO, Geneva. Updated July 2017. Available from: www.who.int/mediacentre/factsheets/fs266/en/

World Health Organization (WHO). (2015). *World report on ageing and health*. Geneva: WHO. Available from: www.who.int/ageing/publications/world-report-2015/en/

9 Ageing in low- and middle-income countries

Ageing against all odds

Andrea Rishworth and Susan J. Elliott

Introduction

Issues related to population ageing – social security and pensions, health financing, long-term care – have been the subject of public and policy debate in high-income countries for decades (Harper, 2014; McCracken and Phillips, 2005). In contrast, limited consideration has focused on ageing in low- and middle-income countries (LMICs), even though the absolute numbers of older adults in many LMICs are expected to double or even triple over the next decades (United Nations [UN], 2015; Phillips and Feng, Chapter 8). While the ageing of global populations has profound implications for individuals and societies at local, national and transnational levels (Kudo, Mutisya and Nagao, 2015; UN, 2015), geographers have yet to comprehensively examine how ageing is shaped by place across a range of spatial scales. For instance, considerable disciplinary focus attends to the dynamics of ageing in varied community spheres (e.g., home, residential, health and care centres) but does not scale up, nor does this work connect processes across spatial scales (Skinner, Cloutier and Andrews, 2015). This neglect of the whole applies to most scholars working within the sub-discipline. Recent works takes us deeper into the area of non-representational theory (Andrews, Evans and Wiles, 2013; Horton and Kraftl, 2008) yet further away from a unified vision of ageing in place across spatial scales. Others in the sub-discipline continue to contribute quantitative analyses of the spatial dynamics of service access and utilisation with a particular emphasis on underserved populations (e.g., McCracken and Phillips, 2005; Rosenberg *et al.*, 2009). Still others have brought to the fore the spatial inequities inherent in 'rural versus urban' comparisons (Skinner *et al.*, 2008; Skinner and Rosenberg, 2005) and the embodied, emotive experiences of ageing in place (Herron and Skinner, 2012; Herron and Skinner, 2013; Herron, Chapter 14). While these researchers have made essential disciplinary as well as policy contributions, substantive conceptual and theoretical gaps remain. Even more so, the focus of geographies of ageing literature is limited to a very small proportion of the world's population.

This chapter demonstrates the role of space and place in the geographies of ageing research in order to build a more comprehensive understanding; one powerful enough to inform both explanation as well as policy. In order to fully address the changing global dynamics of ageing societies, geography of ageing researchers must be able to connect processes (e.g., political, socio-economic, demographic, health, etc.) across multiple spatial scales. This chapter is organized as follows. First, we present an overview of ageing in LMICs, followed with a case study of ageing in sub-Saharan Africa (SSA), home of the fastest-growing ageing population on the planet. We then discuss the critical role geographers can play in extensively addressing the concepts of space and place in relation to issues of ageing from a global perspective, particularly, the need for geographers to take 'place seriously', fully interrogating the interconnected socio-economic, political and cultural processes that unevenly shape ageing experiences across global spaces and within places. Finally, the chapter argues that geographers must theorize the full complexity of ageing in LMICs to properly expose the geospatial dynamics and long-term effects of global ageing. This will spark a resurgent *geographies of ageing* more fully informed by the concepts and approaches of human geography capable of grappling with the geospatial variations in ageing worldwide.

Ageing in LMICs

By 2050, 80 per cent of the world's older population (60+) will reside in the LMIC regions of Asia, India, Latin America and the Caribbean (UN, 2015). The speed and scale of ageing in these contexts have given rise to a growing body of literature that has begun exploring the various dynamics and emerging realities related to ageing societies.

Budding scholarship from South-East Asia, for instance, documents and explores changing cultural values related to respect and filial obligation, limited institutional support and associated challenges for the family and older persons (Knodel, 2014; Phillips and Feng, 2015). Elderly populations (60+) in South-East Asia are projected to more than double from 12 per cent in 2015 to 25 per cent by 2050, with those aged 80 years and over quadrupling from 1 per cent to 4 per cent in the same time (UN, 2015). This shift raises concern over the availability of health and social services for this growing demographic (Phillips and Feng, 2015). Yet the pace of ageing varies across the region depending on various socio-demographic and political geographical characteristics. For instance, stark differences in life expectancy and disability exist between Myanmar and Thai seniors because of the availability of health and social services across the life course (Knodel, 2014). Studies have compared factors related to poverty, HIV/AIDS and ageing in Cambodia and Thailand, revealing how different levels of economic development, standards of living and government support lead to drastically different experiences in each context (Knodel, 2008). Moreover, literature has brought to the fore previously neglected aspects of ageing with HIV/AIDS in both academic literature and global health agendas (Knodel, 2008). Characterising much of this literature is a growing concern for the intergenerational disjunctures between the desire to adhere to filial norms and changing realities of increased female labour participation, increased rural–urban migration and underdeveloped or non-existent support for older adults (Knodel *et al.*, 2013; Knodel, 2014). These emerging realities underscore the need for intense examination of the diverse regional contexts in order to meet the needs of the changing demographic realities.

Within the Indian context, gerontological literature, though in its infancy (Raju, 2011), has begun exposing the complex realities of ageing as they relate to living arrangements, health concerns and gender dynamics (Agarwal *et al.*, 2016). India's older population is projected to transform from 9 per cent in 2015 to 19 per cent by 2050, with the proportion of those aged 80 years and over tripling from 1 per cent to 3 per cent in the same period (UN, 2015). This growth is occurring in contexts of changing family obligations, heightened rural–urban migration of youth and growing old-age poverty (Agarwal *et al.*, 2016). Health and health care are growing concerns among older adults as over half have at least one chronic disease or disability (Raju, 2011), but more than 90 per cent have no form of health insurance (Bhattacharjya and Sapra, 2008). These changing circumstances are resulting in an increased prevalence of mental illness, psychological problems and substance abuse disorders among the aged (Reddy *et al.*, 2013). This is exacerbated given that the majority (80 per cent) live in households lacking running water or sewer systems (Raju, 2011). The feminisation of India's older adults in some segments of the country raises additional questions over how widows will fare, particularly under traditional Hindu law, where women historically suffer social stigmatisation, discrimination and income insecurity because of inheritance traditions that favor sons (Kadoya and Yin, 2012). Moreover, the literature documents increased portions of older populations residing in slum settings, increasing the risk of physical, verbal and sexual abuse among older populations, particularly among older females (Agarwal *et al.*, 2016).

Equally, a growing number of studies from the Latin America and Caribbean region (LAC) have begun illuminating patterns and challenges of intergenerational transfers, disease burden and the social perceptions of ageing (Leeson, 2013; Rosero-Bixby, 2009). The LAC is expected to see profound shifts in age composition, with older adults (60+) projected to more than double from 11 per cent in 2015 to 26 per cent by 2050, with those aged 80 years and over nearly quintupling from 1 per cent to 4.5 per cent (UN, 2015). These trends are occurring in dynamic environments experiencing growing poverty and inequality among older adults as social security programmes are eliminated and replaced by fully or partially privatized systems (Palloni and Souza, 2013). Because of increasingly fragile institutional environments, studies reveal the older adults are more likely to provide financial support to their adult children in some LAC countries (e.g., Costa Rica, Uruguay, Brazil), compared to other LMIC contexts (Rosero-Bixby, 2009). Furthermore, unique health characteristics of LAC's ageing populations have emerged revealing the region is in the midst of a diabetes epidemic, never before documented on such a regional scale among older populations (Palloni and McEniry, 2007). The growing older adult population together with demand for health services, precisely when access to health care is shrinking and costlier under privatized schemes, creates a context of great societal ambiguity (Palloni and McEniry, 2007; Palloni and Souza, 2013).

The foregoing regional ageing dynamics offer unique opportunities to understand the diverse experiences, mechanisms, and activities currently shaping the geographical processes related to ageing. There is a need to fully examine the relationships among demographic ageing, individual experiences and societal responses, as well as to employ techniques and methodologies that account for these processes in order to inform knowledge and policy. Equally, it is fundamental to recognize the great heterogeneity among ageing populations (i.e., rural–urban differences, gender dynamics, intergenerational and age relations, etc.) among and within regions, at the national and subnational scales. The wide variations in levels of development and socio–economic status of people living within and among different regional contexts make it imperative to ascertain how broad macro characteristics and micro specificities of ageing intersect in these different geographies. Given that much of the aforementioned literature remains descriptive and a-theoretical, geographers have great opportunities to engage these emerging geographic realities in their work. While growing attention has recently shifted towards rapidly ageing LMICs, limited attention has been given to the ageing of individuals in relatively young societies. Sub-Saharan Africa represents such a context.

Ageing in SSA

Ageing in SSA is unlike ageing in any other part of the world; perhaps as a result, in contrast to all other global regions, ageing issues on the continent have received marginal attention in public and policy discourses (Aboderin and Beard, 2015; Pillay and Maharaj, 2013). This may be due, in part, to the fact that despite the growing absolute numbers of older people, demographic trends in SSA indicate that it will remain by far the youngest world region. That is, the median age is currently 19.7 years and will rise to just 26.4 years by 2050 compared to approximately 40 or more in all other world regions (United Nations Population Division, 2011). Concomitantly, the population proportion of older persons (60+) in SSA is presently only 5.5 per cent and projected to rise to 9.8 per cent by 2050, compared to 20 to 30 per cent in Europe and North America (UN, 2015). The contrast to other global regions will become even starker if assumed declines in SSA's fertility do not materialise, and there are signs they may not (Machiyama, 2010).

Regardless, there is an indication that SSA will see an overall 13-fold increase in the absolute size of its older population from the current 56 million to 716 million, by the end of this century (UN, 2011). This does not necessarily mean that everyone is living longer; in fact, life expectancy at birth currently ranges from 45 to 75 years. Rather, it simply means that the proportion of the older population is becoming larger (UN, 2015). Those who do survive to the age of 60, however, can expect to live as long, if not longer, than older populations in other global regions (Aboderin, 2009). Furthermore, marked geographic variation exists in the ways ageing is unfolding among and within regions across the continent. Regionally, tremendous diversity in national and societal contexts in terms of language, cultural expression and social organisation, shape the ways ageing is unfolding (Douglas, 2015). For instance, Ghana has implemented social and health services for older adults (e.g., National Health Insurance Scheme, Growth and Livelihood Empowerment Against Poverty plan), allowing them to (hypothetically) access free health and social services (e.g., primary health care, monetary support; Biritwum *et al.*, 2013). Other West African countries are seeing rapidly changing family structures because of significant outmigration resulting from high unemployment rates (Aboderin, 2009). In East Africa (e.g., Uganda, Kenya), we see large numbers of older Africans assuming financial and caretaking responsibilities for HIV/ AIDS orphans within their nuclear and extended families (Musangali, Daire and DeLorenzi, 2016; Schatz and Seeley, 2015).

Across and within this diversity range a number of cross-cutting themes that in many ways characterise the experience of ageing in SSA: growing income inequality, increasing burdens of both chronic and infectious diseases, civil strife and corrupt governments (Aboderin, 2017; Pillay and Maharaj, 2013). Older populations across the continent systematically have less availability of and access to health care and are relatively disadvantaged compared to young adults (29–59 years) in both wealth and health (Aboderin, 2009). The majority are often denied access to social services, with little recourse to any form of social safety net (e.g., pension, formal social security provisions; Aboderin and Beard, 2015; Pillay and Maharaj, 2013). Moreover, older populations are often food-insecure and have limited access to safe water and sanitation (Kimokoti and Hamer, 2008). This, in turn, forces many to depend on the informal economy, most often in subsistence agriculture or family support systems (Aboderin and Beard, 2015; Pillay and Maharaj, 2013). Likewise, marginal government support may propel older adults to engage in transactional sex (de Waal and Whiteside, 2003) and the production of illegal alcohol to maintain some form of subsistence (Mkandawire, Luginaah and Tobias, 2011).

Current ageing researchers are addressing a broad spectrum of topics ranging from living arrangements, family and intergenerational relations, work and migration patterns and experiences of (health/social) care receipt and provision (Aboderin, 2016; Ferreira, 2006; Kuuire *et al.*, 2016), drawing predominantly on World Health Organization Study of Global Ageing and Adult Health Studies (SAGE) data. While inquiry remains geographically confined to approximately 20 of the 54 major countries (Aboderin, 2016), researchers provide varied insight surrounding the changing socio-economic and livelihood situations of the aged. For instance, a growing body of literature highlights the social impacts of changing migratory patterns and spatial mobility of family members on ageing populations in countries of west Africa (Aboderin, 2009; Aboderin and

Hoffman, 2012). Considerable attention also centres on older people's livelihoods in the context of HIV/AIDS in coutnires of east and southern Africa. A rapidly growing body of research in this area has explored older persons' roles as care givers for ailing children and/or orphaned grandchildren, as well as focusing on their own loss of support from younger kin (Ferreira, 2006; Musangali *et al.*, 2016; Schatz and Seeley, 2015).

Yet, to date, most investigation remains overly descriptive and inconsistent in terms of findings and methodology (Aboderin, 2009, 2016). While SAGE surveys have captured pertinent information regarding older adults there is a general lack of nationally representative data allowing robust generalisations into the specific needs of older people. Few data provide insight into causative factors and processes that may act contemporaneously or occur over the life course. Moreover, rarely is empirical evidence provided by older adults themselves in which they describe their lived experiences of ageing (Aboderin and Ferreira, 2008). As a result, most investigation has failed to examine and theorise the actual nature, causes and processes pertaining to age and ageing, nor has it illuminated the diversity within Africa's older populations (Aboderin, 2009, 2016).

To accurately understand and respond to the needs of older people of SSA, literature must move beyond empirical descriptions towards the production of grounded incisive analysis and theoretically informed explanations of older people's circumstances and realities in order to both expand our knowledge and improve our response. Approaches that capture individual micro-level perspectives, the recursive relations within meso-level family-community environments and the broader macro-level social context would substantively enhance explanations and knowledge of ageing in SSA (Aboderin and Ferreira, 2008). The rich diversity and geographical variation of Africa require an understanding of context-specific nuances to analyse the interrelations among and across scales over the life course (Aboderin, 2009, 2016; Giddens, 1991). Hence, elucidating the ways in which space and place broadly affect older persons, their families and the societies in which they are situated (Newbold, 2014) would help foster a more variegated, specific in-depth engagement with policy agendas and debates not to mention new insights, concepts and theories of age and ageing in SSA.

Geographical gerontology is well poised to enhance empirical, conceptual and theoretical knowledge of ageing in SSA. Yet, substantially little attention has been directed towards how older people co-experience ageing and place in LMICs, particularly African (Bloom, Mahal and Rosenberg, 2012). This lack of attention is partially the result of long-standing disciplinary and interdisciplinary connections between geographers and gerontologists. As a result, empirical, conceptual and theoretical challenges inherent in both disciplines have fissured knowledge and broader discussions of global ageing (Andrews *et al.*, 2007).

Accordingly, the need exists to query the role geographers *can* play in extrapolating the dynamics of global ageing with a focus on the fundamental concepts of space and place and how they are used within geographic gerontology.

Current approaches in geographic gerontology

Geographic gerontology has generated substantive insight regarding how older people locate in, experience and negotiate the physical and social worlds, flourishing into a comfortable place to co-develop theories, concepts and empirical knowledge on the relationships among space, place and ageing (Andrews *et al.*, 2007). Despite this growth, much scholarly work remains (Skinner et al., 2015). On one hand, the geographies of ageing have expanded theoretically and empirically, recognising the co-constructed, transactional relations between people and places as they age (Andrews *et al.*, 2013; Skinner et al., 2015). This research demonstrates the 'interrelatedness' of people and places at scales from individual homes, residences or caregiving facilities (Milligan and Wiles, 2010), but the focus remains firmly entrenched in perspectives of macro space or micro place (Andrews and Phillips, 2004; Skinner *et al.*, 2015). Granted, while humanist perspectives provide much-needed attention to ageing experiences beyond simple spatial analysis, the separation of the individual from the larger environment underestimates the complexities of ageing in place.

On the other hand, gerontologists highlight how socio-economic and political processes shape the abilities of older adults to respond to challenges in varied environments (Carrol and Bartlett, 2015). While gerontologists' theories have expanded from the prevailing psychosocial considerations of environmental gerontology (Cutchin, 2009), they have faltered in their recognition of place as a complex set of situated social dynamics occurring within myriad relations (Twigg and Martin, 2015). Emphasis is often given to the role of place *on* the individual without dissecting the relational qualities of people and place, that is, how people also *make* place (Valentine, 2001).

This has resulted in a literature riddled with fractured conceptual frameworks within which to situate ageing, space and place. Geographers have a critical role to play in pushing the boundaries of geographic gerontology to fully engage the complexity of person–place relations, thus exposing the complex dynamics of ageing in LMICs. To do this, geographers must fully engage with the conceptual repertoire of human geography and begin taking 'place seriously'.

The role of geographers in the global ageing context

There are key process-related questions that demand spatial and temporal attention international in scope. These include the changing spatial distribution, segregation and migration patterns of ageing populations (Yi *et al.*, 2014), the making and remaking of social space among ageing populations (Schwanen, Hardill and Lucas, 2012), and the impacts of socio-economic and political relations and inequality along the life course (Kendig and Nazroo, 2016).

Globally, demographic transitions will have profound short- and long-term implications for macroeconomics, social well-being and the long-term health care needs of individuals, societies and nations (Harper and Hamblin, 2014). Cross-national differences in age profiles, for example, raise questions of economic prosperity and international mobility. Will SSA's youthful population be successfully translated into a 'demographic dividend', or will the region's demographic transition lead to a glut of unemployed youth, heightening social risks and tension? (Canning, Raja and Yazbeck, 2015). Similarly, the convergence of changing demographics together with rising rates of chronic disease in both high- *and* low-income contexts suggests an imminent labour-demanding 'silver tsunami' with global implications (Bartels and Naslund, 2013). This raises particular concern as increased demands for skilled workers in ageing societies has already resulted in significant unmet labour needs in LMIC regions with higher outmigration rates (i.e., Africa, India; Connel and Walton-Roberts, 2016). Likewise, ageing populations trigger issues of justice and ethics pertaining to societal expectations of intergenerational responsibilities and (in)equities within and among regional contexts and issues that arise from differential access to education, employment or health care (Harper and Hamblin, 2014), all of which have geopolitical implications (Robbins and Smith, 2016).

In light of these incipient global demographic trends and geospatial variations, geographers have a fundamental role in exposing the interplay between sociostructural processes and context-specific dynamics and experiences in late life (Milligan and Wiles, 2010). Foremost, to advance disciplinary knowledge and explanations, geographers must fully engage with social theory and begin 'taking place seriously'. A thorough reconsideration of the disciplinary tendency to distinguish macro space from micro place (Andrews *et al.*, 2013) is essential to push current approaches focused on welfare, care and daily lives in specific places (Andrews *et al.*, 2007) to broader forms of comprehension. Rather than restrict ageing to variables endogenous to specific places, 'taking place seriously' reconceptualises ageing as a dynamic phenomenon, best understood as exogenous to a broader system. This would not only enhance disciplinary capacity but also provide the power to develop needed empirical, conceptual and theoretical knowledge of ageing in LMICs. For instance, this would provide the ability to interrogate the kinds of political economies and ecological environments created in lieu of declining and moving labour surplus, especially where it influences the geographies of ageing (Robbins and Smith, 2016).

Though global-scale processes are often eschewed in exchange for keen local insight, exposing the complexity of time and space in structuring social processes is precisely the current intellectual challenge of the geographies of ageing. Despite long-standing disciplinary interests in relations among age, space and place, overarching philosophical frameworks that account for structure and agency and society and space are missing. While geographers have investigated issues of age, considering structural processes of space or the characteristics of actors in place from positivist (Warnes, 1982), phenomenological (Rowles, 1978), post-structural (Harper and Laws, 1995), critical (Hopkins and Pain, 2007) or non-representational (Horton and Kraftl, 2008) theoretical perspectives, there has been little effort to join space and place in an overarching disciplinary framework. This common separation of spatial structural factors from place-specific ageing inhibits simultaneous understandings of structure and agency and fundamental notions of structuration (Callinicos, 2004). Although some have advocated for a more self-conscious consideration of structure and agency in the geographies of ageing (Andrews and Phillips, 2002; Skinner *et al.*, 2015), little work to date has sought to bridge this divide.

If the geographies of ageing are to move knowledge forward, the discipline must critically reflect on its current use of space and place. Conceptual frameworks and theoretical perspectives elemental to human geography, that account for the complex spatial processes (socio-economic, political, cultural, environmental) operating across multiple temporal and spatial scales (e.g., subjective life-course experiences, community and environmental contexts, national and international arenas) would allow geographers to conceptualise how the ageing of populations is occurring in tandem with other broad global socio-ecological systems, such as increased migration, changing disease burdens, environmental degradation and economic insecurity (Kudo *et al.*, 2015; Phillips and Feng, Chapter 8).

Looking forward

In 2007, Andrews *et al.* asked, What should the discipline look like in another decade? (Andrews *et al.*, 2007). We are suggesting here that we still do not have an answer. Despite momentous global demographic change, and its rapidity in LMICs, geographic insights into its direct and indirect processes have, to date, been notably absent. Geographers must give new meaning to 'ageing in place', engaging with concepts fundamental to human geography to propel geographical gerontology forward to greater heights of understanding and relevance. We call on geographers to pay critical attention to the geographical variations in global ageing and the immediate and long-term challenges encountered by ageing individuals, communities and societies. While this will take work, we feel geographers are fundamental to exposing how different geographies across time and over space configure dynamics and experiences of ageing (Newbold, 2014). Like all other good geographers, we must look globally, be grounded theoretically and engage with policy if we are to make a difference.

References

Aboderin, I. (2009). Ageing in West Africa, in P. Uhlenberg (ed.), *International handbook of population ageing* (pp. 253–276). Berlin/Heidelberg: Springer Science+Business Media.

Aboderin, I. (2017). Coming into its own? Developments and challenges for research on ageing in Africa. *The Journals of Gerontology Series B*, 72(4), 643–645. http://dx.doi.org/10.1093/geronb/gbw017

Aboderin, I. and Beard, J. R. (2015). Older people's health in sub-Saharan Africa. *The Lancet*, 385(9968), e9–e11. http://dx.doi.org/10.1016/S0140-6736(14)61602-0

Aboderin, I. and Ferreira, M. (2008). Linking ageing to development agendas in sub-Saharan Africa: Challenges and approaches. *Journal of Population Ageing*, 1(1), 51–73.

Aboderin, I. and Hoffman, J. (2012). *Care for older adults in Sub-Saharan Africa: Discourses and realities of family solidarity*. Invited Symposium: Taking Care? Global discourses on intergenerational relationships and family support, 65th Annual Scientific Meeting, Gerontological Society of America (GSA), San Diego, USA, 14–18 November [Abstract in

The Gerontologist, 52(S1), 662]. Available from: https://academic.oup.com/gerontologist/issue/52/S1

Agarwal, A., Lubet, A., Mitgang, E., Mohanty, S. and Bloom, D. E. (2016). *Population ageing in India: Facts, issues, and options*. Discussion Paper #10162, The Institute for the Study of Labor (IZA), Bonn. Available from: http://ftp.iza.org/dp10162.pdf

Andrews, G. J., Cutchin, M., McCracken, K., Phillips, D. R. and Wiles, J. (2007). Geographical gerontology: The constitution of a discipline. *Social Science & Medicine*, 65(1), 151–168. http://dx.doi.org/10.1016/j.socscimed.2007.02.047

Andrews, G. J., Evans, J. and Wiles, J. L. (2013). Re-spacing and re-placing gerontology: Relationality and affect. *Ageing & Society*, 33(8), 1339–1373. http://dx.doi.org/10.1017/S0144686X12000621

Andrews, G. J. and Phillips, D. R. (2002). Changing local geographies of private residential care for older people 1983–1999: Lessons for social policy in England and Wales. *Social Science & Medicine*, 55(1), 63–78. http://dx.doi.org/10.1016/S0277-9536(01)00207-6

Andrews, G. J. and Phillips, D. R. (eds). (2004). *Ageing and place*. London: Routledge.

Bartels, S. and Naslund, J. (2013). The underside of the silver tsunami: Older adults and mental health care. *New England Journal of Medicine*, 368(6), 493–496. http://dx.doi.org/10.1056/NEJMp1211456

Bhattacharjya, A. S. and Sapra, P. K. (2008). Health insurance in China and India: Segmented roles for public and private financing. *Health Affairs*, 27(4), 1005–1015. http://dx.doi.org/10.1377/hlthaff.27.4.1005

Biritwum, R., Mensah, G., Yawson, A. and Minicuci, N. (2013). *Ghana: Study on global Ageing and adult health Wave 1: The Ghana national report*. University of Ghana Medical School, Department of Community Health, WHO, Geneva. Available from: https://apps.who.in/healthinfo/systems/surveydata/index.php/catalog/6/download/1940

Bloom, D. E., Mahal, A. and Rosenberg, L. (2012). Design and operation of health systems in developing countries, in J. R. Beard, S. Biggs, D. E. Bloom, L. P. Fried, P. Hogan, A. Kalache and S. J. Olshansky (eds), *Global population ageing: Peril or promise?* (pp. 65–68). Geneva: World Economic Forum, 2011. Available from: www3.weforum.org/docs/WEF_GAC_GlobalPopulationAgeing_Report_2012.pdf

Callinicos, A. (2004). *Making history: Agency, structure, and change in social theory*. Historical Materialism, Vol. 3. Leiden and Boston, MA: Brill.

Canning, D., Raja, S. and Yazbeck, A. S. (eds). (2015). *Africa's demographic transition: Dividend or disaster?* Washington, DC: World Bank Publications. Available from: https://openknowledge.worldbank.org/bitstream/handle/10986/22036/AfrDemographicTransitionOVERVIEW.pdf

Carrol, M. and Bartlett, H. (2015). Ageing well across cultures, in J. Twigg and W. Martin (eds.), *The Routledge handbook of cultural gerontology*. London: Routledge.

Connell, J. and Walton-Roberts, M. (2016). What about the workers? The missing geographies of health care. *Progress in Human Geography*, 40(2), 158–176. http://dx.doi.org/10.1177/0309132515570513

Cutchin, M. P. (2009). Geographical gerontology: New contributions and spaces for development. *The Gerontologist*, 49(3), 440–444. http://dx.doi.org/10.1093/geront/gnp095

de Waal, A. and Whiteside, A. (2003). New variant famine: AIDS and food crisis in southern Africa. *The Lancet*, 362(9391), 1234–1237. http://dx.doi.org/10.1016/S0140-6736(03)14548-5

Douglass, R. (2015). The ageing of Africa: Challenges to African development. *African Journal of Food, Agriculture, Nutrition and Development*, 16(1), 15pp. Available from: www.ajfand.net/Volume16/No1/Commentary_Douglass.pdf

Ferreira, M. (2006). HIV/AIDS and older people in sub-Saharan Africa: Towards a policy framework. *Global Ageing: Issues and Action*, 4(2), 56–71.

Giddens, A. (1991). *Modernity and self-identity: Self and society in the late modern age*. Stanford, CA: Stanford University Press.

Harper, S. (2014) Economic and social implication of aging societies. *Science*, 346(6209), 587–591. http://dx.doi.org/10.1126/science.1254405

Harper, S. and Hamblin, K. (eds). (2014). *International handbook on ageing and public policy*. Cheltenham: Edward Elgar Publishing.

Harper, S. and Laws, G. (1995). Rethinking the geography of ageing. *Progress in Human Geography*, 19(2), 199–221. http://dx.doi.org/10.1177/030913259501900203

Herron, R. V. and Skinner, M. W. (2012). Farmwomen's emotional geographies of care: A view from rural Ontario. *Gender, Place & Culture*, 19(2), 232–248. http://dx.doi.org/10.1080/0966369X.2011.572432

Herron, R. V. and Skinner, M. W. (2013). The emotional overlay: Older person and carer perspectives on negotiating ageing and care in rural Ontario. *Social Science & Medicine*, 91, 186–193. http://dx.doi.org/10.1016/j.socscimed.2012.08.037

Hopkins, P. and Pain, R. (2007). Geographies of age: Thinking relationally. *Area*, 39(3), 287–294. http://dx.doi.org/10.1111/j.1475-4762.2007.00750.x

Horton, J. and Kraftl, P. (2008). Reflections on geographies of age: A response to Hopkins and Pain. *Area*, 40(2), 274–288. http://dx.doi.org/10.1111/j.1475-4762.2008.00809.x

Kadoya, Y. and Yin, T. (2012). *Widow discrimination and family care-giving in India*. ISER Discussion Paper, Institute of Social and Economic Research (ISER), Osaka University, Osaka. Available from: www.econstor.eu/bitstream/10419/92680/1/730628124.pdf

Kendig, H. and Nazroo, J. (2016). Life course influences on inequalities in later life: Comparative perspectives. *Population Ageing*, 9(1), 1–7. http://dx.doi.org/10.1007/s12062-015-9138-7

Kimokoti, R. W. and Hamer, D. H. (2008). Nutrition, health, and ageing in sub-Saharan Africa. *Nutrition Reviews*, 66(11), 611–623. http://dx.doi.org/10.1111/j.1753-4887.2008.00113.x

Knodel, J. (2008). Poverty and the impact of AIDS on older persons: Evidence from Cambodia and Thailand. *Economic Development and Cultural Change*, 56(2), 441–475. http://dx.doi.org/10.1086/522892

Knodel, J. (2014). *The situation of older persons in Myanmar: Results from the 2012 survey of older persons* (Rev. 2014). Yangon: HelpAge International Myanmar Country Office. Available from: https://deepblue.lib.umich.edu/bitstream/handle/2027.42/100343/Myanmar%20report.pdf?sequence=1&isAllowed=y

Knodel, J., Kespichayawattana, J., Wivatwanich, S. and Saengtienchai, C. (2013). The future of family support for Thai elderly: Views of the populace. *Journal of Population and Social Studies*, 21(2), 110–132. Available from: www.tci-thaijo.org/index.php/jpss/article/view/71987/58162

Kudo, S., Mutisya, E. and Nagao, M. (2015). Population ageing: An emerging research agenda for sustainable development. *Social Sciences*, 4(4), 940–966. http://dx.doi.org/10.3390/socsci4040940

Kuuire, V. Z., Tenkorang, E. Y., Rishworth, A., Luginaah, I. and Yawson, A. E. (2016). Is the pro-poor premium exemption policy of Ghana's NHIS reducing disparities among the elderly? *Population Research and Policy Review*, 1–19. http://dx.doi.org/10.1007/s11113-016-9420-2

Leeson, G. (2013). The demographics of population ageing in Latin America, the Caribbean and the Iberian Peninsula, 1950–2050, in M. de Oca (ed.), *Envejecimiento -en America Latina y el Caribe* (pp. 53–71). Mexico: UNAM.

Machiyama, K. (2010). *A re-examination of recent fertility declines in sub-Saharan Africa*. DHS working Papers, No.68, ICF Macro, Calverton. Available from: www.dhsprogram.com/pubs/pdf/WP68/WP68.pdf

McCracken, K. and Phillips, D. R. (2005). International demographic transitions, in G. J. Andrews and D. R. Phillips (eds), *Ageing and place: Perspectives, policy, practice* (pp. 36–60). New York: Routledge.

Milligan, C. and Wiles, J. (2010). Landscapes of care. *Progress in Human Geography*, 34(6), 736–754. http://dx.doi.org/10.1177/0309132510364556.

Mkandawire, P., Luginaah, I. and Tobias, J. (2011). Landscapes of economic deprivation and locally distilled liquor (Kachasu): An emerging milieu of HIV/AIDS risk in urban Northern Malawi. *Environment and Planning A*, 43(10), 2384–2398. http://dx.doi.org/10.1068/a4414

Musangali, M., Daire, A. P. and DeLorenzi, L. (2016). The impact of caregiver coping strategies and patient level of functioning on perception of caregiver burden among caregivers of persons living with HIV/AIDS in Kenya. *Journal of HIV/AIDS & Social Services*, 15(4), 450–463. http://dx.doi.org/10.1080/15381501.2014.912176

Newbold, B. K. (2014). *Population geography: Tools and issues* (2nd edn). Lanham, MD: Rowman & Littlefield.

Palloni, A. and McEniry, M. (2007). Ageing and health status of elderly in Latin America and the Caribbean: Preliminary findings. *Journal of Cross-Cultural Gerontology*, 22(3), 263–285. http://dx.doi.org/10.1007/s10823-006-9001-7

Palloni, A. and Souza, L. (2013). The fragility of the future and the tug of the past: Longevity in Latin America and the Caribbean. *Demographic Research*, 29, 543–578. http://dx.doi.org/10.4054/DemRes.2013.29.21

Phillips, D. R. and Feng, Z. (2015). Challenges for the ageing family in the People's Republic of China. *Cambridge Journal on Aging*, 34(3), 290–304. http://dx.doi.org/10.1017/S0714980815000203

Pillay, N. K. and Maharaj, P. (2013). Population ageing in Africa, in J. L. Powell and S. Chen (eds), *International perspectives on ageing: Ageing and health in Africa* (Vol. 4, pp. 11–51). New York: Springer.

Raju, S. (2011). *Studies on ageing in India: A review*. BKPAI Working Paper No. 2, United Nations Population Fund (UNFPA), New Delhi. Available from: www.isec.ac.in/BKPAI%20Working%20paper%202.pdf

Reddy, V. B., Gupta, A., Lohiya, A. and Kharya, P. (2013). Mental health issues and challenges in India: A review. *International Journal of Scientific and Research Publications*, 3(2), 2250–3153. Available from: www.ijsrp.org/research-paper-0213.php?rp=P14792

Robbins, P. and Smith, S. H. (2016). Baby bust: Towards political demography. *Progress in Human Geography*, 41(2), 199–219. http://dx.doi.org/10.1177/0309132516633321

Rosenberg, M. W., Wilson, K., Abonyi, S., Wiebe, A., Beach, K. and Lovelace, R. (2009). Older aboriginal peoples in Canada: Demographics, health status and access to health care. SEDAP Research Paper No. 249, 33 pp., SEDAP, Hamilton, ON. Available from: http://socserv.mcmaster.ca/sedap/p/sedap249.pdf

Rosero-Bixbury, L. (2009). Intergenerational economic transfers and population ageing in Latin America, in *Family support networks and population ageing*, 57–60. UNFP, Northwestern University and UNPA Seminar, June 3–4, Doha, Qatar. Available from: www.unfpa.org/sites/default/files/pub-pdf/family_support_networks2009.pdf

Rowles, G. D. (1978). *Prisoners of space? Exploring the geographic experience of older people*. Boulder, CO: Westview.

Schatz, E. and Seeley, J. (2015). Gender, ageing and carework in East and Southern Africa: A review. *Global Public Health*, 10(10), 1185–1200. http://dx.doi.org/10.1080/17441682.2015.1035664

Schwanen, T., Hardill, I. and Lucas, S. (2012). Spatialities of ageing: The co-construction and co-evolution of old age and space. *Geoforum*, 43(6), 1291–1295. http://dx.doi. org/10.1016/j.geoforum.2012.07.002

Skinner, M. W., Cloutier, D. and Andrews, G. J. (2015). Geographies of ageing progress and possibilities after two decades of change. *Progress in Human Geography*, 39(6), 776–799. http://dx.doi.org/10.1177/0309132514558444

Skinner, M. W. and Rosenberg, M. W. (2005). Co-opting voluntarism? Exploring the implications of long-term care reform for the nonprofit sector in Ontario. *Environment and Planning C: Government and Policy*, 23(1), 101–121. http://dx.doi.org/101068/c0434

Skinner, M. W., Rosenberg, M. W., Lovell, S. A., Dunn, J. R., Everitt, J. C., Hanlon, N. and Rathwell, T. A. (2008). Services for seniors in small-town Canada: The paradox of community. *Canadian Journal of Nursing Research*, 40(1), 81–101.

Twigg, J. and Martin, W. (eds). (2015). *Routledge handbook of cultural gerontology*. London and New York: Routledge.

United Nations (UN), Department of Economic and Social Affairs, Population Division. (2015). *World population ageing 2015* (ST/ESA/SER.A/390). Available from: www.un.org/en/development/desa/population/publications/pdf/ageing/WPA2015_ Report.pdf

United Nations (UN) Population Division. (2011). *World population prospects: The 2010 revision*. Available from: https://dl.vecnet.org/downloads/xs55mc11m

Valentine, G. (2001). *Social geographies: Space and society*. New York: Addison-Wesley Longman.

Warnes, A. M. (ed.). (1982). *Geographical perspectives on the elderly*. Chichester: Wiley.

Yi, C., Zang, C., Wu, S., Gao, M. and Liang, H. (2014). Spatial restructuring of senior population in Beijing from 2000–2010. *Urban Development*, 2, 011.

10 Urban ageing

New agendas for geographical gerontology

Tine Buffel and Chris Phillipson

Introduction

Population ageing and urbanisation have in their different ways become the dominant social trends of the twenty-first century, with their interaction raising issues for all types of communities – from the most isolated to the most densely populated. By 2030, two-thirds of the world's population will be residing in cities; by that time the major urban areas of the developed world will have 25 per cent or more of their population people aged 60 and over. Cities, and the metropolitan regions of which they are a part, are themselves changing. Soja and Kanai (2007) use the term 'global city region' to refer to a: "new metropolitan form characterized by sprawling polycentric networks of urban centers clustered around one or more 'historic' urban cores" (Soja and Kanai, 2007: 58). Such networks vary in size from 1 million, at the lower level, to 10 million and beyond, at the upper, covering well over a billion of the world's residents.

Accelerated urbanisation brings advantages and opportunities of different kinds. Soja and Kanai note the extent to which "dense and heterogeneous cities and city regions have become the driving forces of the global economy, generating enormous wealth as well as technological innovation and cultural creativity" (2007: 68). Cities are regarded as central to economic development, attracting waves of migrants and supporting new knowledge-based industries (Organisation for Economic Co-operation and Development [OECD], 2015; Sudjic, 2016). The rebuilding of many cities – notwithstanding economic recession – provides opportunities for innovations in housing and services suitable for a range of age and income groups. However, the extent to which the 'new urban age' will produce what has been termed 'age-friendly' communities (World Health Organization [WHO], 2007) which support the needs of people as they grow older, remains uncertain. Cities produce advantages for older people regarding access to medical services, provision of cultural and leisure facilities, shopping and general necessities for daily living (Phillipson, 2010). However, they are also seen as threatening environments, often creating insecurity and feelings of vulnerability arising from changes to neighbourhoods and communities (Andrews and Milligan, 2009; De Donder *et al.*, 2012; Smith, 2009).

Urbanisation and urban living may present particular challenges for older people and the planners and service providers charged with assisting them.

This chapter examines this issue in four main ways: first, through examining the relationship between urbanisation and age-friendly communities; second, by assessing geographical and sociological research concerning the issue of place attachment; third, by reviewing research and policy issues relating to mobility and transportation, safety and security, and empowering older people in local communities; finally, the chapter develops a new agenda for urban ageing and geographical gerontology.

Urbanisation and the development of 'age-friendly' communities

Developing what has been termed 'age-friendly cities and communities' (Fitzgerald and Caro; 2016; Moulaert and Garon, 2016) has become a key issue driving policies aimed at improving the quality of life in old age. The 'age-friendly' perspective is especially associated with an initiative from the World Health Organization (WHO) in 2006 that studied the experiences of older people living in urban environments. This produced a guide identifying key characteristics of an age-friendly community in terms of service provision (e.g., health services, transportation), the built environment (e.g., housing, out-door spaces and buildings) and social aspects (e.g., civic and social participation; WHO, 2007). Building on this work, in 2010, the WHO launched the 'Global Network of Age-friendly Cities' in an attempt to encourage implementation of policy recommendations from the 2006 project. The network has a member-ship of over 500 cities and communities across countries in the global North and South (2017 figures).[1]

The WHO initiative may be viewed as commensurate with various urban policies, including those associated with 'sustainable' development (Satterth-waite, 1999; UN-Habitat, 2010) and 'harmonious cities' (UN-Habitat, 2008). The former raised questions about managing urban growth in a manner capa-ble of meeting the needs of future as well as current generations. The idea of harmonious cities emphasises values such as "tolerance, fairness, social justice and good governance" (UN-Habitat, 2008: 12) as essential principles of urban planning. Such themes were also reflected in the elaboration of ideas associated with 'lifetime homes' and 'lifetime neighbourhoods' (Department for Com-munities and Local Government, 2008; Atlanta Regional Commission, 2009) which emerged alongside recognition of the need for more systematic inter-ventions to support population ageing at a neighbourhood and community level (Ball and Lawler, 2016).

An additional influence on age-friendly perspectives came with recogni-tion of the development in many localities of what has been termed 'naturally occurring retirement communities' (NORCS), that is, neighbourhoods that, with the migration of younger people, have in effect evolved into commu-nities of older people (Scharlach and Lehning, 2016). The key issue behind the 'lifetime' concept was an understanding that effective support for older people within neighbourhoods would require a range of interventions linking different parts of the urban system – from housing and the design of streets to transportation and improved accessibility to shops and services, as well as promoting safety and security within neighbourhoods.

These varied elements of the urban system were also reflected in the increased emphasis by geographers, sociologists, psychologists and others about the importance of place in the lives of older people (Cutchin, 2009; Rowles and Bernard, 2014; Skinner, Cloutier and Andrews, 2015; see Golant, Chapter 15). The research on this issue, in the context of urban environments, is summarised in the next section, which reviews debates surrounding the importance of place attachment in the lives of older people.

Urban change and place attachment

Debates about place attachment, social integration and the variety of bonds connecting people to their environments, hold a prominent place in the social science literature (Scannel and Gifford, 2010; see Rowles, Chapter 16). The meanings of, and attachment to, place are considered especially significant for older people given the length of time they are likely to have spent in the same locality (Krause, 2004). Despite evidence of the value of place attachment in old age, much less is known about the extent to which older people differ in the way this is experienced. Expressions of attachment may vary considerably between individuals, and change in significance over the life course (Gilleard, Hyde and Higgs, 2007; Livingston, Bailey and Kearns, 2010). However, attachment may also be affected by characteristics of the locality and structural and social changes that have affected the area over time (Phillipson, 2007).

Research on the determinants of attachment has considered the influence of both personal and place characteristics. Quantitative research has focused upon the effect of individual characteristics such as age and length of residence in the community (Gilleard *et al.*, 2007). For example, Livingston *et al.* (2010), in a UK study, found that someone aged over 65 with 16 or more years of residence in an area was nearly four times as likely to be attached as a 16- to 19-year-old with fewer than two years of residence. Similarly, Gilleard *et al.* (2007) demonstrated that age and ageing in place are both associated with increased feelings of attachment to one's area. Being a homeowner and being more highly educated have also been found to positively predict place attachment (Livingston *et al.*, 2010).

Two contextual determinants of place attachment are *physical-spatial* aspects of the environment and *residential (in)stability*. In relation to the former, qualitative studies in deprived urban settings (Smith, 2009; Livingston *et al.*, 2010) have highlighted those factors that limit a sense of attachment. These include physical deterioration of buildings and infrastructure, a lack of services, traffic congestion and poor neighbourhood design. Older people may be especially sensitive to environmental barriers because of the vulnerabilities associated with increased age (Wight *et al.*, 2009). Mobility limitations, for example, may create unique challenges in crossing a busy intersection, dealing with cracked pavements and a lack of benches for rest (Day, 2008). In these instances, the high-demand character of the environment (*environmental press*) is expected to cause disruptions to place attachment, especially when the capabilities of a person to deal with such demands (*personal competence*) are restricted (Wahl and Oswald, 2010).

Day's (2008) study of three neighbourhoods in Edinburgh (two deprived and one affluent) explored how older residents felt their well-being was affected by their local environment. This research examined how neighbourhood inequalities might affect older people, for example, through local environments affecting older people differently compared with other residents, through spatial inequalities and through limited control over decisions affecting the local environment. Participants in the study felt that a good-quality urban environment:"should be clean and free of litter; that it should have low levels of pollution; . . . that it should be walkable, including for those with mild to moderate mobility difficulties; [and] that it should provide for [a] positive visual and sensory experience (Day, 2008: i). The benefits of such environments were viewed in terms of "[facilitating] exercise, social contact and support . . . relaxation and enjoyment and positive community self-image as well as low exposure to pathogens and environmental health hazards" (Day, 2008: i).

Migrant populations may be especially challenged in relation to place attachment, developing what Lager, van Hoven and Meijering, (2012) refer to within the context of a study of older Antillean migrants living in the Netherlands as a sense of 'in-betweenness' (Lager *et al.*, 2012: 92). Participants in their qualitative study experienced attachment to their current places but retained a strong sense of belonging to the places that they had left behind (see, further, Buffel and Phillipson, 2011). Similarly, Buffel (2015a), in a study of first-generation Turkish migrants living in Brussels (Belgium), found that the idea of 'home' was experienced and created as a site of *connection* within and across communities, reflecting a transnational sense of belonging. However, the experience of place also emerged as a central site of daily struggle in the participants' narratives, reflecting financial hardship, poor housing conditions, urban deprivation and experiences of discrimination and exclusion (Buffel, 2017).

Research in urban areas in both the United Kingdom and the Netherlands found that older people who evaluated their neighbourhood negatively were much more likely to be lonely than those who judged their neighbourhood quality to be high (Scharf and De Jong Gierveld, 2008). Reinforcing this point, a study of urban communities in Belgium found that those who evaluated their neighbourhood positively demonstrated higher levels of social activity and formal participation[2] (Buffel *et al.*, 2013). For example, older people who felt involved in their neighbourhood and had frequent social contact with neighbours were more likely to participate socially. In addition, residents who reported sufficient activities organised by and for older people in their neighbourhood showed higher levels of social activity and formal participation. One important finding for policy-makers, however, was that levels of activity were higher among the 'young-old' and among those in better health. The authors concluded from this that

> in terms of promoting *social activity* [authors' emphasis], intervention strategies will need to differentiate more clearly between strategies aimed at

involving the general 'public' and vulnerable groups of older people, such as the oldest old, those with health limitations and those living on limited incomes. For example, broad-based communication methods could be complemented by more intensive empowerment initiatives with specific 'hard to reach' groups of people. These could address the social, structural and economic conditions that act as a barrier to social activity in old age. . . . Building participatory community networks, and specifically empowering socially excluded and vulnerable groups of older people to enhance their involvement in decision-making processes in their localities, will be crucial tasks for policy-makers.

(Buffel *et al.*, 2013: 667)

Drawing on the preceding assessment of the range of issues faced by older people ageing in urban environments, the next section of this chapter develops an agenda for interventions, focusing on work around, first, promoting mobility and transport within cities; second, ensuring safety and security in the neighbourhood; and, third, empowering older people, including the most vulnerable groups, to effect change within communities.

Developing an agenda for urban ageing

Promoting mobility within cities

As people grow older, they become increasingly reliant on both walking and public transport for everyday mobility. Older people's experiences of mobility can be shaped by gender and life events. Although gender differences are changing and differ by country, older women are more reliant on public transport, being less likely to drive and own a car (Ahern and Hine, 2012). Accessible and affordable transportation is vital for older people living in cities for a number of reasons: it supports changing mobility patterns in old age, it enables access to services that support well-being and it encourages more equal access to the resources available in cities (Handler, 2014). While transportation has a crucial role to play in terms of enabling physical access to resources, it also helps maintain independence in later life and supports a feeling of being connected in some form. For instance, research in the United Kingdom (Webb, Laverty and Millett, 2015) reported a range of benefits linked to the provision of free bus travel for older people. The study found that those who hold a free bus pass are 40 per cent more likely to be physically active than those who do not, after taking socio-demographic characteristics and mobility difficulties into account. The research also showed that among women, in particular, the free bus pass is associated with a healthier later life (Webb *et al.*, 2015).

Research has also highlighted the critical role of supportive walking environments in promoting quality of life (Buffel *et al.*, 2013). 'Walkable' neighbourhoods are seen to be particularly important for older people, as walking is a preferred form of physical activity for a majority of this age group. One study found that two thirds of all trips made by older people (mainly on foot) are restricted to their respective neighbourhoods (Handler, 2014). However, Grant *et al.* (2010), in a comparative case study of four Ottawa neighbourhoods, provide evidence of what they termed "inequitable walking environments" (Grant *et al.*, 2010: 1), suggesting that walking conditions are more supportive in higher as opposed to lower-socio-economic-status (SES) neighbourhoods. The study found that older people in lower-SES neighbourhoods were more affected by traffic hazards and were more reliant on public transit compared to their counterparts in higher-SES neighbourhoods. Lower-SES neighbourhoods also faced greater challenges in creating walkable places. The authors concluded that the study findings highlighted "the need for municipal governments to monitor differences in walking conditions among advantaged and disadvantaged neighbourhoods; to be receptive to the needs of disadvantaged neighbourhoods; and to ensure policy decisions are taken to reduce inequitable walking conditions" (Grant *et al.*, 2010: 14).

In addition, policy interventions need to address the day-to-day problems that deter older people from using public transport and walking in their locality (Handler, 2014). Recreational walking and physical activity have consistently been correlated with perceptions of neighbourhood safety and inversely related to neighbourhood problem (Mendes de Leon *et al.*, 2009). Measures which may support older people's mobility for example include providing dropped kerbs on the street, minimising obstructions which present a safety hazard; phasing traffic light signals at road crossings to allow pedestrians a longer time to cross, ensuring that transport vehicles are accessible and installing legible and standardised signage at transport intersections. However, as Handler (2014) argues, it is also important to acknowledge and address the less concrete, less visible factors that inhibit mobility in older age:

> There is growing recognition that transportation needs to be understood not only as a support mechanism for getting people from A to B but that the route – the journey itself – might be considered as a destination in its own right. In this sense, public transportation might be considered as a place in itself – to meet and chat to people (on the bus, or at the bus stop, for instance). For an Age-friendly city it is important to remember that this latent conviviality of public transport carries a particular importance for those who lead more isolated lives in older age.
>
> (Handler, 2014: 53–54)

As the examples illustrate, accessible and affordable transportation may not only support access to resources and help maintain good health, but it also has the ability to preserve independence and support a feeling of social connectedness. A key issue in this respect will be to ensure that transport planners take into account older people's transport use and journeys together with their actual transport needs. Developing a more 'community-based and user-led approach' in transport planning will therefore be a key issue to address in the development of age-friendly cities.

Safety and security in urban neighbourhoods

The second area for intervention concerns issues relating to safety and security within neighbourhoods. Findings from the WHO Survey of Global AGEing and Adult Health (SAGE) show that fear of crime among older people in high-income countries (data for low-income countries is lacking) is significantly associated with low quality of life and reduced social networking outside the home (Minton, 2009; Lloyd-Sherlock and Agrawal, 2014). Older people who reported they felt unsafe when walking down their own street after dark were found to leave their home less frequently to visit friends and relatives. In a European context, this issue has been examined in the Belgian Ageing Studies (BAS), a large-scale investigation covering nearly 25,000 respondents in 85 municipalities across Belgium. The research used a structured questionnaire to examine a range of issues, including older people's attitudes towards their neighbourhoods, neighbourhood characteristics and feelings of safety and well-being. The findings confirmed that satisfaction with the neighbourhood environment was associated with feelings of safety: "the more satisfied older people are with their neighbourhood, the less they feel unsafe" (De Donder *et al.*, 2012: 441).

An unexpected finding from the BAS research was the perception that '*too many*' older people living in the neighbourhood itself contributed to feelings of unsafety. This led to the policy recommendation that: "intergenerational activities are more desirable than activities exclusively for older people" (De Donder *et al.*, 2012: 441; see also Lindenberg and Westendorp, 2015). An additional finding was that experiencing insufficient opportunities for political participation was the most significant factor increasing feelings of unsafety: "Important recommendations in this matter are giving older people a voice and a role in the policy process, providing them with opportunities to bring solutions to specific problems and to communicate successes and failures" (De Donder *et al.*, 2012: 442).

In further analysis of the BAS data set, De Donder *et al.* (2013) considered additional neighbourhood-based factors related to feelings of unsafety in later life. They suggested that issues relating to accessibility and community infrastructure were important dimensions contributing to feelings of unsafety (see also Day, 2008). For example, a neighbourhood that was perceived as physically adapted for older people was found to relate to greater feelings of safety. Older people who were dissatisfied with the number of toilets and the conditions of the sidewalks also felt less safe. The authors concluded that the findings raised a number of issues for policy and practice:

> The physical quality of the local environment can contribute to higher feelings of safety, indicating that environmental design could be encouraged in order to increase feelings of safety. This view is of particular interest to architects, environmental planners, urban developers and geographers. The study has shown that intervention programmes should aim at older

people *feeling at home and safe in their neighbourhood.* [authors' emphasis]. This kind of policy would be extremely relevant, not only at the individual level – to help older people feel safe and heighten the quality of their life – but also on the social level, given that high feelings of unsafety can also have a negative effect on the quality of life in the community.

(De Donder *et al.*, 2013: 932)

Drawing on these findings, a key area for policy work involves the need for partnerships between local authorities, public health professionals, architects, community organisations and older people to tackle feelings of unsafety in later life. Identifying particular target groups (e.g., socially excluded groups of older people) and understanding older people's insecurities in the context of poverty, inequality and community change will be vital to achieving effective change in this area. Enabling older people to feel safe and secure should be a priority for policy-makers, not least because this age group often shows a strong commitment and attachment to their local neighbourhood. Policy interventions aimed at reducing feelings of insecurity in later life could include promoting strong social networks within communities, improving the walkability and accessibility of the environment, combatting the impact of ageism and discrimination, tackling social exclusion and inequality and providing opportunities for older people to have a *voice* in decision-making processes surrounding their neighbourhood (Burns, Lavoie and Rose, 2012; De Donder *et al.*, 2013). This last issue will now be explored in the penultimate section of this chapter.

Empowering older people in local communities

A third area for intervention concerns the need for a close engagement of older people, and those in later mid-life, in urban regeneration and urban planning. Studies in the United Kingdom by Riseborough and Sribjilanin (2000) and Simpson (2010) found that older people were often 'invisible' in urban regeneration policies. The problem here was less the absence of older people in consultations around policies but more an underlying 'ageism' which viewed them only as 'victims' of neighbourhood change. In contrast, the authors make the point that regeneration practice could greatly benefit from the experience of older people, their attachment to neighbourhoods and their involvement in community organisations. Targeting urban regeneration strategies at different groups within the older population will be essential, with awareness of contrasting issues faced by different ethnic groups, those with particular physical/ mental health needs and those living in areas with high levels of economic and social deprivation (Buffel and Phillipson, 2016).

At the same time, interventions should not only focus on changes for current cohorts of older residents but also work towards longer-term neighbourhood change that can benefit successive cohorts of older residents. There is, therefore, an urgent need to reconnect urban regeneration policies with strategies that support resident-led planning for 'lifetime neighbourhoods' (Bevan and Croucher, 2011) or 'ageing-friendly communities' (Scharlach and Lehning, 2013). Such models promote the empowerment of residents of any age to bring about neighbourhood changes which enable people to meet their basic needs, maintain significant relationships and participate in the community in meaningful ways as they grow older (Scharlach and Lehning, 2013). As Menec *et al.* argue "older adults must be involved in identifying areas of need, prioritizing key issues, and ensuring appropriate implementation" (2011: 487).

Whilst progress has been made in identifying some key policies for age-friendly work, there has been much less success in terms of making older people themselves central to the creation and development of policies and age-friendly initiatives. Here, methods drawn from participatory research can advance efforts to engage older residents as leaders and visionaries in identifying features of their neighbourhood in need of improvement (Buffel, 2015). The case for involving older residents as 'co-researchers' in exploring the age-friendliness of their neighbourhood is that it represents a viable method to engage older residents and mobilise their expertise, skills and knowledge, and to stimulate co-production in developing age-friendly initiatives. In addition, it provides benefits to the older co-researchers, community stakeholders and policy-makers involved because it provides a forum for rich and meaningful social engagement and mutual learning and exchange (Buffel, 2015).

However, the need for experimentation to test and learn from participatory and collaborative approaches involving older people in the co-production of urban space remains. The success of communities in becoming more age-friendly will, to a large extent, depend on whether older people, including those facing social exclusion, will be involved as key actors in setting the agenda for future research and policies on age-friendly developments.

Conclusion: future directions for geographical gerontology

This chapter has reviewed a range of issues and challenges facing older people living in urban environments. The discussion highlights the variety of disciplinary perspectives necessary for understanding the evolving relationship between population ageing, on one side, and urbanisation, on the other. Here, geographical gerontology has an important role in understanding the links between changes at a macro level associated with globalisation and their impact on localities and neighbourhoods. Such changes strengthen the need for a geographical gerontology sensitive to changes in older people's relationship to space and place and appreciative of the way in which spatial inequalities can heighten the vulnerability of older populations (Skinner *et al.*, 2015).

Viewing older people as spatial *and* social is essential for a fuller understanding of the unequal way in which ageing is experienced. On one side, there is increasing evidence of people choosing locations congruent with the lifestyles they wish to adopt in later life, manifest in examples such as retirement/gated communities, second homes, co-housing, and rural areas colonised by older people (Phillipson, 2007). In these examples, spatial characteristics both create and reinforce different styles of ageing. In some cases, these spaces seek to establish new age–based identities (Katz, 2005). On the other side, spatial characteristics may confirm ageing as a form of 'social exclusion', illustrated by residence in neighbourhoods undergoing economic decline, forced urban regeneration favouring younger over older age groups (Buffel and Phillipson, 2016) and transitions into socially isolated residential and nursing homes.

The preceding points suggest that a key task for a geographical gerontology will be to apply a *critical spatial perspective to the age-friendly approach* discussed throughout this chapter. To date, this model has been viewed independently of spatial inequalities and processes. An alternative to this will require, first, new theoretical perspectives, especially those which can apply insights from urban geography and sociology to the frameworks developed in geographical gerontology. Second, empirical studies of daily life in urban environments, in particular, those which seek a better understanding of the 'winners' and 'losers' among older people as a result of urban change. Finally, research is needed on the ways in which particular groups of older people – notably those living on low incomes, people with cognitive/mental health problems or those who are socially isolated – can gain the benefits rather than costs of urban living. Developing such work will require strengthening links between human geography, on one hand, and social gerontology, on the other, and determining new priorities for research and urban policy. Based on the agenda for urban ageing reviewed in this chapter, the next decade should see extensive collaboration on projects which apply insights from geography to the study of ageing societies.

Notes

1 For further information about the Global Network of Age-friendly Cities and Communities, see www.who.int/ageing/projects/age_friendly_cities_network/en/.
2 *Formal participation*, in this study, was defined as a voluntary commitment to community organisations on a regular basis. *Social activity*, on the other hand was defined as participation in social activities that connect individuals to others, involving less formal engagement (e.g., shopping, walking or biking, visit a restaurant/pub/café). Both are seen as key forms of *social participation*.

References

Ahern, A. and Hine, J. (2012). Rural transport – Valuing the mobility of older people. *Research in Transportation Economics*, 34(1), 27–34. http://dx.doi.org/10.1016/j.retrec.2011.12.004

Andrews, G. J., Milligan, C., Phillips, D. R. and Skinner, M. W. (2009). Geographical gerontology: Mapping a disciplinary intersection. *Geography Compass*, 3(5), 1641–1659. http://dx.doi.org/10.1111/j.1749-8198.2009.00270.x

Atlanta Regional Commission. (2009). *Lifelong communities: A regional guide to growth and longevity*. Atlanta: ARC.

Ball, M. S. and Lawler, K. (2016). Commentary: Changing practice and policy to move to scale: A framework for age-friendly communities across the United States, in F. Caro and K. Fitzgerald (eds), *International perspectives on age-friendly cities* (pp. 171–183). London: Routledge.

Bevan, M. and Croucher, K. (2011). *Lifetime neighbourhoods*. London: Department for Communities and Local Government.

Buffel, T. (2017). Ageing migrants and the creation of home: Mobility and the maintenance of transnational ties. *Population Space and Place*, 23(5), e2075–e2087. Online first: October 2015. http://dx.doi.org/10.1002/psp.1994

Buffel, T. (2015). *Researching age-friendly communities: Stories from older people as co-investigators*. Manchester: The University of Manchester Library.

Buffel, T., De Donder, L., Phillipson, C., Dury, S., De Witte, N. and Verte, D. (2013). Social participation among older adults living in medium-sized cities in Belgium: The role of neighbourhood perceptions. *Health Promotion International*, 29(4), 655–668. Available from: http://heapro.oxfordjournals.org/content/29/4/655.full.pdf+html

Buffel, T., De Donder, L., Phillipson, C., Dury, S., De Witte, N. and Verte, D. (2013). Social participation among older adults living in medium-sized cities in Belgium: The role of neighbourhood perceptions. *Health Promotion International*, 29(4), 655–668. Available from: http://dx.doi.org/10.1093/heapro/dat009

Buffel, T. and Phillipson, C. (2011). Experiences of place among older migrants living in inner-city neighbourhoods in Belgium and England. *Diversité Urbaine*, 11(1), 13–38. http://dx.doi.org/10.7202/1007742ar

Buffel, T. and Phillipson, C. (2016). Can global cities be 'age-friendly cities'? Urban development and ageing populations. *Cities*, 55, 94–100. http://dx.doi.org/10.1016/j.cities.2016.03.016

Burns, V. F., Lavoie, J. P. and Rose, D. (2012). Revisiting the role of neighbourhood change in social exclusion and inclusion of older people. *Journal of Aging Research*, 2012, 12pp. Online first: http://dx.doi.org/10.1155/2012/148287

Cutchin, M. P. (2009). Geographical gerontology: New contributions and spaces for development. *The Gerontologist*, 49, 440–444. http://dx.doi.org/10.1093/geront/gnp095

Day, R. (2008). *Local urban environments and the well-being of older people*. Glasgow: Scottish Centre for Research on Social Justice.

De Donder, L., De Witte, N., Buffel, T., Dury, S. and Verté, D. (2012). Social capital and feelings of unsafety in later life: A study on influence of social networks, place attachment, and civic participation on perceived safety in Belgium. *Research on Aging*, 34(4), 425–448. http://dx.doi.org/10.1177/0164027511433879

De Donder, L., Buffel, T., Dury, S., De Witte, N. and Verté, D. (2013). Perceptual quality of neighbourhood design and safety. *Ageing & Society*, 33(6), 917–937. http://dx.doi.org/10.1004/S0144686X12000207

Department for Communities and Local Government. (2008). *Lifetime homes, lifetime neighbourhoods: A national strategy for housing in an ageing society*. London: Department of Health, Department for Work and Pensions. Available from: www.housingcare.org/information/detail-2966-lifetime-homes-lifetime-neighbourhoods-a-national-strateg.aspx

Fitzgerald, K. G. and Caro, F. G. (2016). *An overview of age-friendly communities around the world*. New York: Routledge.

Gilleard, C., Hyde, M. and Higgs, P. (2007). The impact of age, place, aging in place, and attachment to place on the well-being of the over 50s in England. *Research into Ageing*, 29(6), 590–605. http://dx.doi.org/10.1177/01640275073050730

Grant, T., Edwards, N., Sveistrup, H., Andrew, C. and Egan, M. (2010). Inequitable walking conditions among older people: Examining the interrelationship of neighbourhood socio-economic status and urban form using a comparative case study. *BMC Public Health*, 2010, 16pp. http://dx.doi.org/10.1186/1471-2458-10-677

Handler, S. (2014). *An alternative age-friendly handbook*. Manchester: UK Urban Ageing Consortium | MICRA.

Katz, S (2005). *Cultural aging, life course, lifestyle, and senior worlds.* Peterborough, Ontario: Broadview Press.

Krause, N. (2004). Neighbourhoods, health and well-being in later life, in H-W. Wahl, R. Scheidt and P. Windley (eds), *Aging in context: Socio-physical environments* (pp. 223–249). New York: Springer.

Lager, D., van Hoven, B. and Meijering, L. (2012). Places that matter: Place attachment and wellbeing of older Antillean migrants in the Netherlands. *European Spatial Research and Policy*, 19(1), 81–94. http://dx.doi.org/10.2478/v10105-012-0007-6

Lindenberg, J. and Westendorp, R.G.J. (2015). Overcoming old in age-friendliness. *Journal of Social Work Practice*, 29(1), 85–98. http://dx.doi.org/10.1080/02650533.2014.993949

Livingston, M., Bailey, N. and Kearns, A. (2010). Neighbourhood attachment in deprived areas: Evidence from the north of England. *Journal of Housing and the Built Environment*, 25(4), 409–427. http://dx.doi.org/10.1007/s10901-010-9196-3

Lloyd-Sherlock, P. and Agrawal, S. (2014). *Fear of crime and its effects on older people: Comparing six countries.* London: Age UK Policy Brief. Available from: www.ageuk.org.uk/Documents/EN-GB/For-professionals/Research/Fear%20of%20Crime.pdf?dtrk=true

Mendes de Leon, C. Cagney, K. A., Bienias, J. L., Barnes, L. L., Skarupski, K. A., Scherr, P. A. and Evans, D. A. (2009). Neighbourhood social cohesion and disorder in relation to walking in community-dwelling older adults: A multilevel analysis. *Journal of Aging and Health*, 21(1), 155–171. http://dx.doi.org/10.1177/0898264308328650

Menec, V. H., Means, R., Keating, N., Parkhurst, G. and Eales, J. (2011). Conceptualizing age-friendly communities. *Canadian Journal on Aging*, 30(3), 479–493. http://dx.doi.org/10.1017/S0714980811000237

Minton, A. (2009). *Ground control: Fear and happiness in the twenty-first century city.* London: Penguin Books.

Moulaert, T. and Garon, S. (2016). *Age-friendly cities and communities in international comparison: Political lessons, scientific avenues, and democratic issues.* New York: Springer.

Organisation for Economic Co-operation and Development (OECD). (2015). *Ageing in cities.* Paris: OECD.

Phillipson, C. (2007). The 'elected' and the 'excluded': Sociological perspectives on the experience of place and community in old age. *Ageing & Society*, 27(3), 321–342. http://dx.doi.org/10.1017/S0144686X06005629

Phillipson, C. (2010). Growing old in the "century of the city", in D. Dannefer and C. Phillipson (eds), *The Sage handbook of social gerontology* (pp. 597–606). London: Sage.

Riseborough, M. and Sribjilanin, A. (2000). *Overlooked and excluded? Older people and regeneration: A review of policy and practice.* London: Age Concern England.

Rowles, G. and Bernard, M. (eds). (2014). *Environmental gerontology: Making meaningful places in old age.* New York: Springer Publishing.

Satterthwaite, D. (ed.). (1999). *Sustainable cities.* London: Earthscan.

Scannell, L. and Gifford, R. (2010). Defining place attachment: A tripartite organizing framework. *Journal of Environmental Psychology*, 30(1), 1–10. http://dx.doi.org/10.1016/j.jenvp.2009.09.006

Scharf, T. and de Jong Gierveld, J. (2008). Loneliness in urban neighbourhoods: An Anglo-Dutch comparison. *European Journal of Ageing*, 5, 103–115. http://dx.doi.org/10.1007/s10433-008-0080-x

Scharlach, A. E., and Lehning, A. J. (2013). Ageing-friendly communities and social inclusion in the United States of America. *Ageing & Society*, 33(1), 110–136. http://dx.doi.org/10.1017/S0144686X12000578

Scharlach, A. and Lehning, A. (2016). *Creating age-friendly communities.* Oxford: Oxford University Press.

Simpson, C. (2010). *Older people and engagement in neighbourhood renewal: A qualitative study of stoke-on-trent.* Ph.D. diss., Keele University, Keele.

Skinner, M. W., Cloutier, D. and Andrews, G. (2015). Geographies of ageing: Progress and possibilities after two decades of change. *Progress in Human Geography*, 39(6), 776–799. http://dx.doi.org/10.1177/0309132514558444

Smith, A. (2009). *Ageing in urban neighbourhoods.* Bristol: Policy Press.

Soja, E. and Kanai, M. (2007). The urbanisation of the world, in R. Burdett and D. Sudjic (eds), *The endless city* (pp. 54–69). London: Phaidon,

Sudjic, D. (2016). *The language of cities.* London: Allen Lane.

UN-Habitat. (2008). *State of the world's cities 2008/2009: Harmonious cities.* London: Earthscan. Available from: http://mirror.unhabitat.org/pmss/listItemDetails.aspx?publication ID=2562&AspxAutoDetectCookieSupport=1

UN-Habitat. (2010). *State of the world's cities 2010/2011: Bridging the urban divide – overview and key findings.* London: Earthscan. Available from: https://sustainabledevelopment. un.org/index.php?page=view&type=400&nr=1114&menu=35

Wahl, H-W. and Oswald, F. (2010). Environmental perspectives on ageing, in D. Dannefer and C. Phillipson (eds), *The SAGE handbook of social gerontology* (pp. 111–124). London: Sage Publications.

Webb, E., Laverty, A. and Millett, C. (2015). Who has a bus pass, and is it good for their health? *Journal of Transport & Health*, 2(suppl 2), S58–S59. http://dx.doi.org/10.1016/ jth.2015.04.591

Wight, R. G., Cummings, J. R., Karlamangla, A. S. and Aneshensel, C. S. (2009). Urban neighbourhood context and change in depressive symptoms in later life. *Journal of Gerontology: Psychological Sciences*, 64B(2), 247–251. http://dx.doi.org/10.1093/geronb/gbn016

World Health Organization. (2007). *Global age-friendly cities: A guide.* Geneva: WHO. Available from: www.who.int/ageing/publications/Global_age_friendly_cities_Guide_ English.pdf?ua=1

11 Rural ageing

Contested spaces, dynamic places

Mark W. Skinner and Rachel Winterton

Introduction

Around the world, population ageing has become a defining attribute of the twenty-first century and nowhere is this happening more rapidly than in rural regions. Global rates of rural population ageing (i.e., the increasing proportion of rural residents in older age cohorts) are outpacing that of urban and metropolitan populations (World Health Organization, 2015). Rural populations are generally understood to live in a wide range of smaller, dispersed, underserviced settlements that lie outside urban and metropolitan regions, often at a great distance (Woods, 2010). As illustrated in the Australian context, for instance, older adults are less likely to live in urban centres, with one-third of older adults living outside of major cities in regional and remote settings (Australian Institute of Health and Welfare, 2017). These trends are also observed across Europe, North America and in countries located in the global South (see Phillips and Feng, Chapter 8; Rishworth and Elliott, Chapter 9; Rosenberg and Wilson, Chapter 5). They are driven by the rural outmigration of younger cohorts for employment and education, the ageing in place of older rural residents and, in some cases, the in-migration of retirees into rural areas, with significant health care system, socio-economic and cultural implications for individuals, households and communities (Keating, 2008). The result is a scenario poignantly described more than a decade ago as a 'double-jeopardy' (Joseph and Cloutier-Fisher, 2005), whereby increasing numbers of older rural adults throughout the world, many of whom have age-related health issues, are growing older in increasingly vulnerable rural places, many of which are struggling to provide services and supports, and are often reliant on volunteer and informal networks.

Rural ageing is not just about population trends and their implications. In examining the contexts in which people grow older, rural places are assuming greater significance for understanding the processes, outcomes and representations of ageing (see contributions to Glasgow and Berry, 2013; Hagan Hennessy, Means and Burholt, 2014; Hash *et al.*, 2014; Keating, 2008; Milbourne, 2012; Skinner and Hanlon, 2016). There is consensus that rural locales present different challenges and opportunities to health and wellbeing in later life to those evident in urban settings. Despite these differences, the rural ageing context remains relatively under-examined within research and largely unaccounted for within prevailing policy (Scharf, Walsh and O'Shea, 2016). As this chapter shows, however, there is a burgeoning body of geographically informed scholarship – *a rural geographical gerontology* – that has contributed to the depth of our understanding of rural ageing processes and outcomes, experiences and representations. In particular, we review the work of geographers and allied rural gerontology scholars who have enabled explorations of how older people and their communities experience and transform population change across scales, throughout contested spaces and within the dynamics of place. In doing so, we elucidate the scope of rural geographical gerontology scholarship, and informed by a critical turn in gerontology, we reflect on the overarching importance of developing a critical perspective within rural ageing studies more broadly.

Rural geographical gerontology

Geographers and allied scholars working in geographical gerontology have a long-standing interest in older rural populations and understanding the settings, environments and communities where they live (Skinner, Cloutier and Andrews, 2015). From Graham Rowles's groundbreaking geographical studies of the lives of older Appalachians in rural Kentucky in the United States (1983) and Alun Joseph's landmark work on ageing in rural communities (Joseph and Chalmers, 1995; Joseph and Martin-Matthews, 1993), to the work of geographers who have followed including Denise Cloutier, Malcolm Cutchin, Neil Hanlon, Mark Skinner and Janine Wiles, there has been a growing and increasingly coherent body of rural ageing scholarship within geography. Geographers have pioneered the application of place-based approaches for understanding the implications of rural population ageing at various scales (Hanlon and Halseth, 2005), the importance of space and place as constructs within rural gerontology (Cutchin, 2009; Wiles, 2005) and have refined the theorization of the spatial dynamics of ageing rural communities (Joseph and Skinner, 2012), particularly in relation to service delivery (Cloutier-Fisher and Skinner, 2006), volunteerism (Skinner and Joseph, 2011) and informal care (Herron and Skinner, 2013).

The growing importance of geographical concepts and approaches within broader interdisciplinary scholarship, however, was propelled by the concurrent work of leading social gerontology scholars, notably Vanessa Burholt, Nina Glasgow, Norah Keating, Judith Phillips, Thomas Scharf, Kieran Walsh, Jeni Warburton and Rachel Winterton, who have led to a greater understanding and appreciation of the diverse experiences of older rural people and complexity of change in twenty-first-century ageing rural places. Their work has connected rural ageing studies to broader debates about age-friendly communities, place attachment, retirement migration, social inclusion and volunteerism, among other themes. The influence on geographers and gerontologists of Keating's (2008) provocatively titled human ecology book *Rural Ageing: A Good Place to Grow Old?* – not to mention her leadership in calling for interrogations of the interactions between people and place for building our understanding of rural ageing (Keating, Swindle and Fletcher, 2011) – cannot be overstated (see Skinner and Hanlon, 2016). Indeed, the evolution of what now could be called rural geographical gerontology is an excellent example of the co-developments that often occur among disciplines (geography) and broader interdisciplinary studies (social gerontology) as observed by Andrews *et al.* (2007, 2009). This is exemplified by the various interdisciplinary and international contributions to the recent *Journal of Rural Studies* special issue, edited by Milbourne (2012), on growing old in rural places (Bailey and Biggs, 2012; Burholt and Dobbs, 2012; Curry and Fisher, 2012; Davis *et al.*, 2012; Glasgow and Brown, 2012; Joseph and Skinner, 2012; Keeling, 2012; Manthorpe *et al.*, 2012; Milbourne and Doheny, 2012; Shergold and Parkhurst, 2012; Walsh *et al.*, 2012; Winterton and Warburton, 2012; Zanjani and Rowles, 2012).

One such co-development, that of critical rural gerontology (Scharf *et al.*, 2016), has the potential to further influence the application of geographical concepts and approaches within rural ageing studies. Work within culturally informed critical gerontology is illuminating the ways in which gender, sexuality, race, class and ability relate to processes and outcomes, experiences and representations of ageing (see contributions to Twigg and Martin, 2015). Implications for rural geographical gerontology include the need to understand and challenge the construction (and contestation) of spaces, places and environments of rural ageing and the myths and stereotypes associated with prevailing conceptions of ageing and rurality (Skinner and Winterton, 2017). As the next section illustrates, there are many ways in which geographical research can contribute to the critically informed 'rural turn' underway in contemporary gerontology.

What is 'rural' about rural geographical gerontology?

As Scharf, Walsh and O'Shea (2016) observed, there is a long-standing assumption that ageing in rural places is distinct from 'urban' ageing, with research over time continuing to question what is 'rural' about rural ageing (Rowles, 1988; Wahl, 2005). While the nascent geographical gerontology literature has lamented this comparative dichotomy in understanding the experiences of older adults (Andrews *et al.*, 2007, 2009), it has nevertheless provided a lens through which to consider the environmental contexts in which people age (Cutchin, 2009). However, if we interrogate contemporary definitions of geographical gerontology, there are specific aspects that are relevant to understanding the emergence of a specific 'rural' sub-strand of the discipline.

First, the mutually constitutive relationship between ageing and rural sustainability is critical to understanding the settings in which people age. Over time, prominent geographical gerontologists have highlighted the importance of recognizing not only how environmental settings influence older adults but also how older adults simultaneously influence the settings within which they interact (Cutchin, 2009; Wiles, 2005). This is particularly relevant in rural environments, with international research highlighting the complex relationship between rural community sustainability, socio-demographic ageing trends and the agentic actions of older adults (Hermanova, Brown, Goins and Briggs, 2001). While ageing has been identified as a perceived threat to rural community sustainability in terms of the economic trajectories of these settings, research has simultaneously highlighted the pivotal role that older adults play in supplementing declining rural social infrastructure through voluntarism and civic engagement (Davis *et al.*, 2012; Winterton and Warburton, 2014). Consequently, recent contributions to the field advocate developing contextualized studies of rural aging spaces that view the experiences of older adults and the responses of their communities as both mutually constituted and mutually transformative (Skinner and Hanlon, 2016; Skinner and Winterton, 2017).

Second, rural community contexts of ageing are being transformed more radically than their urban equivalents (Joseph and Cloutier-Fisher, 2005). As Wiles (2005) notes, places are subject to ongoing negotiation and – across international contexts – certain macro-level trends are having a significant impact on the experience of rural ageing. Much of this is linked with processes associated with globalization. International research has noted the influence that greater levels of population mobility, welfare state retrenchment, suburbanization of rural settings, rural economic restructuring and the advent of new information and communications technologies are having on older adults' health and well-being in rural settings (Keating *et al.*, 2011; Scharf *et al.*, 2016; Wahl, 2005; Winterton and Warburton, 2012).

Importantly, it is trends related to ageing that are in some cases prompting these rural transformations. While rural population ageing is a consequence of large-scale economic restructuring because of working-age populations migrating to urban areas in search of employment (Glasgow and Brown, 2012), this population ageing is simultaneously changing the social and economic trajectories of rural communities (Rowles and Watkins, 1993). At the same time, macro-level trends associated with the changing notions of retirement are also instigating rural change, with increasing levels of rural retirement migration prompting greater diversity within rural ageing populations (Brown and Glasgow, 2008; Stockdale and MacLeod, 2013). Consequently, rural change is both facilitated by, and contributing to, rural population ageing. As such, there is a significant role for geographical gerontology to interrogate how this rural change is simultaneously negotiated and facilitated by older adults (Chalmers and Joseph, 2006). This is pressing when one considers that these endemic changes are, in many regions, challenging pervasive definitions of what it means to be 'rural' – socially, spatially and environmentally (Rowles, 1988). In the context of research demonstrating the importance of rural place identity to ageing well in rural settings (Burholt, 2006; Rowles, 1983; Winterton and Warburton, 2012), this understanding is critical. Wiles (2005) argues that places are highly contested and a recent paper by Skinner and Winterton (2017) has sought to explicate these contested spaces of rural ageing. In line with contemporary discourses on rural citizenship (Woods, 2006), they argue that conflict in rural ageing spaces emerges over the right to the symbolic, political and cultural spaces of the rural in older age, and the right to age in place within the rural. Consequently, it is important to understand the implications of this conflict for diverse older people and rural communities.

As this chapter suggests, the diverse scales at which rural ageing occurs also justify the emergence of a distinct rural geographical gerontology. While 'rural' is broadly conceptualized as a 'context' within which people age, geographers and social gerontologists have warned against an oversimplified, generalized notion of rurality in understanding the experience of rural ageing (Cutchin, 2009; Rowles, 1988; Scharf *et al.*, 2016). In the subsequent section, we outline the various spaces, places and scales within which rural older adults age, and the implications of this complexity for rural geographical gerontology. We argue that it is the varied experiences within these spaces and places, in terms of how they are constructed and negotiated within and across rural places and at different scales, that are critical in understanding what is uniquely 'rural' about rural ageing.

Geographical scales and spaces of rural ageing

Macro environments

Constructions and classifications of rurality at various scales and within diverse spaces and places influence the lives of older adults. Specifically, the construction of the rural, macro-level context in this regard has been subject to debate. Discussions of the 'rural' context within ageing research have commonly employed macro-level definitions and scales (Andrews *et al.*, 2009); a context where the rural merely acts as the ecological setting within which people age (Rowles, 1988). In this context, the rural reflects an arbitrary set of demographic, physical and spatial characteristics (Warnes, 1990), and is relational in that it is continually being defined against other spatial settings that are more, or less, rural (Andrews *et al.*, 2007; Scharf *et al.*, 2016) Alternatively, a more limited scale of inquiry has sought to examine the rural as a sociocultural context (Rowles, 1988; Keating and Phillips, 2008), one which explores how social and cultural constructions of the rural, such as norms, values and attitudes, influence older adults. However, these constructions are often influenced by, and similarly influence, macro-level changes and processes, which also impact rural older adults in different ways. While some of these processes are exclusive to rural settings, others reflect broader changes that are having specific and unique impacts on rural areas, such as in the arenas of health and ageing policy reforms, increasing population mobility, economic change and restructuring (Andrews *et al.*, 2009; Cutchin, 2009; Skinner and Winterton, 2017). Given the dynamic nature of rural places (Williams and Kulig, 2012), the importance of rural identities in defining the self in older age (Winterton and Warburton, 2012) and the limited capacity of some older adults to respond to environmental change (Golant, 2003), these issues raise critical questions for rural gerontologists regarding how macro-level processes contribute to contested conceptualizations of the rural. For example, places may become less or more 'rural' over time based on arbitrary characteristics, but little is known relative to how this impacts the sociocultural and phenomenological construction of the rural among older adults.

For geographical gerontologists, critical questions also arise in relation to how these varied macro-level scales at which the rural is defined (locality vs. social construction) and how the macro-level processes that impact the rural, interact with and influence community and individual levels of context. While earlier critiques of rural–urban classifications outlined a lack of evidence of causality as a significant issue in understanding this phenomenon (Rowles, 1988), much research since then has sought to explain how macro-level characteristics and processes impact the rural, by directly influencing either the lives of older adults or the community contexts within which they are ageing (e.g., Keating, Eales and Phillips, 2013; Walsh *et al.*, 2012). To explain how macro-level processes and characteristics influence rural community and neighbourhood settings, rural gerontologists have employed human ecology theory, which emphasizes the interactions between different environmental contexts (Keating and Phillips, 2008; Winterton *et al.*, 2016). Much of this research has highlighted not only the role of macro-level factors – low population density, declining populations and geographical isolation – on the rural resource environment (Andrews *et al.*, 2009; Rowles, 1988; Walsh *et al.*, 2012) but also the social construction of rural older people as highly independent and resilient (Keating *et al.*, 2011; Walker *et al.*, 2013).

Local, community and neighbourhood spaces

There is a comprehensive body of literature exploring how rural community level characteristics impact older adults' health, well-being and quality of life (see Menec and Nowicki, 2014; Winterton *et al.*, 2016). In exploring locality-based spaces, a key sphere of inquiry has been the identification of the different spaces that simultaneously impact and are influenced by older adults within rural areas. The wider rural age-friendly literature highlights the various opportunities and constraints that diverse community spaces (e.g., social, political, physical, service-related) pose for diverse older adults (Keating *et al.*, 2013; Spina and Menec, 2015). However, the community environment is not entirely physical and a nascent body of work details how rural older adults' 'imagined' local spaces, in relation to perceptions of local place identity, person–environment fit and belonging, are being affected by macro-level changes and processes and how these are continually being negotiated (Burholt, 2006; Winterton and Warburton, 2012; Keating *et al.*, 2013; Novek and Menec, 2014). While the importance of a more generalised rural identity has been discussed, work from the United Kingdom and Australia has identified the importance of a 'local', territorial, rural culture in fostering attachment to place and well-being in older age (Burholt and Naylor, 2005; Winterton and Warburton, 2012).

Within both material and imagined spaces, older adults are simultaneously engaged in both active and passive negotiations over access to community spaces, and what they should look like (Skinner and Winterton, 2017). Additionally, other community members, volunteers, organizations and policymakers are engaged within these spaces either in support of, or in opposition to, rural older adults' rights and entitlements (Farmer *et al.*, 2012; Joseph and Skinner, 2012; Spina and Menec, 2015). This complex negotiation of the physical and the imagined rural in older age, in the context of broader macro-level characteristics and processes, raises some critical questions. Rural geographical gerontology is well poised to interrogate how different rural communities experience, challenge or acquiesce with macro-level rural processes and characteristics that impact ageing and to examine the role of older people within this process. Moreover, critical inquiry is urgently needed in relation to how diverse forms of population ageing are altering the notion of local, rural identities and the impact of this on individual experiences of ageing.

Micro-level spaces and places

While ageing takes place within rural socio-spatial and community environments against a broader backdrop of macro-level characteristics and processes, these spaces (and the negotiations that occur within them) give rise to the experience of ageing rurally at the individual level. However, early research has questioned to what extent the micro-environments of older adults – their immediate spaces of interaction and care – are actually 'rural' (Rowles, 1988), in terms of how they are distinctly different from those experienced in urban settings. This can be interrogated in relation to two key factors: the varied composition and proximity of rural older people's microenvironments in comparison to urban dwellers and the relative importance assigned to various elements of rural older people's microenvironments.

In response to service restructuring, rural older adults' spaces of care are more likely to encompass voluntary sector or informal supports, delivered by friends and/or family members (Keating *et al.*, 2011; Skinner and Joseph, 2011), with many older adults exhibiting a preference for local, informal care (Bacsu *et al.*, 2012). However, in response to the continued out-migration of working-age populations (Glasgow and Brown, 2012), they are also less likely to have proximal family members available to assist with informal caregiving. Therefore, numerous studies have highlighted the considerable importance of local social connectivity and support networks to rural older adults (Bacsu *et al.*, 2012; Burholt, 2006). In this context, we also must consider that spaces of care may not necessarily be local, particularly in the context of wider service restructuring, resulting in many rural older adults being forced to travel outside of their communities to access medical services and residential care (Bernoth, Dietsch and Davies, 2012; Goins *et al.*, 2005). The risk of residential home care admission also increases with level of rurality (Gilbert *et al.*, 2010), suggesting that rural seniors may be more likely to age in residential care, rather than home settings. In the face of geographic isolation, rural spaces of care are also likely to be virtual, with increasing levels of rural services being delivered by telephone and the Internet (Herron and Waldbrook, 2016). Consequently, it could be argued that the microenvironments of rural older adults are going to be increasingly 'non-local', which is a topic worthy of exploration by rural geographical gerontologists. A key concept also worth examining is the extent to which norms, attitudes and values associated with the rural are incorporated into older adults' spaces of care, social interactions and the home environments. For example, Burholt (2006) has suggested that rural informal care structures may be partly driven by community or macro-level rural social norms.

Towards a critical rural geographical gerontology?

While the literature on geographical gerontology explicitly notes that rural spaces and places of ageing are both dynamic and contested (Skinner and Winterton, 2017), this can, in part, be explained by the heterogeneity of both rural places, and the life-course experiences of, and resources available to, their older residents (Keating and Phillips, 2008; Scharf *et al.*, 2016). Consequently, to understand how rural spaces and places are constructed and contested in older age, many geographical gerontologists are increasingly drawing on approaches from critical gerontology. This critical lens acknowledges the heterogeneity of older adults – in terms of age, race, gender, class, education, personal resources and life-course experiences – and how this can alternatively empower or disenfranchise older adults in different circumstances (Scharf *et al.*, 2001; Wahl, 2005). This is particularly relevant given that changing demographic and family structures, variations in lifestyles, and opportunities across the life course are increasing the diversity among older adults (Scharf *et al.*, 2016). These changes are being felt acutely in rural regions, regions whose ageing populations are also becoming less homogenous as a consequence of increasing population mobility and counter-urbanization trends (Keating and Phillips, 2008).

As Scharf, Walsh and O'Shea (2016) have recently noted, there is much to be gained by combining critical gerontology and geographical gerontology in exploring issues associated with rural ageing. From a geographical perspective, there is scope to further investigate how different rural places and spaces empower or disadvantage older people at different levels of context and how trends associated with diverse forms of population ageing challenge or change power differentials within rural spaces and places. In recognition of the diversity of rural places, there is also a need to interrogate how different rural spaces or places – whether these are spatial, social, physical, political or imagined – impact the ageing experience for diverse older adults. While there is a nascent body of work beginning to examine these issues (see Keating *et al.*, 2013; Walsh *et al.*, 2014; Winterton and Warburton, 2012), continued work in this area will enable us to better understand the complexity of the mutually constitutive relationship between older people and rural places (Skinner and Hanlon, 2016).

To conclude, in reviewing the body of geographically informed scholarship on rural ageing, this chapter provides a foundation for understanding the scope and depth of *rural geographical gerontology*; an increasingly robust field of inquiry that enables explorations of how older people and their communities experience and transform processes and outcomes of population change across scales, throughout contested spaces and within the dynamics of place. Looking forward, we advocate careful attention to developing critical perspectives that are sensitive to the interactions between older people and ageing places, not to mention inclusive of approaches that engage with the diversity and complexity of older persons experiences across time (life course), setting (household, community, institution) and scale (from the corporeal to the global). Rural ageing will remain at the forefront of population change in the twenty-first century; where better than in rural regions, small towns and villages to learn how to best support older people and sustain ageing places?

Reference

Andrews, G. J., Cutchin, M., McCracken, K., Phillips, D. R. and Wiles, J. (2007). Geographical gerontology: The constitution of a discipline. *Social Science & Medicine*, 65(1), 151–168. http://dx.doi.org/10.1016/j.scoscimed.2007.02.047

Andrews, G. J., Milligan, C., Phillips, D. R. and Skinner, M. W. (2009). Geographical gerontology: Mapping a disciplinary intersection. *Geography Compass*, 3(5), 1641–1659. http://dx.doi.org/10.1111/j.1749-8198.2009.00270.x

Australian Institute for Health and Welfare. (2017). *Older Australians at a glance – regional and remote communities*. Canberra: Australian Government.

Bacsu, J. R., Jeffery, B., Johnson, S., Martz, D., Novik, N. and Abonyi, S. (2012). Healthy aging in place: Supporting rural seniors' health needs. *Online Journal of Rural Nursing and Health Care*, 12(2), 77–87.

Bailey, J. and Biggs, I. (2012). "Either Side of Delphy Bridge": A deep mapping project evoking and engaging the lives of older adults in rural North Cornwall. *Journal of Rural Studies*, 28(4), 318–328. http://dx.doi.org.2012.01.001

Bernoth, M., Dietsch, E. and Davies, C. (2012). Forced into exile: The traumatising impact of rural aged care service inaccessibility. *Rural and Remote Health*, 12(1), 1–8. Available from: www.rrh.org.au/publishedarticles/article_print_1924.pdf

Brown, D. and Glasgow, N. (2008). *Rural retirement migration*. New York: Springer.

Burholt, V. (2006). Adref: Theoretical contexts of attachment to place for mature and older people in rural North Wales. *Environment and Planning A*, 38(6), 1095–1114. http://dx.doi.org/10.1068/a3767

Burholt, V. and Dobbs, C. (2012). Research on rural ageing: Where have we got to and where are we going in Europe? *Journal of Rural Studies*, 28(4), 432–446. http://dx.doi.org/10.1016/j.jrurstud.2012.01.009

Burholt, V. and Naylor, D. (2005). The relationship between rural community type and attachment to place for older people living in North Wales, UK. *European Journal of Ageing*, 2(2), 109–119. http://dx.doi.org/10.1007/s10433-005-0028-3

Chalmers, A. and Joseph, A. (2006). Rural change and the production of otherness: The elderly in New Zealand, in P. Cloke, T. Marsden and P. Mooney (eds), *Handbook of rural studies* (pp. 388–400). London: Sage.

Cloutier-Fisher, D. and Skinner, M. W. (2006). Levelling the playing field? Exploring the implications of managed competition for voluntary sector providers of long-term care in small town Ontario. *Health & Place*, 12(1), 97–109. http://dx.doi.org/10.1016/j.healthplace.2004.10.012

Curry, N. and Fisher, R. (2012). The role of trust in the development of connectivities amongst rural elders in England and Wales. *Journal of Rural Studies*, 28(4), 358–370. http://dx.doi.org/10.1016/j.jrurstud.2012.06.005

Cutchin, M. (2009). Geographical gerontology: New contributions and spaces for development. *Gerontologist*, 49(3), 440–444. http://dx.doi.org/10.1093/geront/gnp095

Davis, S., Crothers, N., Grant, J., Young, S. and Smith, K. (2012). Being involved in the country: Productive ageing in different types of rural communities. *Journal of Rural Studies*, 28(4), 338–346. http://dx.doi.org/10.1016/j.jrurstud.2012.01.008

Farmer, J., Nimegeer, A., Farrington, J. H. and Rodger, G. (2012). Rural citizens' rights to accessible health services: An exploration. *Sociologia Ruralis*, 52(1), 134–144. http://dx.doi.org/10.1111/j.1467-9523.2011.00549.x

Gilbert, R., Todd, C., May, M., Yardley, L. and Ben-Shlomo, Y. (2010). Socio-demographic factors predict the likelihood of not returning home after hospital admission following a fall. *Journal of Public Health*, 32(1), 117–124. http://dx.doi.org/10.1093/pubmed/fdp077

Glasgow, N. and Berry, E. H. (eds). (2013). *Rural aging in 21st century America*. Dordrecht, The Netherlands: Springer.

Glasgow, N. and Brown, D. L. (2012). Rural ageing in the United States: Trends and contexts. *Journal of Rural Studies*, 28(4), 422–431. http://dx.doi.org/10.1016/j.rurstud.2012.01.002

Goins, R. T., Williams, K. A., Carter, M. W., Spencer, S. M. and Solovieva, T. (2005). Perceived barriers to health care access among rural older adults: A qualitative study. *The Journal of Rural Health*, 21(3), 206–213. http://dx.doi.org/10.1111/j.1748-0361.2005.tb00084.x

Golant, S. (2003). Conceptualizing time and behavior in environmental gerontology: A pair of old issues deserving new thought. *Gerontologist*, 43(5), 638–648. http://dx.doi.org/10.1093/geront/43.5.638

Hagan Hennessy, C., Means, R. and Burholt, V. (eds). (2014). *Countryside connections: Older people, community and place in rural Britain*. Bristol: Policy Press.

Hanlon, N. and Halseth, G. (2005). The greying of resource communities in northern British Columbia: Implications for health care delivery in already-underserviced communities. *The Canadian Geographer*, 49(1), 1–24. http://dx.dio.org/10.1111/j.0008-3658.2005.00077.x

Hash, K. M., Jurkowski, E. T. and Krout, J. A. (2014). *Aging in rural places: Programs, policies, and professional practice*. New York: Springer.

Hermanova, H., Brown, D. K., Goins, R. T. and Briggs, R. (2001). The first international conference on rural aging: A global challenge. *Journal of Rural Health*, 17(4), 303–304. http://dx.doi.org/10.1111/j.1748-0361.2001.tb00275.x

Herron, R. V. and Skinner, M. W. (2013). The emotional overlay: Older person and carer perspectives on negotiating aging and care in rural Ontario. *Social Science & Medicine*, 91, 186–193. http://dx.doi.org/10.1016/j.socscimed.2012.08.037

Herron, R. and Waldbrook, N. (2016). Emerging issues in ageing resource communities, in M. Skinner and N. Hanlon (eds), *Ageing resource communities: New frontiers of rural population change, community development and voluntarism* (pp. 195–205). New York: Routledge.

Joseph, A. E. and Chalmers, A. I. (1995). Growing old in place: A view from rural New Zealand. *Health & Place*, 1(2), 79–90. http://dx.doi.org/10.1016/1353-8292(95)00011-A

Joseph, A. and Cloutier-Fisher, D. (2005). Ageing in rural communities: Vulnerable people, vulnerable places, in G. Andrews and D. Phillips (eds), *Ageing and place: Perspectives, policy and practice* (pp. 133–146). London: Routledge.

Joseph, A. E. and Martin-Matthews, A. (1993). Growing old in aging communities. *Journal of Canadian Studies*, 28(1), 14–29. http://dx.doi.org/10.3138/jcs.28.1.14

Joseph, A. E. and Skinner, M. W. (2012). Voluntarism as a mediator of the experience of growing old in evolving rural spaces and changing rural places. *Journal of Rural Studies*, 28(4), 380–388. http://dx.doi.org/10.1016/j.jrurstud.2012.07.007

Keating, N. (2008). *Rural ageing: A good place to grow old?* Bristol: Policy Press.

Keating, N., Eales, J. and Phillips, J. E. (2013). Age-friendly rural communities: Conceptualizing 'best-fit'. *Canadian Journal on Aging/La Revue canadienne du vieillissement*, 32(4), 319–332. http://dx.doi.org/10.1017/S0714980813000408

Keating, N. and Phillips, J. (2008). A critical human ecology perspective on rural ageing, in N. Keating (ed.), *Rural ageing: A good place to grow old?* (pp. 1–10). Bristol: Policy Press.

Keating, N., Swindle, J. and Fletcher, S. (2011). Aging in rural Canada: A retrospective and review. *Canadian Journal on Aging*, 30(3), 323–338. http://dx.doi.org/10.1017/S0714980811000250

Keeling, S. (2012). Grandchildren's perspectives on grandparents in rural New Zealand. *Journal of Rural Studies*, 28(4), 371–379. http://dx.doi.org/10.1016/j.jrurstud.2012.04.002

Manthorpe, J., Moriarty, J., Stevens, M., Hussein, S. and Sharif, N. (2012). Promoting the mental well-being of older people from black and minority ethnic communities in United Kingdom rural areas: Findings from an interview study. *Journal of Rural Studies*, 28(4), 406–411. http://dx.doi.org/j.jrurstud.2012.01.006

Menec, V. H. and Nowicki, S. (2014). Examining the relationship between communities'"age-friendliness" and life satisfaction and self-perceived health in rural Manitoba, Canada. *Rural and Remote Health*, 14(2594). Available from: www.rrh.org.au/articles/subviewnew.asp?ArticleID=2594

Milbourne, P. (2012). Growing old in rural places. *Journal of Rural Studies*, 28(4), 315–317. http://dx.doi.org/10.1016/j.jrurstud.2012.10.001

Milbourne, P. and Doheny, S. (2012). Older people and poverty in rural Britain: Material hardships, cultural denials and social inclusions. *Journal of Rural Studies*, 28(4), 389–397. http://dx.doi.org/10.1016/j.jrurstud.2012.06.007

Novek, S. and Menec, V. H. (2014). Older adults' perceptions of age-friendly communities in Canada: A photovoice study. *Ageing and Society*, 34(6), 1052–1072. http://dx.doi.org/10.1017/S0144686X1200150X

Rowles, G. (1983). Place and personal identity in old age: Observations from Appalachia. *Journal of Environmental Psychology*, 3(4), 299–313. http://dx.doi.org/10.1016/S0272-4944(83)80033-4

Rowles, G. (1988). What's rural about rural aging? An Appalachian perspective. *Journal of Rural Studies*, 4(2), 115–124. http://dx.doi.org/10.1016/0743-0167(88)90029-0

Rowles, G. D. and Watkins, J. F. (1993). Elderly migration and development in small communities. *Growth and Change*, 24(4), 509–538. http://dx.doi.org/10.1111/j.1468-2257.tb00136.x

Scharf, T., Phillipson, C., Kingston, P. and Smith, A. E. (2001). Social exclusion and older people: Exploring the connections. *Education and Ageing*, 16(3), 303–320. Available from: www.researchgate.net/profile/Paul_Kingston/publication/228808448_Social_Exclusion_and_Older_People_exploring_the_connections/links/542fcf400cf27e39fa997997.pdf

Scharf, T., Walsh, K. and O'Shea, E. (2016). Ageing in rural places, in M. Shucksmith and D. Brown (eds), *Routledge international handbook of rural studies* (Chapter 4). Oxon: Routledge.

Shergold, I. and Parkhurst, G. (2012). Transport-related social exclusion amongst older people in rural Southwest England and Wales. *Journal of Rural Studies*, 28(4), 412–421. http://dx.doi.org/10.1016/j.jrurstud.2012.01.010

Skinner, M. W., Cloutier, D. and Andrews, G. J. (2015). Geographies of ageing Progress and possibilities after two decades of change. *Progress in Human Geography*, 39(6), 776–799. http://dx.doi.org/10.1177/0309132514558444

Skinner, M. and Hanlon, N. (eds). (2016). *Ageing resource communities: New frontiers of rural population change, community development and voluntarism*. New York: Routledge.

Skinner, M. W. and Joseph, A. E. (2011). Placing voluntarism within evolving spaces of care in ageing rural communities. *GeoJournal*, 76(2), 151–162. http://dx.doi.org/10.1007/s107088-009-9283-8

Skinner, M. and Winterton, R. (2017). Interrogating the contested spaces of rural aging: Implications for research, policy and practice. *Gerontologist*. http://dx.doi.org/10.1093/geront/gnx094

Spina, J. and Menec, V. H. (2015). What community characteristics help or hinder rural communities in becoming age-friendly? Perspectives from a Canadian prairie province. *Journal of Applied Gerontology*, 34(4), 444–464. http://dx.doi.org/10.1177/0733464813496164

Stockdale, A. and MacLeod, M. (2013). Pre-retirement age migration to remote rural areas. *Journal of Rural Studies*, 32, 80–92. http://dx.doi.org/10.1016/j.jrurstud.2013.04.009

Twigg, J. and Martin, W. (eds). (2015). *Routledge handbook of cultural gerontology*. London: Routledge.

Wahl, H-W. (2005). Ageing research along the urban-rural distinction: Old questions and new potential. *European Journal of Ageing*, 2(2), 131–136. http://dx.doi.org10.1007/s10433-005-0030-9

Walker, J., Orpin, P., Baynes, H., Stratford, E., Boyer, K., Mahjouri, N., Patterson, C., Robinson, A. and Carty, J. (2013). Insights and principles for supporting social engagement in rural older people. *Ageing and Society*, 33(6), 938–963. http://dx.doi.org/10.1017/S0144686X12002402

Walsh, K., O'Shea, E., Scharf, T. and Murray, M. (2012). Ageing in changing community contexts: Cross-border perspectives from rural Ireland and Northern Ireland. *Journal of Rural Studies*, 28(4), 347–357. http://dx.doi.org/10.1016/j.jrurstud.2012.01.012

Walsh, K., Scharf, T. and Shucksmith, M. (2014). Exploring the impact of informal practices on social exclusion and age-friendliness for older people in rural communities. *Journal of Community & Applied Social Psychology*, 24(1), 37–49. http://dx.doi.org/10.1002/casp.2176

Warnes, A. M. (1990). Geographical questions in gerontology: Needed directions for research. *Progress in Human Geography*, 14(1), 24–56. http://dx.doi.org/10.1177/030913259001400103

Wiles, J. (2005). Conceptualizing place in the care of older people: The contributions of geographical gerontology. *Journal of Clinical Nursing*, 14(s2), 100–108. http://dx.doi.org/10.1111/j1365-2702-200501281.x

Williams, A. and Kulig, J. (2012). Health and place in rural Canada, in J. Kulig and A. Williams (eds), *Health in rural Canada* (pp. 2–19). Vancouver/Toronto: UBC Press.

Winterton, R. and Warburton, J. (2012). Ageing in the bush: The role of rural places in maintaining identity for long term rural residents and retirement migrants in North-East Victoria, Australia. *Journal of Rural Studies*, 28(4), 329–337. http://dx.doi.org/10.1016/j.jrurstud2012.0.005

Winterton, R. and Warburton, J. (2014). Healthy ageing in Australia's rural places: The contribution of older volunteers. *Voluntary Sector Review*, 5(2), 181–201. http://dx.doi.org/10.1332/204080514X14020463739601

Winterton, R., Warburton, J., Keating, N., Petersen, M., Berg, T. and Wilson, J. (2016). Understanding the influence of community characteristics on wellness for rural older adults: A meta-synthesis. *Journal of Rural Studies*, 45, 320–327. http://dx.doi.org/10.1016/j.jrurstud.2015.12.001

Woods, M. (2006). Political articulation: The modalities of new critical politics of rural citizenship, in P. Cloke, T. Marsden and P. Mooney (eds), *Handbook of rural studies* (pp. 457–471). London: Sage.

Woods, M. (2010). *Rural: Key ideas in geography*. London: Taylor and Francis.

World Health Organization (WHO). (2015). *World report on ageing and health*. Geneva: WHO. Available from: http://apps.who.int/iris/bitstream/10665/186463/1/9789240694811_eng.pdf?ua=1

Zanjani, F. and Rowles, G. D. (2012). "We don't want to talk about that": Overcoming barriers to rural aging research and interventions on sensitive topics. *Journal of Rural Studies*, 28(4), 398–405. http://dx.doi.org/10.1016/j.jrurstud2012.03.005

12 Ageing communities

Sarah A. Lovell

Introduction

The town of Roseto in Pennsylvania, USA became renowned when evidence emerged that its strong social cohesion appeared to have a sustained, positive impact on the community's health outcomes (Egolf *et al.*, 1992; Wolf and Bruhn, 1998). The town was founded by immigrants from a southern Italian city where close intergenerational ties saw extended family living together. Roseto was unique for its stability and high levels of community support (Egolf *et al.*, 1992). The experience of Roseto residents speaks to the important role of community in sustaining social norms and, in turn, health outcomes. Older adults were valued, respected and continued to play an important role in family decision-making; traditional family structures were evident with men as the clear head of the household (Wolf and Bruhn, 1998). The town itself had high levels of civic pride; close ties to neighbours were an important feature of the community, and local initiatives were well supported. Roseto, at this time, had a myocardial infarction rate that was significantly lower than the more diverse neighbouring town (Egolf *et al.*, 1992). Ultimately, social change is reported to have taken place in Roseto. Younger adults moved away for jobs, the community cohesion that discouraged conspicuous consumption faded and lifestyles changed (Wolf and Bruhn, 1998). Over 30 years, the lower myocardial infarction rate in Roseto had, by 1965, caught up to neighbouring towns (Egolf *et al.*, 1992). The authors attributed this to changing familial and community ties:

> The evidence at hand confirms a familiar belief that whether a community lives in social equilibrium or social disruption depends on the way people treat each other as they cope with everyday challenges. It also strongly suggests that the way individuals perceive themselves and their role in society may palpably influence their health and that of their community.
>
> (Wolf and Bruhn, 1998: 128)

Possessing a sense of community is akin to perceiving that one belongs to a supportive group (Pretty, Chipuer and Bramston, 2003). When communities are effective, this sense of group membership enables reciprocal exchanges that may be considered a form of caring (Pretty, Chipuer and Bramston, 2003). The benefits of belonging to a supportive community are self-evident to many older adults and were an important element in my grandmother's renewed commitment to her church as a widow. Research has consistently illustrated that a positive sense of identity can emerge from strong connections to community (Glendinning *et al.*, 2003; Stedman, 2002), and the longer one lives in a community, the greater the likelihood of involvement (Davis *et al.*, 2012). However, a problem of 'civic malaise' has emerged across Western nations in which individualism appears to have collectively undermined our social ties (Deijerm and Uslaner, 2001; Fukuyama, 1999). In the following sections we critically examine how social changes have reshaped understandings of community over time, identify how communities support healthy ageing, and consider how ageing communities can build capacity.

Neoliberalism, individualism and challenges for 'community'

Building a place-based sense of community offers a potentially inclusive means by which ageing populations can be supported and sense of identity maintained. Such an approach relies on shared norms; thus, Rose characterises 'good' citizenship as "guided by common virtues and a commitment to the common good, whose active engagement in the life of the polis and the affairs of the community would revitalize civil society" (Rose, 1999: 169). Fukuyama similarly insists that a community is more than just a group of people; he argues that "true communities are bound together by the values, norms and experiences shared among their members. The deeper and more strongly held those common values are, the stronger the sense of community is" (Fukuyama, 1999: 14). Importantly, however, such an understanding does not place a shared geography at the heart of his definition of community. Place-based ideals of community rely on the assumption of a 'natural' sense of shared identity and often fail to address the increasingly pluralistic – indeed, even individualistic – nature of society (Rose, 1999). The decline of the collective and rise of the autonomous individual is considered part of a broader shift to an ethic of care for the 'self' and the loss of traditional values (Etzioni, 1995; White and Hunt, 2000). Such criticisms may explain why the geographical gerontology literature appears leery of 'community', instead engaging concepts such as 'place' and 'home' to understand how context impacts experiences of ageing (e.g., Wiles, 2005; Andrews *et al.*, 2007). From a geographical perspective, 'community' broadly refers to our connections within – and to – the places we live. As with communities formed through common interests, residential communities rely on the strengths of social networks in addition to physical dimensions, such as attachment to place and local amenities, to shape the behavioural, material and affective realities of residents.

In the past, those carrying out community studies were criticised for applying an "enlarged social imagination" to their work, particularly through idealized notions of communitarianism, and this led many geographers to distance themselves from the subject in the 1960s and 1970s (Phillips, 1998: 130). The late 1990s saw a burgeoning body of geographic research assessing the impact of neoliberal policies on rural populations. Low population density, the downturn of key industries, and the outmigration of younger working-age populations from rural areas were factors in rural population ageing (see Skinner and Winterton, Chapter 11). Rural communities were, in turn, lauded as fostering high levels of community participation, and older adults appear to play an important role in the sustainability of these places (Davis *et al.*, 2012). Rich voluntary networks made rural communities targets for offloading state services or withdrawing social investment with implications for the ageing populations in such locations. Age-friendly communities, for instance, are recognised as possessing robust physical infrastructure such as accessible public transport, affordable housing, and adequate health and social services to ensure older adults feel supported and confident to age in place (Scharlach and Lehning, 2013; World Health Organisation: WHO, 2007; see Phillips and Feng, Chapter 8).

In the face of service withdrawal, communities have resisted, for example, by protesting the closure of state schools and hospitals (Barnett and Barnett, 2003; Witten *et al.*, 2003). In other cases, the withdrawal of the state has presented an opportunity to reinvigorate a community's voluntary sector through the establishment of new partnerships, funding contracts, and the opportunity to build leadership and volunteer capacity (Skinner, 2008). Skinner and Joseph go further to suggest these changes have enhanced the agency of some rural communities: "It is in meeting the challenge of sustaining services and communities that the voluntary sector has become an active and direct agent of rural change" (Skinner and Joseph, 2007: 122). Such analyses shed light on the assemblage of political, economic, and social factors that shape a community and, as such, are key lifelines for the concept within the mainstream geographical literature (see also Lovell, Gray and Boucher, 2015; Walsh and O'Shea, 2008).

Communities are often the scale at which the voluntary sector is mobilised, opportunities for social participation identified, or a lack of social networks lamented. Critical to the idealised place-based 'community' is a vibrant voluntary sector that serves as a source of social connectivity by bridging diversity and having a liberalising effect that increases tolerance (Putnam, 2000). As such, offloading service delivery to the community is easily rationalised as a means of increasing community determinism (Fyfe, 2005). Yet some commentators raise concerns that expecting citizens to address their own needs plays into the state's neoliberal cost-saving agenda by transferring governance activities to the community and further eroding the welfare state (Mohan and Mohan, 2002; Whitehead and Diderichsen, 2001). Too often, the rural voluntary sector is 'threadbare', and the success of voluntary initiatives is reliant on public-sector support to overcome the challenges of limited resources, difficulty navigating the increasingly professional third sector to secure funds, and an overreliance on older adults as volunteers (Winterton *et al.*, 2014).

The concept of 'community', neglected through much of the 1980s and 1990s, was reinvigorated in the academic imagination with the publication of

Putnam's (2000) *Bowling Alone* and subsequent research on social networks as potential sources through which resources might be accessed to ease a range of health and social problems (Carpiano, 2006). Extending into the political realm, a subtle shift in discourse was observed from the mid-2000s as 'socialised neoliberalism' (Gauld, 2009) took hold. Administrations such as the Blair government reframed communities as more than simply a 'shadow state', or a tool for cost savings, but as a potential resource in the battle against social ills such as civil disobedience (Fyfe, 2005; Geiger and Wolch, 1986). Much of the 'neo-communitarianism' of the 2000s idealised the transformative potential of communities (Fyfe, 2005); for example, Bowles and Gintis (2002) state that "communities can sometimes do what governments and markets fail to do because their members, but not outsiders, have crucial information about other members' behaviours, capacities and needs" (Bowles and Gintis, 2002: F423). Yet variability in the capacity of communities can increase geographic inequities. It is within this context that geographic gerontologists have increasingly adopted the language of 'community' to examine macro-level influences on communities and their potential to support healthy ageing (Knapp *et al.*, 2012; Skinner *et al.*, 2014; Winterton *et al.*, 2014). Such a perspective recognises that communities are dynamic places in which actors intersect with the social, political and economic context of places to shape collective outcomes.

Healthy ageing and a sense of community

Western societies are seeing populations live to older ages with concerns over levels of isolation beginning to enter the policy arena (Bartlett *et al.*, 2013). Communities provide opportunities for social participation, the development of supportive networks and can reduce isolation. Importantly, these factors are predictors of higher self-rated health and stronger mental health outcomes (Muhajarine *et al.*, 2008; Nummela, Seppanen and Uutela, 2011; Pollack and von dem Knesebeck, 2004). Holmén and Furukawa (2002), for example, report that those who are unsatisfied with their social relationships experience higher levels of loneliness. Participating in one's community is thought to benefit health by providing opportunities for physical activity and mental stimulation (Pollack and von dem Knesebeck, 2004), as well as a broader sense of well-being and purpose (Wiles and Jayasinha, 2013). 'Giving back' to the community provides older adults with a sense of purpose that can also facilitate social connections providing support and continuity throughout life changes (Black, Dobbs and Young, 2012). An important caveat is that poor health may be a factor in reducing community interactions.

Social exclusion is a concept synonymous with citizenship; that is, it refers to the non-participation of particular individuals in the everyday activities of society due to unsupportive settings. Older adults are more reliant on their local environment for the resources and support that help them function in their day-to-day lives (Buffel, Phillipson and Scharf, 2012). As a result, those who feel unsafe in traffic perceive high levels of crime, and those who struggle with poverty may experience higher levels of social exclusion (Buffel, Phillipson and Scharf, 2012). While experiences of disablement and financial exclusion are common among older adults, attitudes can be a further source of social exclusion (Scharlach and Lehning, 2013). Research with older adults reminds us that ageism and a lack of respect are common experiences and can undermine one's willingness and opportunity to contribute to the community (Black, Dobbs and Young, 2012; Warburton, Ng and Shardlow, 2013).

Ageing in place emerged as a concept that captured both the social connections and appreciation for the physical environment; these are important elements of older adults' desire to age in their own community (Wiles *et al.*, 2011; Golant, Chapter 15). Rather than a purely individual concept, a sense of place is often co-constituted through the social and cultural practices of a group. Thus a community may be a powerful site of belonging (Scannell and Gifford, 2010). Amongst older adults, qualitative research has highlighted that perceptions of a community as 'safe' and 'vibrant' are important, but co-location among one's friends is critical. The authors elaborate:

> Being greeted by numerous people as they walked down the street provided daily evidence of belonging, and a sense of security is derived from familiarity with the wider community, both in terms of people (such as neighbours who comprise 'your own little community') and places (the "little details" of knowing the local supermarket or health service well).
>
> (Wiles *et al.*, 2011: 365)

The importance of neighbourhoods as a source of social connection is echoed in quantitative research illustrating that those who interact with their immediate neighbours on a near daily basis are less likely to feel lonely (Kearns *et al.*, 2015). Social connections have been found to elicit greater place attachment than the physical features of place such as the home or neighbourhood (Hidalgo and Hernandez, 2001).

As with social ills, social capital is not evenly distributed across society. Higher levels of loneliness and lower levels of social engagement and trust have been observed among lower-socio-economic groups, divorced people, and people with lower levels of education (Fukuyama, 1995; Kearns *et al.*, 2015; Subramanian, Lochner and Kawachi, 2003). Research in the United States has found that homogeneity (both ethnic and economic) is an important predictor of higher rates of social capital at the state, as well as the neighbourhood, level (Alesina and La Ferrara, 2000). In contrast to social capital research, the literature on age-friendly communities espouses the importance of heterogeneity (Gonyea and Hudson, 2015), in alignment with Jacobs (1961) and others who argue for inclusive neighbourhoods. Alesina and La Ferrara suggest that "more homogeneous communities have a higher level of social interaction leading to more social capital" (2000: 849), whereas community involvement is lowest in ethnically diverse communities. This research suggests that notions of vibrant, heterogeneous communities are more often an ideal than a reality.

Community capacity building

Theories of social capital and the growth of ecological perspectives in health have furthered thinking around communities as dynamic systems embedded in, and shaped by, broader processes (Raeburn *et al.*, 2006). Asset-based approaches to community development capitalise on this dynamism by enabling communities to identify and build on their collective assets while proving acceptable to communities. Such approaches have the potential to address Skinner *et al.'s* (2014) call to consider the contribution that older adults make to building resilient communities. Despite their popularity, asset-based approaches remain largely untested (Friedli, 2013).

Community capacity building is a strength-based approach to community development that considers enhancing the skills, networks, and resources of a community as critical to collective action and gaining a sense of control over local problems (Kar, Pascual and Chickering, 1999). While evidence that community action can impact the health and social well-being of a community is thin (Lovell and Rosenberg, 2009), the success and particularly the sustainability of a broad range of interventions and policies often hinges on the development of strong partnerships and community participation in decision-making (Hawe *et al.*, 1997; Jackson *et al.*, 2006). This emphasis is reflected in many contemporary ageing initiatives that consider communities as critical to transformation (Buffel *et al.*, 2012; Menehan, 2011; Partners for Livable Communities, 2016). I turn now to a case study of integrated housing to explore community capacity building in practice and the implications of diversity.

Community capacity building case study: age-integrated housing

Housing for older adults is frequently organised to facilitate social interaction and foster a sense of community (Howden-Chapman, Signal and Crane, 1999). Age-integrated communities, exemplified by initiatives such as establishing a preschool inside a nursing home to enable older adult residents to watch and interact with their daytime visitors (Jansen, 2016), are thought to provide older adults with greater cognitive stimulation and reduced loneliness. Research on age-integrated housing highlights the importance of community context to feelings of security and victimisation among older adults. For example, in a large US study across 42 housing sites in 15 US cities, Normoyle and Foley (1988) found higher rates of fearfulness of crime among age-segregated seniors living in high-rises when compared to their age-integrated counterparts living alongside younger families in public housing. Integrating seniors with troubled young persons has proven much more problematic. When frail seniors were age-integrated with younger persons with chronic mental illness and substance abuse histories, the quality of life and care for the frail seniors were diminished (Heumann, 1996). The author found that seniors spent more time in their apartments because of fear of younger, volatile residents, and as a result, social opportunities were reduced, building safety suffered, external visitors declined, and resident stress and turnover increased (Heumann, 1996). Similarly, Lawton and Yaffe (1980) found fear of crime was higher among seniors residing in the same building as teenaged children of problem families. Such concerns are backed up by evidence from longitudinal research, which suggests that mid-late adolescents exhibit the highest crime rates and that peers serve as a critical delinquency trigger among a cohort of adolescents at these ages (Piquero and Moffitt, 2014).

My own experiences working in an age-integrated apartment building to implement community-based participatory research highlighted the importance of shared values and experiences to community capacity building. In the age-integrated apartment building, older adults lived in harmony with a smaller number of disabled adults (aged 40–65). I approached the community to see if it could identify a problem that would mobilise them with the goal of taking transformative action-building levels of social capital. Reflecting high levels of communication and problem identification within the community, participants quickly identified several priority areas: (1) the poor services and limited funding for the disability transit service, (2) a lack of activities for seniors and (3) the poor behaviour of local youth. At the heart of participatory research are principles of community-led action and philosophies of working *with* the community. Less acknowledged is the importance of readiness for action as a prerequisite to capacity building. Health promoters have noted that communities with no history of collective action often require the investment of considerable time before they will accept and/or participate in change (Lovell, Kearns and Rosenberg, 2011).

The community I engaged with appeared ready for action: residents were particularly upset at the quality of accessible transport and this was key to their decision to advocate for a more reliable parallel transit service. While this problem primarily affected the quality of life of the younger, disabled residents, the older residents initially voiced indignation on their neighbours' behalf and concern that they would lose mobility and become similarly reliant on the services. Yet stakeholder involvement in the process went through a number of changes over the course of the project and those who were not users of the accessible transit service grew restless at meetings and ceased attending.

Over the course of the participatory research project, participants carried out focus groups with users of the transit service from across their city and used the results to advocate for service improvement. At the project's conclusion, stakeholders felt their work had led to some positive change and felt satisfied their voices had been heard by the transit provider. Yet participants did not see the same positive effects within the social housing setting; the expressed frustration at the lack of wider support for their work, and felt they had instead established stronger connections with the disabled community for whom they were advocating. Despite being grounded in a geographic community with a strong social network this research illustrates the limits of community and, potentially, echoes contemporary criticisms of the 'natural' community, and highlights the importance of a sense of shared identity to community mobilisation.

Conclusion

Returning to Roseto, it appears that 50 years after social cohesion was observed to be important to longevity, we have a better understanding of how a sense of community belonging can facilitate reciprocal relationships that provide support, companionship, and purpose. We have also come to understand the socially nuanced nature of 'community' and its importance for health; this includes a recognition that *perceptions* of one's community are among the most important influences on older adults' use of their neighbourhood and that participation in local affairs appears to be a more consistent predictor of health outcomes than objective measures of neighbourhood quality (Wen, Hawkley and Cacioppo, 2006). Yet alongside these relationships between health and place-based communities is a recognition of the increasingly pluralised societies in which we live. The increasingly subjective and culturally mediated nature of community has important implications for how groups in society are mobilised and the causes which they mobilise around or against.

The neoliberal turn of the 1990s saw the state rollback the provision of health and social services, particularly in rural communities. At a time when our social connections beyond our neighbourhood were strengthening, a gap emerged for local communities, in their capacity as individual volunteers or non-profit service providers, to fill the hole left by the state. Collectively, the geographical gerontology literature illustrates that such communities are left to negotiate the challenges of recruiting staff, navigating the contractual landmine of the third sector, managing an ageing volunteer workforce and meeting the needs of vulnerable older adults (Kearns, 1997; Kearns and Joseph, 1997; Skinner and Joseph, 2007; Skinner *et al.*, 2014). It is little surprise that some communities have proven capable of responding to local issues and have a vibrant third sector to rely on while others are struggling.

Turning to the health literature, strength-based approaches to community development seek to recognise and work with the tools communities already have at their disposal. The effectiveness of capacity building often relies on the density of social networks and whether a community has been able to identify its needs. Health promoters report investing considerable time – sometimes years – into building relationships before capacity-building work can take place within a given community (Lovell *et al.*, 2011). As this chapter illustrates, even within relatively cohesive communities, there is no guarantee that capacity-building initiatives will be widely embraced or that a community will mobilise independently.

Moving forward, building communities that support healthy ageing must involve both grassroots capacity building initiatives alongside top-down support for strategic infrastructure such as public transportation, home care, affordable housing and amenities. Across all levels, older adult participation is necessary to design appropriate services and recognise how communities can be accessible and inclusive across the life course. This means reversing the trend toward marginalising older voices in discussions of urban regeneration intended to build age-friendly cities, towns and communities (Buffel *et al.*, 2012), recognising the diversity within ageing communities and providing socially and culturally specific opportunities for engagement.

References

Alesina, A. and La Ferrara, E. (2000). Participation in heterogeneous communities. *The Quarterly Journal of Economics*, 115(3), 847–904. http://dx.doi.org/10.1162/003355300554935

Andrews, G. J., Cutchin, M., McCracken, K., Phillips, D. R. and Wiles, J. (2007). Geographical gerontology: The constitution of a discipline. *Social Science & Medicine*, 65(1), 151–168. http://dx.doi.org/10.1016/j.socscimed.2007.02.047

Barnett, R. and Barnett, P. (2003). "If you want to sit on your butts you'll get nothing!" Community activism in response to threats of rural hospital closure in southern New Zealand. *Health & Place*, 9, 59–71. http://dx.doi.org/10.1016S1353-8292(02)00019-9

Bartlett, H., Warburton, J., Luii, C., Peach, L. and Matthew, C. (2013). Preventing social isolation in later life: Findings and insights from a pilot Queensland intervention study. *Ageing & Society*, 33, 1167–1189. http://dx.doi.org/10.1017/S01477686X12000463

Black, K., Dobbs, D. and Young, T. (2012). Aging in community: Mobilizing a new paradigm of older adults as a core social resource. *Journal of Applied Gerontology*, 32, 219–234. http://dx.doi.org/10.1177/0733464812463984

Bowles, S. and Gintis, H. (2002). Social capital and community governance. *The Economic Journal*, 112(483), F419–F436. http://dx.doi.org/10.1111/1468-0297.00077

Buffel, T., Phillipson, C. and Scharf, T. (2012). Ageing in urban environments: Developing 'age-friendly' cities. *Critical Social Policy*, 32(4), 597–617. http://dx.doi.org/10.1177/0261018311430457

Carpiano, R. M. (2006). Toward a neighbourhood resource-based theory of social capital for health: Can Bourdieu and sociology help? *Social Science & Medicine*, 62(1), 165–175. http://dx.doi.org/10.1016/jsocscimed.2005.05.020

Davis, S., Crothers, N., Grant, J., Young, S. and Smith, K. (2012). Being involved in the country: Productive ageing in different types of rural communities. *Journal of Rural Studies*, 28(4), 338–346. http://dx.doi.org/10.1016.jrurstud.2012.01.008

Deijerm, P. and Uslaner, E. (2001). Introduction, in P. Dekker and E. Uslaner (eds), *Social capital and participation in everyday life* (pp. 3–8). London: Routledge.

Egolf, B., Lasker, J., Wolf, S. and Potvin, L. (1992). The Roseto effect: A 50-year comparison of mortality rates. *American Journal of Public Health*, 82, 1089–1092.

Etzioni, A. (1995). *The spirit of community*. New York: Crown.

Friedli, L. (2013). 'What we've tried, hasn't worked': The politics of assests based public health. *Critical Public Health*, 23, 131–145. http://dx.doilorg/10.1080/09581596.2012.748882

Fukuyama, F. (1995). *Trust: The social virtues and the creation of prosperity*. New York: The Free Press.

Fukuyama, F. (1999). *The great disruption: Human nature and the reconstitution of social order*. New York: The Free Press.

Fyfe, N. R. (2005). Making space for "Neo-communitarianism"? The third sector, state and civil society in the UK. *Antipode*, 37(3), 536–557. http://dx.doi.org/10.1111/j.00664812.2005.00510x

Gauld, R. (2009). *New health policy* (1st edn). Berkshire: McGraw-Hill Education.

Geiger, R. K. and Wolch, J. R. (1986). A shadow state? Voluntarism in metropolitan Los Angeles. *Environment and Planning D: Society and Space*, 4(3), 351–366. http://dx.doi.org/10.1068/d040351

Glendinning, A., Nuttall, M., Hendry, L., Kloep, M. and Wood, S. (2003). Rural communities and well-being: A good place to grow up? *The Sociological Review*, 51(1), 129–156. http://dx.doi.org/10.1111/1467-954X00411

Gonyea, J. G. and Hudson, R. B. (2015). Emerging models of age-friendly communities: A framework for understanding inclusion. *Public Policy & Aging Report*, 25(1), 9–14. http://dx.doi.org/10.1093/ppar/pru056

Hawe, P., Noort, M., King, L. and Jordens, C. (1997). Multiplying health gains: The critical role of capacity-building within health promotion programs. *Health Policy*, 39(1), 29–42. http://dx.doi.org/10.1016/S0168-8510(96)00847-0

Heumann, L. (1996). Assisted living in public housing? A case study of mixing frail older adults and younger persons with chronic mental illness and substance abuse histories. *Housing Policy Debate*, 7(3), 447–471. http://dx.doi.org/10.1080/10511482.1996.9521229

Hidalgo, M. and Hernandez, B. (2001). Place attachment: Conceptual and empirical questions. *Journal of Environmental Psychology*, 21(3), 273–281. http://dx.doi.org/10.1006/jevp.2001.0221

Holmén, K. and Furukawa, H. (2002). Loneliness, health and social network among older adults people – a follow-up study. *Archives of Gerontology and Geriatrics*, 35(3), 261–274. http://dx.doi.org/10.1016/S0167-4943(02)00049-3

Howden-Chapman, P., Signal, L. and Crane, J. (1999). Housing and health in older people: Ageing in place. *Social Policy Journal of New Zealand*, 13, 14–30. Available from: www.healthyhousing.org.nz/wp-content/uploads/2010/01/Housing-and-health-in-older-people-aging-in-place.pdf

Jackson, S. F., Perkins, F., Khandor, E., Cordwell, L., Hamann, S. and Buasai, S. (2006). Integrated health promotion strategies: A contribution to tackling current and future health challenges. *Health Promotion International*, 21(S1), 75–83. http://dx.doi.org/10.1093/heapro/dal054

Jacobs, J. (1961). *The death and life of great American cities*. New York: Random House.

Jansen, T. (2016). The preschool inside a nursing home. *The Atlantic*, 20 January.

Kar, S. B., Pascual, C. A. and Chickering, K. L. (1999). Empowerment of women for health promotion: A meta-analysis. *Social Science & Medicine*, 49(11), 1431–1460. http://dx.doi.org/10.1016/S0277-9536(99)00200-2

Kearns, A., Whitley, E., Tannahill, C. and Ellaway, A. (2015). Loneliness, social relations and health and well-being in deprived communities. *Psychology, Health & Medicine*, 20(3), 332–344. http://dx.doi.org/10.1080/13548506.2014.940354

Kearns, R. A. (1997). Guest editorial: Third sector stories – taking community seriously. *Health and Social Care in the Community*, 6(4), 221–223. http://dx.doi.org/10.1046/j.1365-2524.1998.00122.x

Kearns, R. A. and Joseph, A. E. (1997). Restructuring health and rural communities in New Zealand. *Progress in Human Geography*, 21(1), 18–32. http://dx.doi.org/10.1191/030913297666611118

Knapp, M., Bauer, A., Perkins, M. and Snell, T. (2012). Building community capital in social care: Is there an economic case? *Community Development Journal*, 48(2), 312–331. http://dx.doi.org/10.1093/cdj/bss021

Lawton, M. P. and Yaffe, S. (1980). Victimization and fear of crime in elderly public housing tenants. *Journal of Gerontology*, 35(5), 768–779. http://dx.doi.org/10.1093/geronj/35.5.768

Lovell, S., Gray, A. and Boucher, S. E. (2015). Experiences of economic change in small town New Zealand: Implications for voluntarism and community capacity, in M. W. Skinner and N. Hanlon (eds), *Ageing resource communities: New frontiers of rural population change, community development and voluntarism*. Abingdon: Routledge.

Lovell, S. A., Kearns, R. A. and Rosenberg, M. W. (2011). Community capacity building in practice: Constructing its meaning and relevance to health promoters. *Health and Social Care in the Community*, 19(5), 531–540. http://dx.doi.org/10.1111/j.1365-2524.2011.01000.x

Lovell, S. and Rosenberg, M. (2009). Community capacity amongst people living with HIV/AIDS. *GeoJournal*, 76(2), 111–121. http://dx.doi.org/10.1007/s10708-009-9289-2

Menehan, K. (2011). *Community partnerships for older adult: Programs* [Online]. Robert Wood Johnson Foundation. Available from: www.rwjf.org/en/library/research/2011/12/community-partnerships-for-older-adults.html [Accessed 26 July 2016]

Mohan, G. and Mohan, J. (2002). Placing social capital. *Progress in Human Geography*, 26(2), 191–210. http://dx.doi.org/10.1191/0309132502ph364ra

Muhajarine, N., Labonte, R., Williams, A. and Randall, J. (2008). Person, perception, and place: What matters to health and quality of life. *Social Indicators Research*, 85(1), 53–80. http://dx.doi.org/10.1007/s11205-007-9124-4

Normoyle, J. B. and Foley, J. M. (1988). The defensible space model of fear and older adults public housing residents. *Environment and Behavior*, 20(1), 50–74. http://dx.doi.org/10.1177/0013916588501003

Nummela, O., Seppanen, M. and Uutela, A. (2011). The effect of loneliness and change in loneliness on Self-Rated Health (SRH): A longitudinal study among aging people. *Archives of Gerontology and Geriatrics*, 53(2), 163–167. http://dx.doi.org/10.1016/j.archger.2010.10.023

Partners for Livable Communities. (2016). *Ageing in place initiative: Developing liveable communities for all ages*. Available from: www.livable.org/program-areas/livable-communities-for-all-ages-a-aging-in-place/the-aging-in-place-initiative [Accessed 26 July 2016]

Phillips, M. (1998). The restructuring of social imaginations in rural geography. *Journal of Rural Studies*, 14(2), 121–153. http://dx.doi.org/10.1016/S0743-0167(97)00056-9

Piquero, A. R. and Moffitt, T. E. (2014). Moffitt's developmental taxonomy of antisocial behavior, in G. Bruinsma and D. Wiesburd (eds), *Encyclopedia of criminology and criminal justice*. New York: Springer New York.

Pollack, C. E. and von dem Knesebeck, O. (2004). Social capital and health among the aged: Comparisons between the United States and Germany. *Health & Place*, 10(4), 383–391. http://dx.doi.org/10.1016/j.healthplace.2004.08.008

Pretty, G. H., Chipuer, H. M. and Bramston, P. (2003). Sense of place amongst adolescents and adults in two rural Australian towns: The discriminating features of place attachment, sense of community and place dependence in relation to place identity. *Journal of Environmental Psychology*, 23(3), 273–287. http://dx.doi.org/10.1016/S0272-4944(02)00079-8

Putnam, R. (2000). *Bowling alone: The collapse and revival of American community*. New York: Simon & Schuster.

Raeburn, J., Akerman, M., Chuengsatiansup, K., Mejia, F. and Oladepo, O. (2006). Community capacity building and health promotion in a globalized world. *Health Promotion International*, 21(S1), 84–90. http://dx.doi.org/10.1093/heapro/dal055

Rose, N. (1999). *Powers of freedom: Reframing political thought*. Cambridge: Cambridge University Press.

Scannell, L. and Gifford, R. (2010). Defining place attachment: A tripartite organizing framework. *Journal of Environmental Psychology*, 30(1), 1–10. http://dx.doi.org/10.1016/j.envp.2009.09.006

Scharlach, A. and Lehning, A. (2013). Ageing-friendly communities and social inclusion in the United States of America. *Ageing & Society*, 33(1), 110–136. http://dx.doi.org/10.1017/S0144686X12000578

Skinner, M. W. (2008). Voluntarism and long-term care in the countryside: The paradox of a threadbare sector. *The Canadian Geographer/Le Géographe canadien*, 52(2), 188–203. http://dx.doi.org/10.1111/j.1541-0064.2008.00208.x

Skinner, M. W. and Joseph, A. E. (2007). The evolving role of voluntarism in ageing rural communities. *New Zealand Geographer*, 63(2), 119–129. http://dx.doi.org/10.1111/j.1745-7939.2007.00096.x

Skinner, M. W., Joseph, A. E., Hanlon, N., Halseth, G. and Ryser, L. (2014). Growing old in resource communities: Exploring the links among voluntarism, aging, and community development. *The Canadian Geographer/Le Géographe canadien*, 58(4), 418–428. http://dx.doi.org/10.1111/cag.12087

Stedman, R. C. (2002). Toward a social psychology of place: Predicting behavior from place-based cognitions, attitude, and identity. *Environment and Behavior*, 34(5), 561–581. http://dx.doi.org/10.1177/0013916502034005001

Subramanian, S., Lochner, K. and Kawachi, I. (2003). Neighborhood differences in social capital: A compositional artifact or a contextual construct? *Health & Place*, 9(1), 33–44. http://dx.doi.org/10.1016/S1353-8292(02)00028-X

Walsh, K. and O'Shea, E. (2008). Responding to rural social care needs: Older people empowering themselves, others and their community. *Health & Place*, 14(4), 795–805. http://dx.doi.org/10.1016/j.healthplace.2007.12.006

Warburton, J., Ng, S. and Shardlow, S. (2013). Social inclusion in an ageing world: Introduction to the special issue. *Ageing & Society*, 33(1), 1–15. http://dx.doi.org/10.1017/S0144686X12000980

Wen, M., Hawkley, L. C. and Cacioppo, J. T. (2006). Objective and perceived neighborhood environment, individual SES and psychosocial factors, and self-rated health: An analysis of older adults in Cook County, Illinois. *Social Science & Medicine*, 63(10), 2575–2590. http://dx.doi.org/10.1016/j.socscimed.2006.06.025

White, M. and Hunt, A. (2000). Citizenship: Care of the self, character and personality. *Citizenship Studies*, 4(2), 93–116. http://dx.doi.org/0.1080/13621020050078041

Whitehead, M. and Diderichsen, F. (2001). Social capital and health: Tip-toeing through the minefield of evidence. *Lancet*, 358(9277), 165–166. http://dx.doi.org/10.1016/S0140-6736(01)05398-3

Wiles, J. (2005). Conceptualizing place in the care of older people: The contributions of geographical gerontology. *International Journal of Older People Nursing*, 14(2), 100–108. http://dx.doi.org/10.1111/j.1365-2702.2005.01281.x

Wiles, J. L. and Jayasinha, R. (2013). Care for place: The contributions older people make to their communities. *Journal of Aging Studies*, 27(2), 93–101. http://dx.doi.org/10.1016/j.aging.2012.12.001

Wiles, J. L., Leibing, A., Guberman, N., Reeve, J. and Allen, R.E.S. (2011). The meaning of "ageing in place" to older people. *The Gerontologist*, 52(3), 357–366. http://dx.doi.org/10.1093/geront/gnr098

Winders, J. (2014). Society, in R. Lee, N. Castree, R. Kitchin, V. Lawson, A. Paasi, C. Philo, S. Radcliffe, S. Roberts and C. Withers (eds), *The Sage handbook of human geography* (pp. 478–504). London: Sage.

Winterton, R., Warburton, J., Clune, S. and Martin, J. (2014). Building community and organisational capacity to enable social participation for ageing Australian rural populations: A resource-based perspective. *Ageing International*, 39(2), 163–179. http://dx.doi.org/10.1007/s12126-013-9187-2

Witten, K., Kearns, R., Lewis, N., Coster, H. and McCreanor, T. (2003). Educational restructuring from a community viewpoint: A case study of school closure from Invercargill. *Environment and Planning C: Government and Policy*, 21(2), 203–223. http://dx.doi.org/10.1068/c05r

Wolf, S. and Bruhn, J. (1998). *The power of clan: The influence of human relationships on heart disease*. New Brunswick, NJ: Transaction Publishers.

World Health Organisation (WHO). (2007). *Global age-friendly cities: A guide. Aging and life course*. Geneva: World Health Organisation. Available from: www.who.int/ageing/age_friendly_cities_guide/en/

13 Household spaces of ageing

When care comes home

Anne Martin-Matthews and Denise S. Cloutier

Introduction

As social geographers acknowledge, home is perhaps the "most emotive of geographical concepts, inextricable from that of self, family ..., sense of place ..." (Duncan and Lambert as cited in Gregory *et al.*, 2009: 340). While the language of 'household' denotes material dwelling and domestic sphere, the language of 'home' reflects "a space of belonging and alienation, intimacy and violence, desire and fear, ... invested with emotions, experiences, practices and relationships that lie at the heart of human life" (ibid., 339). The meaning of home includes the social, physical and emotional home (Oswald and Wahl, 2005). At all life stages, the evocation of home encompasses both the lived experiences of everyday, domestic life and a wider sense of being and belonging in the world.

Geographies of ageing consider the spatial contexts in which ageing occurs, recognising that older adults prefer to live at home and to 'age in place' for as long as possible (Skinner, Cloutier and Andrews, 2015). Geographies of care consider the spatial distribution of care provision to vulnerable older populations, including challenges and experiences of both giving and receiving care. Increasingly, home is the expected primary site for delivery and receipt of care 'services'. Indeed, home care for frail elderly people represents "a rare convergence between the preferences of older clients and their families to age in place, and the intentions of public policy to reduce institutionalisation and promote functional and residential independence" (Larsson, Silverstein and Thorslund, 2005: 636). Home care services support this goal and function as an essential pillar of health care systems in many industrialised nations (Doyle and Timonen, 2008).

Cloutier *et al.* (2015) explored how the relational space of the home was perceived as an environment for care, and acted upon by care recipients and care providers in relationship building. That study examined how a relational ethics framework (Bergum and Dossetor, 2005) advances understanding of relationship building and the conceptualisation of home space as the principal environment for care. This chapter extends that analysis and considers how social forces shape experiences of ageing and caring 'at home'. The objective is to advance a critical but constructive understanding of social factors that contextualise household – and home – dynamics of relational ethics (which includes relational space). We consider meanings of home in later life, the nature of home care and the relevance of relational ethics and relational space as examined by Cloutier *et al.* (2015). We adapt the concept of relational properties, as used by Marshall, Matthews and Rosenthal (1993) to characterise power or privilege associated with gender and birth order in families. Here, the concept is constructed quite differently, as a heuristic device to facilitate the examination of key social factors shaping experiences of home as a site of care. These include collectivity, contingency and cultural diversity. In response to these forces, home care space can be collaborative and/or contested. Also, by invoking the language of relational ethics and relational space as used by Bergum and Dossetor (2005), and Cloutier *et al.* (2015), the construct of relational properties emphasises the link to – but also the extension beyond – these more 'micro' and dyadic elements in home care.

Home and home care in later life

In social gerontology, home is an important site of inquiry as it relates to ageing, imperatives of ageing in place and the provision and receipt of care. Researchers, social commentators and policymakers recognise the salience of home for elders in the Third Age and, ever more frequently, the Fourth. Worldwide, older people increasingly adopt independent households, with an estimated 14 per cent of people age 60 and over living alone – rising to 25 per cent of older people in Europe, Oceania and North America (Victor, 2010).

Home space may have heightened meaning for older people, as homes are often a principal possession and resource. The preference to 'age in place' is shared around the world, until it is "no longer safe for [older people] to live in their homes due to physical frailty or cognitive disorders" (Larsson, Silverstein and Thorslund, 2005: 630).

Space is not neutral, and place is more than a passive backdrop or container for experience (Andrews *et al.*, 2007; Cloutier *et al.*, 2015). Control and privacy may be threatened when home space becomes care space. The home environment shifts from a private to a more public space when it becomes the site of both paid and unpaid care work (Martin-Matthews, 2007; Martin-Matthews and Phillips, 2008). Some geographical research moves beyond conceptualising home as a private sphere separate from the public sphere of work (Rowles and Chaudhury, 2005). However, for many older people, the home has been a private domain over their entire life course. That changes when, in the words of Armstrong and Armstrong (2001), "care comes home". The constriction of social and geographical worlds with declining health (e.g., functional impairment and chronic illness) alters the meaning of home space.

Home care itself has changed in response to global sociopolitical shifts and to the press of resources. Consequently, characteristics of recipients have changed as resources are targeted to more frail and dependent older persons, particularly those living alone. The nature of service has changed from household chores to personal and even medical care in people's homes. Hours of service are reduced for many, and the labour force (especially for the personal care assistance of interest here) is largely composed of low-paid, casual labour, often immigrant and marginalised. In writing about Sweden, Larsson *et al.* (2005: 635) suggest that "previous ambitions to develop more socially oriented elder care have not been possible to maintain", especially with further reductions in "the time for contact between the elderly person and the home helper" (*ibid.*, 635).

The home care work force includes professional workers such as nurses, case managers, or physiotherapists, and unregulated, non- or para-professional home support workers. The latter are, primarily, the 'workers' referenced in this chapter. They provide assistance with the 'personal care' of bathing, feeding, toileting and dressing. In many jurisdictions, home care services are under-funded, and, consequently, fragmented and uncoordinated (Martin-Matthews *et al.*, 2013). Service provision is driven by efficiency rather than quality of care. The result is multiple care providers coming into the home as dictated by scheduling, rather than strong worker–client relationships developed through regular and consistent care providers coming into the home.

Relational ethics and relational space: informing further inquiry

A decade-long research programme focused on the home at the nexus of the public and the private spheres (the Nexus project) examined the receipt of home care services by older people. Home care was examined from multiple perspectives (older clients, care workers, family carers and policymakers), in diverse jurisdictions (three Canadian provinces) and through (inter)disciplinary lenses including sociology, gerontology, social work and geography. Dozens of publications, empirically informed by the Nexus studies, examine the gendered, shared, contingent, temporal and dynamic nature of home care from these varied perspectives (e.g., Martin-Matthews and Sims-Gould, 2008; Sims-Gould and Martin-Matthews, 2010a; Byrne *et al.*, 2011; Martin-Matthews, Sims-Gould and Tong, 2013; Cloutier *et al.*, 2015).

Cloutier *et al.* (2015) analysed data from the Nexus project to explore strategies for relationship building used by home support workers and older residents according to a 'relational ethics' framework enacted across relational space (Bergum and Dossetor, 2005). This framework was developed in the acute care sector to examine relationships between nurses and patients. It centres on four principles: engagement, embodiment, mutual respect and environment – each deemed necessary for ethical care. Cloutier *et al.* (2015) sought to understand both how the territory of home space itself influences the nature of relationship building between older adult care recipients and care providers and the utility of the relational ethics framework in the home care context.

Bergum and Dossetor (2005) define relational ethics as a way of being and acting. Relational space is the space between moral agents acting together in a relational way and a social and policy space for discovering knowledge about one another through sensitive interaction. Cloutier *et al.* (2015) applied the relational ethics framework (from health care) to insights about meanings of home and the complexity of home space (from social geography). This application advanced the understanding of home as relational space, how this is enacted by clients and workers, and the ways in which home care space is both multilayered and complex.

In this chapter, we widen the lens to examine the broader social forces and institutions that contribute to these layers and complexity. The focus is on how relational ethics and relational space are impacted by what are characterised here as relational properties of the home care context: collectivity, contingency and cultural diversity. These inform an understanding of how home care functions as collaborative and/or contested. Throughout this chapter, the 'data' sources are derived from a selection of relevant publications from the Nexus project.

Home space as relational space: social forces framing the context

Sociological inquiry considers the role of social forces in shaping individual lives. The work of Bergum and Dossetor is useful in explicating the nature of relational ethics and relational space, and the application of this perspective to home care is important (Cloutier *et al.*, 2015). But it is equally important to consider what frames this 'relational context'. Might the primary focus on relational dynamics reflect, in Dannefer's words, "[t]he strong appeal of reductionist, individual-level approaches . . . to contain the explanatory potentials of social forces while continuing to privilege individual-level explanation" (2013: 794)? How is the home care experience of older people impacted by larger structural forces that shape the everyday dynamics within which care is delivered and received, beyond the impact of proximate contextual experiences (Hagestad and Dannefer, 2001)?

Thus, the focus becomes the relationship between the micro ethics of the home care experience and the larger social forces that influence them (Bergum and Dossetor do acknowledge macro ethics, but they appear more policy-oriented). How is relationship building – and other dynamics of home as the site of care – to be understood in the context of these social forces? For example, it is recognised that different health care workers entering home space blur distinctions between private and public spheres (Cloutier *et al.*, 2015). A wider lens that considers social forces as 'relational properties' enables the consideration of which "structural and cultural dynamics shape the contour and content" (Settersten and Hagestad, 2015: 29) of this blurring.

Many broad social factors (including issues of 'macro-ethics' as identified by Bergum and Dossetor, 2005) impact the context of care at home. These, as adapted from Stone (2006: 413), include the value placed on caregiving; local labour market conditions; regulatory and reimbursement policies; global, national, and local workforce resources; and immigration policies. As Stone (2006: 413) argues, "[t]he confluence of these factors and individual employer and employee decisions are played out in workplaces that are clients' homespaces."

The focus here is on contexts that influence relational ethics in home space. Perspectives from sociology draw attention to the ways in which social forces (here defined as collectivity, contingency and cultural diversity) – characterised as relational properties – impact home space as care space. They enable us to take full account of ways in which social forces shape the dynamics of home care as collaborative relationship building toward common goals and/or as contested terrain within households.

Relational properties when home is the site of care

Home space as collectivity

Central to notions of home and home life is the concept of collectivity. Marshall, Matthews and Rosenthal (1993: 39) note the importance of

> [a]nalytical distinctions among social structure, social system, cultural system, and personal meaning to address some of the inherent obstacles that distance family life lived from the way it is described in research reports. . . . Social system refers to the patterned behavior linking members of a collectivity . . . [which] . . . could be a society, an organisation, a family or some other set of individuals.

Here the sociological concept of the collectivity is applied to findings from the Nexus project.

In this context, the relational space of home care may potentially be crowded space. Although the client may live alone, the collectivity typically includes a succession of workers and multiple family members – and may include other co-residents. The size and complexity of the collectivity are highly variable, comprising both strangers and 'familiars'. Such relational properties can extend well beyond, but also frame, the specifics of relational ethics of the dyadic client–worker relationship. Family and client depictions of workers as adept or inept, and their scheduled arrival as predictable or not (Sims-Gould *et al.*, 2015), constitute important features of the collectivity. Much home care literature is monolithic in its characterisation of 'workers', whereas clients (and family members) emphasise distinctions (Byrne *et al.*, 2011).

In addition, where caregiver 'respite' is an expectation of home care service, the collectivity of the home space is directly altered by the worker's presence. The 'designated' family member is expected to leave the home setting when the worker is present. Some workers do recognise that "the family members' ability to receive respite was predicated on rapport with and trust of the worker" (Sims-Gould *et al.*, 2015). This, in itself, alters the 'relational space' of the home.

Home care is also gendered. Because of greater longevity, and more years of unhealthy life expectancy, most recipients of home care are women – and especially those living alone. Most providers of informal care are also women: wives and daughters. Similarly, most paid care workers are women. Gender dynamics are at the heart of home care (Sims-Gould and Martin-Matthews, 2010a, b).

Home care as contingent

Contingency as a relational property of home care operates in three primary ways: the Care Plan, home space, and variability of workers. The document guiding what, how much and within what time frame services are to be provided by workers is the Care Plan. This plan is based on an assessment of a given client's needs at a single point in time. Thus, what workers 'do' in providing care in the relational space of home is largely contingent upon agency rules and regulations as specified in the Care Plan. While ideally it is to be updated regularly, this is not always the case. In Nexus research, over three quarters of family members noted the bureaucratic nature of service and expressed frustration with the restrictive nature of the Care Plan (Sims-Gould et al., 2015). Workers who have a relationship with a particular client may "bend the rules" and provide assistance beyond that assigned or deviate from the Plan when clients have other needs (Sims-Gould and Martin-Matthews, 2010b:102). Conversely, clients described care needs as "non-negotiable" when requests were characterised by workers as "not allowed" or "that's the rules" (Byrne *et al.*, 2010: 389).

The nature of care is also contingent on characteristics of home space. The Nexus research project was initiated on the recognition that the provision of care in residential care settings differs fundamentally from care provision in the private space of one's home. Cleanliness and clutter are material characteristics of place that challenge relational space (Sims-Gould and Martin-Matthews, 2010b). Each household is unique in its physical set-up, condition and sanitation (Sims-Gould *et al.*, 2015). Care work is thus contingent, as workers are required to undertake tasks beyond those assigned, because of the physical characteristics of household environments.

Home care is also highly contingent on the variability of skills, capacities, schedules and training of individual workers (Byrne *et al.*, 2010). Some, but not all, workers possess the empathic awareness so important to working effectively with many older clients (Sims-Gould *et al.*, 2015). Contingency is further exacerbated by scope of practice, training, wage equity and casualisation of labour in home support (Martin-Matthews, Sims-Gould and Tong, 2013) – all factors that contribute to worker turnover and the lamented 'revolving door' of home care workers. These structural features of the labour force fundamentally challenge relational space and relational ethics: workers rotating into unfamiliar environments and care situations cannot possess the knowledge, understanding and respect for each client's circumstances and preferences. Encounters between workers and family members are highly variable: they can range from providing respite for an exhausted caregiver, to confrontation with family members critical of meal preparation. Workers average at least four clients per day, necessitating adaptability in interactions with clients and families.

A final contingency, given the episodic nature of home care (brief periods on specified weekdays), is that it may run quite counter to the 'typical' rhythm of home life for many older people. Scheduled visits can contribute to tenuous situations for clients and families as they adjust the routine and rhythm of their lives to 'fit in' with home care services at inconvenient or undesirable times. Relational space is contingent on temporal concerns, as in the case of the older home care client who explained, "I'm satisfied with all but the time of waking up and getting up" (Byrne *et al.*, 2011: 169).

Cultural diversity in home care space

The collectivity of individuals who inhabit (and circulate through) home care space and the contingent nature of activities and dynamics in that space are further complicated by the ethno-cultural diversity of clients and families, and of workers rotating through home space within that collectivity. Here, social forces represented by national immigration policies, labour laws and housing affordability all impact home care contexts. Massey (1992: 6) acknowledges the "truly major re-shaping of the spatial organization of social relations at every level, from local to global . . . [with] . . . different geographical scales . . . less easy to separate – rather they constitute each other: the global, the local, and vice versa". In this context, "ethnicity and gender . . . are also deeply implicated in the ways in which we inhabit and experience space and place" (*ibid.*, 9).

As noted, home care work is highly gendered, is considered low-paid domestic labour, and is dominated by immigrant workers (Doyle and Timonen, 2008). As Dannefer (2013: 798) suggests, complexity here is associated not with "migration *per se* but [with] transnational lives . . .". Thus, (multiple) worker(s), clients and families, often distinguished by different cultures, reflect frameworks "of beliefs, expressive symbols, and values [by] which individuals define their world, express their feelings, and make their judgments" (Marshall *et al.*, 1993: 41).

Because of the proliferation of transnational lives, shifts in the ethno-cultural diversity of societies, and the recurring de-skilling of immigrants into low-paid 'care work', ethno-cultural diversity remains an enduring relational property of the care of older people (Martin-Matthews, Sims-Gould and Naslund, 2010). At best, negotiation is required. Examples include the following: employment rules requiring footwear practices inconsistent with workers' religious and cultural proscriptions; preparing of 'unfamiliar', perhaps taboo, types of foods; and language and communication. This is the 'mutual respect' of Bergum and Dossetor's (2005) relational ethics framework. But cultural diversity often gives rise to a contested terrain in home care, as discussed shortly.

Contested and/or collaborative home care space

Home care space as collaborative

Provision of care services to older people in their home space is a dynamic negotiated process that includes contributions from workers and family members themselves. Formal paid care and informal unpaid care do not run parallel to one another as two separate 'systems' but – certainly in home care – are inextricably interconnected. While workers, directed by a Care Plan, provide care to their older clients, they also provide assistance to, and receive assistance from, their clients' family members. The boundaries between what families and workers provide are often not as clear as agency policies dictate they should be. Family caregivers describe how they and paid workers "share the care", either by collaborating to accomplish a task or by setting the stage for one another's tasks (Sims-Gould and Martin-Matthews, 2010a).

Some older clients and more 'regular' workers celebrate their ethno-cultural diversity as an opportunity (for the worker) to learn new English words or the ways of 'real Canadians' or (for the client) to learn new ways of food preparation or about customs (Martin-Matthews and Sims-Gould, 2008). Older clients and their families also especially value workers 'knowing', understanding and 'seeing' the client as a person (Byrne *et al.*, 2010; Sims-Gould and Martin-Matthews, 2010a), a goal fundamental to Bergum and Dossetor's (2005) relational ethics framework.

Workers characterise their interactions with family members (co-resident with the client or not) in terms of *complementary labour* (family members providing informational and instrumental support to workers), and, as noted previously, *disrupted labour* (family members creating emotion work and additional instrumental work for workers). They *complement one another's labour* by sharing vital techniques necessary for safe care provision. Moreover, *complementary instrumental support* provided by families balances the additional 'work' often generated when family members are involved (Sims-Gould *et al.*, 2015).

The last decade has seen a notable shift in policy discourse towards the 'sharing of care' between paid health care professionals and unpaid family caregivers and the promotion of partnerships between formal (paid) and informal (family) caregivers for older adults receiving home care. Indeed, the care system now presumes and indeed relies on this collaboration, whereas previously, the 'public and the private' were different spheres.

Home care space as contested

Home care space can become a contested domain when any of the relational properties give rise to conflicts: within the collectivity of home, families and care; in the contingencies of home care services; or when diverse cultures clash. Relational space is contested or even conflicted when social forces impact private home space. Conflict can manifest as confrontations between individuals or among groups, over scarce resources or incompatible goals. Typically, "in all relationships, conflict is something to be dealt with and managed; it is not something to be 'solved'" (Sprey, 1979: 133).

Relational properties of collectivity reflect the multiplicity and variability of 'players' in families and in home care. A common scenario is one wherein clients' habits conflict with workers' values and preferences, such as clients' smoking, the presence of pets and so on (Sims-Gould and Martin-Matthews, 2010b). Family expectations of workers' individualised or idiosyncratic knowledge of clients also present challenges. Workers cannot or do not always possess knowledge of individual client preferences, of how to "follow their way" with aspects of care work, cleaning, and food preparation (Sims-Gould *et al.*, 2015: 5).

Home space exacerbates this conflict, with home becoming a contested territory, reflecting a public–private dichotomy when it is also the place of work (Rowles and Chaudhury, 2005: 9). In that private sphere, workers may be caught between family disputes and family demands, expectations and conflicts, adding contestation and complexity to their work experience. Relational ethics have relevance well beyond the client–worker relationship and interaction alone.

The contingent nature of home care, especially as manifest in the Care Plan, impacts worker–family interactions. Relational space becomes fraught when family members add extra duties beyond what is on the Care Plan. Home care is time-compressed, and repeated clarifications and explanations of the Care Plan are "time consuming". This domain is contested when requested tasks are not only not listed in the clients' personal Care Plan but also "not even legitimate functions in worker's job descriptions" (Sims-Gould *et al.*, 2015: 6). In such cases, family members *disrupt the labour* of the workers, adding to their emotional workload as well.

Cultural differences are, however, the prime contested domains within home care. Nexus study transcripts capture worker perceptions of deeply held, and at times highly racialised, beliefs among some clients and/or family members, with presumptions of the dominance of one culture (or cultures) over another. When care is provided in home space, workers and clients (and families, if present) experience differences between themselves in ethnicity, culture and class. Much research frames this issue in terms of the immigrant status and ethnicity of workers. However, in the cultural mosaic of Canada, a more appropriate focus is on ethno-cultural diversity *between* workers and clients.

A significant minority of workers (43 per cent) in the Nexus project had encountered hostility from clients over specific issues (footwear, cooking, bathing) or through outright racism and discrimination. Clients were described by workers as not trusting, being quite "mean", refusing help (particularly with personal care, being touched), phoning the agency to specify no "ethnic" person in their home or, on hearing a foreign-born worker's accent, refusing entry at the door (Martin-Matthews, Sims-Gould and Naslund, 2010: 89). Clients perspectives, particularly when living alone, may focus on "differences in others . . . [are] . . . recognized as troublesome or fearful" (Cruikshank, 2008: 150). Relational ethics take on a different meaning in these circumstances.

Summary and conclusion

Older people prefer to age in place in their own homes. This expectation assumes that, with advancing age, home will become a site of care. However, characteristics of clients, families, households and workers and the institutional structures that organise the delivery and frame the receipt of home support services all provide context for relational ethics and relational space in home care. This chapter advances a critical but constructive understanding of social forces impacting household spaces of ageing when care 'comes home'. It builds on Cloutier *et al.*'s (2015) application of Bergum and Dossetor's (2005) concepts of *relational ethics* and *relational space* to home care.

It modifies the sociological concept of *relational properties* as a heuristic device to examine ways in which social structures, institutional imperatives and cultural factors shape the experience of care at home. This focus reflects an aim of this book to advance concepts and approaches in geographical gerontology. Social forces, framed here as *relational properties*, provide important context for home care. These properties include the generative processes used to achieve 'care at home', interactions associated with these processes and policies and practices that shape them. Select findings from a decade of research on home care in Canada (the Nexus project) informed the identification of three relational properties that characterise home as a site of care: collectivity, contingency and cultural diversity.

The concept of collectivity advances the framing of home care space as more than dyadic (between older client and worker). Recognising this, geographies of ageing and of care at home cannot take the individual as the unit of analysis, even if the individual is, from a provision/provider perspective, the 'unit of service' in home care. The collectivity of home and family life is fundamental to the understanding of 'the social' in the context of home care, and thus home space (and the individuals who populate and/or associate with it) must be understood in the broader context of social structure, social system and culture.

Contingency as a relational property of home care is experienced by older clients and workers in individual and intersecting ways: through the regulatory authority of agency Care Plans, through the materiality and characteristics of the home space itself and through variability in the skills, expertise and characteristics of a succession of workers. Cultural diversity is also a relational property of home care. Much research frames this issue in terms of the immigrant status and ethnicity of a poorly paid, often casualised labour force of marginalised home support workers. However, the key issue is how ethno-cultural diversity *between* workers and clients supports or challenges relational ethics and relational space. Discordant viewpoints arising from issues of race, culture, religion or class can characterise home care. These relational properties – collectivity, contingency and cultural diversity – enable us to take full account of ways in which social forces shape the dynamics of home care along a continuum from highly collaborative to contested space.

The focus here has been on the ways in which social forces shape home care today. However, with the decline of the welfare state and the rise of neoliberal policies, home care is a highly variable and ever-changing form of service delivery for older people. The rise of technological innovation is also changing home space, and along with it, "the terms under which frail older people manage in their own homes" (Larsson, Silverstein and Thorslund, 2005: 634). As well, the profile of (publicly funded) home care clients is changing, increasingly restricted to people who are sicker and frailer and, not uncommonly, with dementia. The 'who', 'what' and 'how' of home spaces of ageing will continue to be challenging when 'care comes home'.

References

Andrews, G. J., Cutchin, M., McCracken, K., Phillips, D. R. and Wiles, J. (2007). Geographical gerontology: The constitution of a discipline. *Social Science and Medicine*, 65(1), 151–168. http://dx.doi.org/10.1016/j.socscimed.2007.02-047

Armstrong, P. and Armstrong, H. (2001). The context for health care reform in Canada, in P. Armstrong, C. Amaratunga, J. Bernier, K. Grant, A. Pederson and K. Wilson (eds), *Exposing privatization: Women and health care reform in Canada* (pp. 11–48). Toronto: University of Toronto Press.

Bergum, V. and Dossetor, J. (2005). *Relational ethics: The full meaning of respect*. Hagerstown: University Publishing Group.

Byrne, K., Frazee, K., Sims-Gould, J. and Martin-Matthews, A. (2010). Valuing the older person in the context of delivery and receipt of home support: Client perspectives. *Journal of Applied Gerontology*, 31(3), 377–401. http://dx.doi.org/10.1177/0733464810387578

Byrne, K., Frazee, K., Sims-Gould, J. and Martin-Matthews, A. (2011). 'I'm satisfied … but': Clients' and families' contingent responses about home care. *Home Health Care Services Quarterly*, 30(4), 161–177. http://dx.doi.org/10.1080/01621424.2011.622242

Cloutier, D. S., Martin-Matthews, A., Byrne, K. and Wolse, F. (2015). The space between: Using 'relational ethics' and 'relational space' to explore relationship building between care providers and care recipients in the home space. *Social and Cultural Geography*, 16(7), 764–782. http://dx.doi.org/10.1080/14649365.2015.1020336

Cruikshank, M. (2008). Aging and identity politics. *Journal of Aging Studies*, 22(2), 147–151. http://10.1016/j.jaging.2007.12.011

Dannefer, D. (2013). Age and sociological explanation: Expanding horizons in the study of aging and the life course. *Contemporary Sociology*, 42(6), 793–800. http://dx.doi.org/10.1177/0094306113506870

Doyle, M. and Timonen, V. (2008). *Home care for ageing populations: A comparative analysis of domiciliary care in Denmark, Germany, and the United States*. Northampton: Edward Elgar Publishing.

Gregory, D., Johnston, R., Pratt, G., Watts, M. and Whatmore, S. (eds). (2009). *The dictionary of human geography* (5th edn). Malden, MA: John Wiley and Sons.

Hagestad, G. O. and Dannefer, D. (2001). Concepts and theories of aging: Beyond microfication in social science approaches, in R. H. Binstock and L. K. George (eds), *Handbook of aging and the social sciences* (5th edn., pp. 3–21). Cambridge, MA: Academic Press.

Larsson, K., Silverstein, M. and Thorslund, M. (2005). Delivering care to older people at home, in M. L. Johnson (ed.), *The Cambridge handbook of age and ageing* (pp. 630–637). Cambridge and New York: Cambridge University Press.

Marshall, V. W., Matthews, S. H. and Rosenthal, C. J. (1993). Elusiveness of family life: A challenge for the sociology of aging, in G. L. Maddox and M. P. Lawton (eds), *Annual review of gerontology and geriatrics* (Vol. 13, pp. 39–72). New York: Springer Publishing Company.

Martin-Matthews, A. (2007). Situating 'home' at the nexus of the public and private spheres: Ageing, gender and home support work in Canada. *Current Sociology*, 55(2), 229–249. http://dx.doi.org/10.1177/0011392107073305

Martin-Matthews, A. and Phillips, J. E. (eds). (2008). *Aging at the intersection of work and home life: Blurring the boundaries*. New York: Taylor and Francis.

Martin-Matthews, A. and Sims-Gould, J. (2008). Employers, home support workers and elderly clients: Identifying key issues in delivery and receipt of home support. *Healthcare Quarterly*, 11, 69–75.

Martin-Matthews, A., Sims-Gould, J. and Naslund, J. (2010). Ethno-cultural diversity in home care work in Canada: Issues confronted, strategies employed. *International Journal of Ageing and Later Life*, 5(2), 77–101.

Martin-Matthews, A., Sims-Gould, J. and Tong, C. E. (2013). Canada's complex and fractionalized home care context: Perspectives of workers, elderly clients, family carers, and home care managers. *Canadian Review of Social Policy*, 68/69, 55–74.

Massey, D. (1992). A place called home. *New Formations: Journal of Culture, Theory, Politics*, 17(7), 3–15.

Oswald, F. and Wahl, H. W. (2005). Dimensions of the meaning of home, in G. D. Rowles and H. Chaudhury (eds), *Home and identity in late life: International perspectives* (pp. 21–46). New York: Springer.

Rowles, G. D. and Chaudhury, H. (eds). (2005). *Home and identity in late life: International perspectives*. New York: Springer.

Settersten, R. A. and Hagestad, G. O. (2015). Subjective ageing and new complexities of the life course, in M. Diehl and H-W. Wahl (eds), *Annual review of gerontology and geriatrics* (Vol. 35, pp. 29–53). New York: Springer.

Sims-Gould, J., Byrne, K., Tong, C. and Martin-Matthews, A. (2015). Home support workers' perceptions of family members of their older clients: A qualitative study. *BMC Geriatrics*, 15(165), 9 pp. http://dx.doi.org/10.1186/s12877-015-0163-4

Sims-Gould, J. and Martin-Matthews, A. (2010a). 'We share the care': Family caregivers' experiences of their older relative receiving home support services. *Health and Social Care in the Community*, 18(4), 415–423. http://dx.doi.org/10.1111/j.1365-2524.2010.00913.x

Sims-Gould, J. and Martin-Matthews, A. (2010b). Strategies used by home support workers in the delivery of care to elderly clients. *Canadian Journal on Aging*, 29(1), 97–107. http://dx.doi.org/10.1017/S0714980809990353

Skinner, M. W., Cloutier, D. and Andrews, G. J. (2015). Geographies of ageing: Progress and possibilities after two decades of change. *Progress in Human Geography*, 39(6), 776–799. http://dx.doi.org/10.1177/0309132514558444

Sprey, J. (1979). Conflict theory and the study of marriage and the family, in W. R. Burr, R. Hill, F. I. Nye and I. L. Reiss (eds), *Contemporary theories about the family: Research-based theories* (Vol. 1). Elmsford: Pergamon Press.

Stone, R. I. (2006). Emerging issues in long term care, in R. Binstock and George, L. (eds), *Handbook of aging and the social sciences* (6th edn, pp. 397–418). Cambridge, MA: Academic Press.

Victor, C. (2010). The demography of ageing, in D. Dannefer and C. Phillipson (eds), *The Sage handbook of social gerontology* (pp. 61–74). Thousand Oaks, CA: Sage Publishing.

14 Embodiment and emotion in later life

Ageing from the inside out

Rachel Herron

Introduction

Our bodies and our emotions are integral to our understanding of the world around us. Our bodies are our closest environment (Rich, 1986), and our emotions, whether pre-reflexive or carefully reflected on, reveal much about our relationships to these environments as well as our sense of well-being within them. We respond to, and feel, environmental stimuli (e.g., heat, cold, music or the voice of a familiar friend) through the body. However, the bulk of geographic research on the body and emotions has not paid enough attention to the experiences of older people (for notable exceptions see Herron and Skinner, 2013; Milligan, 2005; Tarrant, 2010). For instance, in the last five years the journal *Emotion, Space and Society* has published only two articles loosely related to ageing (Baldassar, 2015; Jacobson, 2014). While this represents merely a small subsection of work on emotion in geography, I contend that there has been a lack of engagement among researchers exploring emotional geographies and researchers studying ageing. To reach its full potential, geographical gerontology must pay more attention to the micro scale embodied and emotional experiences in a broad range of settings.

In the sections that follow, I do not attempt to provide an exhaustive review of geographic and social science literature on emotion and embodiment; rather, I outline key forces behind the growing interest in emotion and the body. First, I point to everyday spaces where interest in embodiment and emotion has grown, albeit largely in line with a neoliberal agenda and not without reinforcing ageist stereotypes. Next, I outline key developments and debates in humanist, feminist and non-representational geographies that influence and place limitations on how we view emotions (for a more detailed review see Bondi, 2005). With this background in place, I examine how emotion and affect can enhance our understanding of older people's being and well-being in the world. Using research on dementia as an example, I argue that careful attention to embodied experiences and feelings may help us get closer to understanding complex experiences of ageing from the inside as well as in relation to other people and places.

Growing interest in bodies and emotions

Western society is increasingly interested in the body and emotions. In the doctor's office, the supermarket, the gym and the public washroom, we are bombarded with messages that stress the importance of "listening to the body" and managing our emotions. While such messages acknowledge the significance of our bodies and emotions in relation to well-being, they suggest that bodies and emotions are sites of individual and private responsibility disconnected from broader social processes and relationships. This is problematic because individuals experience their bodies and feelings in relation to the particular places and people surrounding them. For instance, a person may experience feelings of sadness while sitting on a park bench *and* while sitting in the hospital; however, how we understand the roots and responses to sadness may be very different in each of these contexts. As Davidson and Milligan (2004) put it, "emotions are understandable – 'sensible' – only in the context of particular places. Likewise, place must be felt to make sense" (Davidson and Milligan, 2004: 524). Moreover, emotions matter beyond the immediate environment across a broad range of sites and scales. Wider social trends (i.e., ideas about gender, class, race, age and sexuality) influence how we view both emotions and their embodiment.

Implicit and explicit ageism shapes prevailing views of older people's bodies and feelings in limiting ways. In Western society, ageing bodies are typically viewed as undesirable, something to be prevented through daily exercise, the right diet or even a pill (Katz, 2000). The ageing body is associated with loss of strength, control and autonomy. Following these associations, later life is generally thought to be depressing as well, even though studies suggest that self-rated mental health improves with age and that the prevalence of common mental health conditions decreases (Reynolds *et al.*, 2015). Negative assumptions about older people are not isolated to lay discourses; they have long pervaded scholarship on ageing as well. Many studies begin with a focus on the *problem* of population ageing exploring the implications of declining labour force participation, increasing dependency ratios and the costs of service provision for an ageing population (Burholt and Dobbs, 2012; Gee and Gutman, 2000). Such deficit models of ageing fail to acknowledge the contributions of older adults and can cause unnecessary stigma and suffering for older people (Dupuis, Wiersma and Loiselle, 2012). Arguably, if we pay attention to the embodied, emotional experiences of older people, we will find that there are many more positive things to be said about ageing. However, the identification and examination of problems are at the root of particular scientific traditions, which inform what topics are worthy of study and what methods are appropriate for investigation.

Toward "reasonable" emotional and embodied research

Emotion has been excluded from "reasonable" scientific research and writing throughout much of the last century because it is often thought of as the opposite of reason (Thien, 2005). However, this is a false binary. A second binary, the mind/body binary, has also shaped scientific thinking. The body is commonly constructed as subordinate to the mind; it is something to be examined objectively rather than understood subjectively (Parr, 2002). Criticism of such objective rationality has come from many different corners in the last forty years. Among the first geographic gerontologists, Rowles (1978) drew attention to the need to consider the subjective dimensions of human life in his ethnographic study of the nature of older people's involvement within their everyday spaces. At the time, however, his book was reviewed with some scepticism. Ley (1978) wrote, "I felt myself to be an intruder in the highly personal and private world of an elderly person. It was sacred space to which the outsider should not, perhaps, have been given such privileged access" (Ley, 1978: 356). Ley questioned whether such personally revealing research could produce enough data and whether the ethical goals of the research could be reached. These are enduring questions (for some geographers), to which I return later in the chapter. What is important to note here is that Rowles not only challenged objective rationality; he also challenged the nature of how research should be done by developing deep relationships with older people, privileging interpersonal knowing and writing narratively. Ultimately, humanistic geographers pushed the discipline to recognise the value of understanding meanings, perceptions and emotions because they relate to so much of what people do and say in the world. Although humanistic geographers drew attention to the subjective realities that evoke emotion in individuals, it did not acknowledge the politics and differences between these individual bodies.

Feminist geographers challenged the individualised (white male) subjectivity inherent in humanistic geography as well as the emotional/rational and mind/body binaries at the heart of positivist scientific thinking (Longhurst, 1995, 1997; Parr, 2002). Drawing on poststructuralism, they sought to deconstruct prevailing conceptions that link femininity, body and emotions while positioning them as subordinate to masculinity, rationality and the mind (Bondi, 2005). They argued that examining and writing about bodies and emotion are necessary and legitimate scientific research (Haraway, 1996; McDowell, 1992; Sharp, 2009). For example, research on geographies of fear and intimate partner violence demonstrated that feelings in and about places can influence how, whether and in what ways women move across space (Pain, 2014). Research on political movements illustrated the importance of emotion as a force behind social justice movements (Wright, 2010). Research on caregiving has illuminated the importance of emotion in care work across a range of settings (Dyck, 2005; Milligan, 2005). Such feminist scholarship has contributed substantially to geographic understandings of emotion as something that moves people, moves with people and changes in relation to particular contexts (Ahmed, 2004; Bondi, 2005; Sharp, 2009). Feminist scholars established a strong tradition of giving voice to emotion through the use of verbatim quotations from qualitative interviews (Moss, 2002), and their work has given rise to emotional geographies – the study of the relationship between emotions and the contexts in which they arise (Anderson and Smith, 2001; Davidson and Milligan, 2004). However, some feminist geographers have been criticised for presenting these narratives uncritically and for their over-reliance on language as the primary means of understanding people's feelings (Bondi, 2005; Thien, 2005).

Before emotion: Non-representational theory and affect

In the last two decades, there has been a lively and ongoing debate between feminist geographers interested in emotion and non-representational theorists interested in affect about the need to move beyond emotion (Bondi, 2005; Pile, 2010; Thien, 2005). Non-Representational Theory (NRT) stresses the importance of understanding inexpressible affects (Pile, 2010). The term *affect* has a long history in psychology but has only recently been taken up in geography (Thrift, 1996). Cresswell defines affect as a "pre-cognitive sensation resulting from an encounter with a person or thing" (Cresswell, 2013: 275). Affect occurs before it is sensed and it takes place in encounters between bodies including non-human objects. It is "the transition of the body and the process whereby the body is affected, that modifies and affects other bodies" (Andrews and Grenier, 2015: 1525). In contrast, emotion refers to the cognitive, intersubjective feelings that are expressed through language and other modes of representation (e.g., a hug). Proponents of NRT argue that before we can examine the feelings and attachments that we have to, and in, particular places, these feelings are already in the process of forming or "showing up" (Andrews, 2015: 338). Rather than focus exclusively on the end product, we should look at how, when and where this sense happens (Andrews, Chen and Myers, 2014). Moreover, NRT posits that how we live and act in the world is equally or even more important than how we reason our actions and attach meaning to them (Pile, 2010).

NRT raises important questions about how we can capture and write about less-than-fully-conscious feelings (Andrews, 2015). Certainly, relationships to places as well as objects in places are not always fully conscious. People act without a sense of the meaning of their actions – perhaps most of the time. Interviews, for example, can capture valid reflections on what has happened, often in hindsight, but they may not get at the *how* or the less-conscious actions that constitute our lives across time and space. NRT challenges researchers to observe more-than-human influences on our actions and the more-than-fully conscious to produce a more nuanced understanding of being in the world. Although a focus on affect can deepen our understanding of how and what bodies do in the world, it is not without its critics. Thien (2005) and Bondi (2005) suggest that research on affect is too abstract and distant from the human relationships and meanings that complicate our living in the world. In attempting to be trans-human, they argue that research on affect has become almost inhuman. Other scholars, such as Hanlon (2014) and Kearns (2014), question how researchers can observe and measure affect. Colls (2012) addresses a number of these critiques by suggesting that geographers adopt a "nomadic consciousness", where they continue to be critical of universalism and gender-blindness while remaining open to what affect offers. In addition, Andrews and Grenier (2015) clarify that ethnographic methods, as well as a range of video and photographic methods, already present the actions and movements of people in time and space. Ultimately, NRT offers geographical gerontologists a new consciousness to enhance existing methods and approaches.

Research on affect and emotion both offer a relational approach to the study of feelings, viewing them as fluid and changing in response to situated encounters. They employ a variety of qualitative, especially ethnographic, methods with a focus on the body and its immediate environment rather than distant places. Taken together, studies of affect and emotion present conceptual strengths (i.e. relationality), avenues for methodological innovation (i.e. variation in ethnographic and qualitative approaches) and areas for further exploration (i.e. diversity and social difference). Rather than focus on emotion alone or affect alone, I add to the growing chorus of scholars that suggest we need to appreciate the full range of embodied responses to the world around us – the less-than-fully conscious and the conscious (Andrews and Grenier, 2015; Colls, 2012).

Where do we go with emotions and affect?

Geographical gerontology can extend work on embodiment, emotion, and affect by examining older people's body–self, body–other, and body–environment relationships. To illustrate what theoretical and empirical work in this vein can contribute to our understanding of ageing, I focus on dementia research in the social sciences. Although dementia is often thought of as a disease of the mind, the last decade has seen the emergence of a growing body of literature on embodiment and dementia (Downs, 2013). Research in this area continues to challenge assumptions about how it feels to have dementia, particularly pervasive popular discourses around loss of self. Ultimately, this work provides a more inclusive perspective of what it means to live well with dementia by examining day-to-day issues from the inside out.

Examining body–self relationships

Examining body–self relationships with attention to emotion and affect is essential to recognising the dignity and citizenship for people with dementia and older people in general. Indeed, the body is a site of individual expression where rights, freedoms and injustices are felt but not always easily, or entirely, expressed through language; thus, an affective and emotional approach to such embodied experiences is important. For example, as countries such as Canada move to legalise physician assisted death for those who have a grievous and irremediable condition, research might examine pain as a bodily sensation to which people respond affectively, often before thinking through or attaching any sort of meaning to that response (Bissell, 2009). While this is an extreme case for the importance of embodiment and affect, it illustrates a very poignant link between affects and emotion and choice in maintaining a dignified life.

Choice is often taken for granted in the mundane acts that constitute self-care such as bathing and dressing. However, Twigg (2010) contends that dress, as our most intimate environment, has a significant impact on an individual's well-being. If it is rough, it can make a person feel irritated. If it is soft and loose it may make a person feel comfortable. Clothing also situates us within our social relations and culture. If we wear the "wrong thing" we may be made to feel out of place. It has a performative quality in that it can remind people where they are and where they come from. Twigg (2010) argues that enabling people with dementia to maintain their mode of dress is essential to supporting their identity at a pre-reflexive cognitive level. In doing so, she highlights the important affective role of dress at the scale of the body and in relation to broader material and cultural environments. For example, people in long-term care are often made to wear clothes that make dressing, toileting and personal care more efficient for staff. Such dress can undermine the autonomy of the individual, and it can also be disturbing for family members and friends. Overall, attention to embodiment and feelings offers a framework for scholars to challenge the tendency toward rationalisation of supports and the biomedical view of the body (Kontos and Naglie, 2007).

Alternative treatments for dementia and other conditions associated with later life could benefit significantly from focusing more attention on embodied experiences across different settings. Studies on music therapy, dance therapy and reminiscence have drawn attention to the potential benefits of non-pharmacological treatments. For example, research on the role of music in the lives of older people with dementia has demonstrated its ability to reduce agitation, enhance relationships among older adults, and improve memory (Elliot and Gardner, 2016). However, much of this literature lacks attention to the insider perspective which might best be captured by examining affect and embodied expression. Kontos and colleagues (2015) provide an example, of how this might be done in their research on elder-clowning. The study focused on non-verbal communication as a means of examining the practice and experience of elder-clowning for people with moderate to severe dementia. They combined interviews with the clowns with video-recorded ethnographic observation revealing the intrinsic capacities of people with dementia to be playful and creative. Importantly, they note that the clowns in the study not only facilitated laughter; they also supported people with dementia through their sadness. Elder-clowning presents an innovative approach to validating and supporting the active, embodied, emotional lives of people with dementia while recognising that they still have rights to expression and self-fulfillment. Rather than focusing on the problem of managing different types of emotions (i.e., depression), the study examines how a diverse range of emotions can be supported. Geographical gerontology has much to contribute to this type of work by further grounding such embodied experiences in a range of relational contexts (material, performative and trans-scaled; Andrews and Grenier, 2015).

Examining interpersonal relationships

Many of the preceding examples illustrate the importance of the body in its own right as well as the difficulty of understanding personal embodied experiences without considering the relationships in which they occur. Emotions have been most present in ageing and dementia research focused on care and caregiving. This is likely because caring is often cited as a feeling in and of itself. However, care is seldom a singular feeling. It involves a range of emotions, some of which are recognised as being productive or consistent with the act itself while others are not. Concepts such as "emotion work" (Dyck, 2005), "emotional labour" (Bailey *et al.*, 2015) and "emotion rules" (Hochschild, 1983) have been used in sociology and geography to examine the role of emotions in care work. For example, Milligan (2005) examined how emotions in care relationships are managed as embodied inner experiences that may remain internalised or they may be expressed outwardly in the act of caring. Drawing on this work, Herron and Skinner (2013) examined how emotions are expressed or repressed in caring relationships, in the home, and in other community settings. Both of these analyses emphasise the suppression of emotions which are perceived to be negative or counter-productive to caring (e.g., anger, frustration, guilt, resentment). Like earlier feminist geographers, they make a case for the need to recognise and give voice to emotions that have not previously had a place to be expressed. However, more attention could be paid to the performance of emotion in this work as well as researcher's readings of emotion to address scepticism about the superficiality of emotion.

The work of Bailey and colleagues (2015) provides an example of how research on dementia care can give voice to emotion as well as take into account performance and action. Although not explicitly framed by NRT, their ethnographic approach combining participant observation and interviews with the health care assistants captured a complex range of reactions from non-cognitive to cognitive. They examined the ambiguity of "feeling rules" between health care assistants and people in the later stages of dementia. Feeling rules refer to the norms and beliefs about what types of feelings are to be displayed in particular kinds of work (Hochschild, 1983). Bailey and colleagues challenged the notion that emotional meanings and expectations are shared between the carer and the person being cared for, suggesting that feeling rules are the product of conflict more often than the product of consent. As they examined day-to-day care practices, they raised questions about the taken-for-granted role of touch in care work. Who initiates the touch and how do we read embodied responses to touch? Is a raised eyebrow or a smile adequate consent to a hug that is being offered? This work is incredibly important in informing care practice for those working with people with limited verbal capacity. Recognising the embodied and spoken tensions between carers and the people with dementia for whom they care is essential to providing quality care, as well as quality of work life, across care settings.

Choices about who cares, where care should take place, and when or what changes need to be made are profoundly shaped by emotions. Herron and Rosenberg (2017) make a case for these issues in their work on reactive behaviours (e.g., hitting, spitting, swearing and kicking). As dementia progresses and verbal communication becomes more difficult, it is not uncommon for people with dementia to react to personal, social and environmental challenges in ways that might be interpreted as "violent" or "aggressive". Not unlike Bailey and colleagues (2013), Herron and Rosenberg suggest that the meanings and feelings associated with these behaviours can be quite different for both partners in the caring relationship. Care partners can be particularly reticent to voice their fear and discomfort because they feel the need to protect the person with dementia. Other relationships of care are threatened by fear of violence and ultimately, so, too, are the continuity and quality of care. Partners in care from the study reported ceasing home care or having trouble keeping the same home care worker because of fear of aggression in the home. Other participants in the study reported neglect in hospital settings because of the fear of staff. Research on reactive behaviours might be enriched further with more direct attention to the embodied and affective dimensions of these expressions. Current research in this area focuses more on the feelings and actions of carers, partially because these behaviours are characteristic of the latter stages of dementia. A multimethod study involving some participant observation and reflection would provide a deeper understanding of the needs of persons with dementia. Most importantly, all the preceding studies emphasise that feelings can maintain or destabilise a sense of care across a range of settings.

Examining places of care

Caring relationships shape, and are shaped by, the physical and social environment. Similarly, an individual's sense of self is also linked to place. A critical area of inquiry for geographical gerontologists has been exploring what constitutes a good place to grow old for different people at different stages of later life. Work in this area has examined the experiences of older people in long-term care (Rijnaard *et al.*, 2016), the meaning of ageing in place for older people (Wiles *et al.*, 2012) and the role of alternative settings such as camps, woodlots and gardens as therapeutic places (Wiersma, 2008). Most of this research has focused on expressed emotions rather than the "showing up" of sense of place and well-being in place. The lack of attention to how emotions take place is partially the product of the slow uptake of NRT in health geography and geographical gerontology (Andrews and Grenier, 2015). It also involves addressing long-standing concerns around privacy and respect, as Ley's comments suggested some forty years ago. Although University ethics committees have grown to address such concerns, examining the personal experiences, movements, and forces in older people's lives demands more time and care throughout the research process (e.g., ethics review, relationship building, etc.) and may act as a deterrent to this type of work. To date, there is only a very small body of research examining the importance of place and the situated feelings of people with dementia; however, there are certainly a number of areas of research that might benefit from this perspective.

Research on dementia-friendly communities has examined aspects of the physical and social environment which enhance the lived experiences of people with dementia (Clarke and Bailey, 2016; Herron and Rosenberg, 2017). Such experiences are not typically understood as embodied and emotional, but the physical environment and objects within it can produce comfort or discomfort for people living with dementia. For example, poor signage can confuse a person with dementia while family photographs and personal objects can remind the person with dementia of their past and potentially enhance their sense of belonging in the present. Understanding the successes and failures in achieving dementia-friendly communities requires that research examine how people with dementia feel in and about the places they live. The development of more caring physical and social environments is dependent on acknowledging attachment to place and freedom of expression within place.

A growing body of dementia research in the social sciences illustrates that a focus on embodiment and emotion can help us challenge the narrative of loss that pervades popular discourse around ageing, in general, and dementia, in particular. Understanding embodiment and emotion in later life can lead us to more innovative treatments and therapies for conditions associated with later life, it can help us to recognise and support the work involved in caregiving and it can help identify and build more caring environments. It also challenges us to engage with persistent philosophical questions related to identity, self-worth and our changing relationships with people and places. Recognising embodiment and emotion in later life is central to respecting individuals' dignity and values.

Concluding comments

Our bodies are our most intimate geographies, and our feelings help us make sense of the world around us. To avoid or only superficially engage with them as researchers is to fail to understand a good life and a good place to grow old. The concept of affect can help us understand embodied experiences beyond cognition and self-reflection, which is particularly important in the study of ageing as these faculties diminish for many people in later life. However, engaging with a full range of emotions in geographical gerontology requires us to reflect on how we know and present feelings in the research process. What methods should we use? Whose emotions do we choose to present? Are there embodied experiences and emotions that are too sensitive to know about? I would suggest that it is only by examining some of the more private and contentious embodied experiences that we can come to terms with the discomfort these encounters produce and respond to them appropriately. Ultimately, careful attention to diverse bodies and feelings will move us further toward understanding ageing from the inside out.

References

Ahmed, S. (2004). *The cultural politics of emotion*. London: Routledge.

Anderson, K. and Smith, S. J. (2001). Editorial: Emotional geographies. *Transactions of the Institute of British Geographers*, 26(1), 7–10. http://dx.doi.org/doi:10.1111/1475-5661.00002

Andrews, G. J. (2015). The lively challenges and opportunities of non-representational theory: A reply to Hanlon and Kearns. *Social Science & Medicine*, 128, 338–341. http://dx.doi.org/10.1016/j.socscimed.2014.09.004

Andrews, G. J., Chen, S. and Myers, S. (2014). The 'taking place' of health and wellbeing: Towards non-representational theory. *Social Science & Medicine*, 108, 210–222. http://dx.doi.org/10.1016/j.socscimed.2014.02.037

Andrews, G. J. and Grenier, J. M. (2015). Ageing movement as space-time: Introducing non-representational theory to the geography of ageing. *Progress in Geography*, 34(12), 1512–1534. http://dx.doi.org/10.18306/dlkxjz.2015.12.003

Bailey, S., Scales, K., Lloyd, J., Schneider, J. and Jones, R. (2015). The emotional labour of health-care assistants in inpatient dementia care. *Ageing & Society*, 35(2), 246–269. Online first: August 2013. http://dx.doi.org/10.1017/S0144686X13000573

Baldassar, L. (2015). Guilty feelings and the guilt trip: Emotions and motivation in migration and transnational caregiving. *Emotion, Space and Society*, 16, 81–89. http://dx.doi.org/10.1016/j.emospa.2014.09.003

Bissell, D. (2009). Obdurate pains, transient intensities: Affect and the chronically pained body. *Environment and Planning A*, 41(4), 911–928. http://dx.doi.org/10.1016/a40309

Bondi, L. (2005). Making connections and thinking through emotions: Between geography and psychotherapy. *Transactions of the Institute of British Geographers*, 30(4), 433–448. http://dx.doi.org/10.1111/j.1475-5661.2005.00183.x

Burholt, V. and Dobbs, C. (2012). Research on rural ageing: Where have we got to and where are we going in Europe? *Journal of Rural Studies*, 28(4), 432–446. http://dx.doi.org/10.1016/j.jrurstud.2012.01.009

Clarke, C. L. and Bailey, C. (2016). Narrative citizenship, resilience and inclusion with dementia: On the inside or on the outside of physical and social places. *Dementia*, 15(3), 434–452. http://dx.doi.org/10.1177/1471301216639736

Colls, R. (2012). Feminism, bodily difference and non-representational geographies. *Transactions of the Institute of British Geographers*, 37(3), 430–445. http://dx.doi.org/10.1111/j.1475-5661.2011.00477x

Cresswell, T. (2013). *Geographic thought: A critical introduction*. New York: John Wiley & Sons.

Davidson, J. and Milligan, C. (2004). Embodying emotion sensing space: Introducing emotional geographies. *Social & Cultural Geography*, 5(4), 523–532. http://dx.doi.org/10.1080/1464936042000317677

Downs, M. (2013). Embodiment: The implications for living well with dementia. *Dementia*, 12(3), 368–374. http://dx.doi.org/10.1177/1471301213487465

Dupuis, S. L., Wiersma, E. and Loiselle, L. (2012). Pathologizing behavior: Meanings of behaviors in dementia care. *Journal of Aging Studies*, 26(2), 162–173. http://dx.doi.org/10.1016/j.aging.2011.12.001

Dyck, I. (2005). Susan Mackenzie memorial lecture feminist Geography, the 'everyday', and local-global relations: Hidden spaces for place-making. *The Canadian Geographer*, 49(3), 233–43. http://dx.doi.org/10.1111/j.0008-3658.2005.00092.x

Elliott, M. and Gardner, P. (2016). The role of music in the lives of older adults with dementia ageing in place: A scoping review. *Dementia.* Online first: March 2016. http://dx.doi.org/10.1177/1471301216639424

Gee, E. and Gutman, G. (eds). (2000). *The overselling of population ageing: Apocalyptic demography, intergenerational challenges, and social policy.* Toronto: Oxford University Press.

Hanlon, N. (2014). Doing health geography with feeling. *Social Science & Medicine,* 115, 144–146. http://dx.doi.org/10.1016/jsocscimed.2014.05.039

Haraway, D. (1996). Situated knowledges: The science question in feminism, in J. Agnew, D. Livingstone and A. Rogers (eds), *Human geography: An essential anthology* (pp. 108–128). Oxford: Blackwells Publishers.

Herron, R. V. and Rosenberg, M. W. (2017). "Not there yet": Examining community support from the perspective of people with dementia and their partners in care. *Social Science & Medicine,* 173, 81–87. http://dx.doi.org/10.1016/j.socscimed.2016.11.041

Herron, R. V. and Rosenberg, M. W. (2017). Responding to aggression and reactive behaviours in the home. *Dementia.* http://dx.doi.org/10.1177/1471301217699676.

Herron, R. V. and Skinner, M. W. (2013). The emotional overlay: Older person and carer perspectives on negotiating aging and care in rural Ontario. *Social Science & Medicine,* 91, 186–193. http://dx.doi.org/10.1016/jsocscimed.2012.08.037

Hochschild, A. R. (1983). *The managed heart: Commercialization of human feeling.* Berkeley, CA: University of California Press.

Jacobson, K. (2014). The temporality of intimacy: Promise, world, and death. *Emotion, Space and Society,* 13, 103–110. http://dx.doi.org/10.1016/j.emospa.2013.08.006

Katz, S. (2000). Busy bodies: Activity, aging, and the management of everyday life. *Journal of Aging Studies,* 14(2), 135–152. http://dx.doi.org/10.1016/S0890-4065(00)80008-0

Kearns, R. A. (2014). The health in "life's infinite doings": A response to Andrews et al. *Social Science & Medicine,* 115, 147–149. http://dx.doi.org/10.1016/j.socscimed.2014.05.040

Kontos, P., Miller, K. L., Mitchell, G. J. and Stirling-Twist, J. (2015). Presence redefined: The reciprocal nature of engagement between elder-clowns and persons with dementia. *Dementia.* 16(1), 46–66. http://dx.doi.org/10.1177/1471301215580895

Kontos, P. C. and Naglie, G. (2007). Bridging theory and practice Imagination, the body, and person-centred dementia care. *Dementia,* 6(4), 549–569. http://dx.doi.org/10.1177/1471301207084394

Ley, D. (1978). Review of book prisoners of space? Exploring the geographical experience of older people, by G. D. Rowles. *Economic Geography,* 54(4) (October), 355–356. http://dx.doi.org/10.2307/143284

Longhurst, R. (1995). The body and geography. *Gender, Place & Culture,* 2(1), 97–106. http://dx.doi.org/10.1080/09663699550022134

Longhurst, R. (1997). (Dis)embodied geographies. *Progress in Human Geography,* 21(4), 486–501. http://dx.doi.org/10.1191/03091329

McDowell, L. (1992). Doing gender: Feminism, feminists and research methods in human geography. *Transactions of the Institute of British Geographers,* 17(4), 399–416. http://dx.doi.org/10.2307/622707

Milligan, C. (2005). From home to 'home': Situating emotions within the caregiving experience. *Environment and Planning A,* 37(12), 2105–20. http://dx.doi.org/10.1068/a37419

Moss, P. (2002). Taking on, thinking about, and doing feminist research in Geography, in P. Moss (ed.), *Feminist geography in practice: Research and methods* (pp. 1–20). Malden, MA: Blackwell.

Pain, R. (2014). Everyday terrorism: Connecting domestic violence and global terrorism. *Progress in Human Geography,* 38(4), 531–550. http://dx.doi.org/10.1177/0309132513512231

Parr, H. (2002). Medical geography: Diagnosing the body in medical and health geography, 1999–2000. *Progress in Human Geography*, 26(2), 240–251. http://dx.doi./org/10.1191/0309132502ph367pr

Pile, S. (2010). Emotions and affect in recent human geography. *Transactions of the Institute of British Geographers*, 35(1), 5–20. http://dx.doi.org/10.1111/j.1475-5661.2009.00368.x

Reynolds, K., Pietrzak, R. H., El-Gabalawy, R., Mackenzie, C. S. and Sareen, J. (2015). Prevalence of psychiatric disorders in US older adults: Findings from a nationally representative survey. *World Psychiatry*, 14(1), 74–81. http://dx.doi.org/10.1002/wps.20193

Rich, A. (1986). *Notes towards a politics of location*. New York: W.W. Norton & Company.

Rijnaard, M. D., van Hoof, J., Janssen, B. M., Verbeek, H., Pocornie, W., Eijkelenboom, A., Beerens, H. C., Molony, S. L. and Wouters, E.J.M. (2016). The factors influencing the sense of home in nursing homes: A systematic review from the perspective of residents. *Journal of Aging Research*, 2016, 16 pp. http://dx.doi.org/10.1155/2016/6143645

Rowles, G. D. (1978). *Prisoners of space? Exploring the geographical experience of older people*. Boulder, CO: Westview Press.

Sharp, J. (2009). Geography and gender: What belongs to feminist geography? Emotion, power and change. *Progress in Human Geography*, 33(1), 74–80. http://dx.doi.org/10.1177/0309132508090440

Tarrant, A. (2010). Constructing a social geography of grandparenthood: A new focus for intergenerationality. *Area*, 42(2), 190–197. http://dx.doi.org/10.1111/j.1475-4762.2009.00920.x

Thien, D. (2005). After or beyond feeling? A consideration of affect and emotion in geography. *Area*, 37(4), 450–454. http://dx.doi.org/10.1111/j.1475-4762.2005.00643a.x

Thrift, N. (1996). *Spatial formations*. London: Sage.

Twigg, J. (2010). Clothing and dementia: A neglected dimension? *Journal of Aging Studies*, 24(4), 223–230. http://dx.doi.org/10.1016/j.jaging.2010.05.002

Wiersma, E. C. (2008). The experiences of place: Veterans with dementia making meaning of their environments. *Health & Place*, 14(4), 779–794. http://dx.doi.org/10.1016/j.healthplace.2008.01.001

Wiles, J. L., Leibing, A., Guberman, N., Reeve, J. and Allen, R. E. (2012). The meaning of "aging in place" to older people. *The Gerontologist*, 52(3), 357. http://dx.doi.org/10.1093/geront/gnr098

Wright, M. W. (2010). Geography and gender: Feminism and a feeling of justice. *Progress in Human Geography*, 34(6), 818–827. http://dx.doi.org/10.1177/0309132510362931

Part IV

Key issues in geographical gerontology

15 Explaining the ageing in place realities of older adults

Stephen M. Golant

Introduction

An international body of research from such diverse academic and professional specialties as geography, environmental gerontology, occupational therapy, urban planning and architecture finds that older people prefer to age in place or stay put in their current residential settings rather than to move or migrate to another destination. These studies, mainly from the United States, Canada, the United Kingdom, Sweden, China, Japan, New Zealand, Australia, Malaysia and Taiwan, also show that compared to younger people, ageing adults, especially homeowners, remain occupants of their current dwellings for much longer durations (Vasunilashorn *et al.*, 2012).

Importantly, these investigations typically treat older people's ageing in place behaviours as more than just demographic events that identify who moves or not (Golant, 1975). Rather, in their efforts to understand what ageing in place means to older adults, they have focused on three different areas of inquiry (Golant, 2015a; Wahl, Iwarsson and Oswald, 2012; Wiles *et al.*, 2012). First, they evaluate whether older adults who age in place are occupying places that are congruent with their self-identities, lifestyles and capabilities, as indicated by their self-reported subjective experiences or the objective assessments of academics or policymakers (Koss and Ekerdt, 2017; Golant, 2003; Heatwole Shank and Cutchin, 2016; Smetcoren *et al.*, 2015). Typically, these studies conclude that the reluctance of older individuals to move, particularly homeowners, is linked to their strong emotional attachments to both their built and social environments (Pinquart and Burmedi, 2003; Rowles and Ravdal, 2002; Rubenstein, 2003).

Second, they interpret the ageing in place behaviours of older adults as adaptive strategies by which they avoid relocating to nursing homes and other long-term care facilities, even as they require care and assistance to cope with their debilitating physical and cognitive health declines or difficulties performing their everyday activities (Golant, 2008a; Wiles *et al.*, 2011). Consequently, these studies consider the dwellings in which older people age in place not just as residential environments but also as care environments that enable them to live independently and autonomously as long as possible (Barrett, Hale and Gauld, 2012; Sixsmith and Sixsmith, 2008).

Third, they evaluate if public-sector policies that are designed to make it more feasible for older adults to age in place are both beneficial and cost-effective. Health and long-term care providers argue that when lower-income older individuals stay put in their own dwellings, government programmes can offer them community- and home-based health and long-term services and supports less expensively than if comparable care was provided to them as occupants of long-term care facilities (Cao *et al.*, 2014). Likewise, urban and civic leaders argue that older persons can age in place more successfully if they occupy more livable or age-friendly communities (Golant, 2014; Heatwole Shank and Cutchin, 2016; Skinner and Power, 2011). Exemplary is the World Health Organization's (WHO) Global Age-friendly Cities Project started in 2006 that promotes active ageing, "the process of optimizing opportunities for health, participation and security in order to enhance quality of life as people age" (WHO, 2007: 5). These initiatives are designed to improve shopping, health care, home care, leisure, volunteer, recreational, and social opportunities – actions that are "prerequisites for ageing in place" (Van Dijk *et al.*, 2015: 1772).

Goals and rationale

This chapter seeks to further our conceptual understanding of the realities of ageing in place. It outlines an emotion-based theory to judge whether older adults feel they occupy residential (or care) environments congruent with their needs and goals and explains why they do not always depend on moving as a strategy to cope with their residential and care deficiencies and challenges.

We should be concerned about whether older people stay put or move from their current dwellings because of overwhelming evidence that the quality of their residential and care environments influence whether they will age successfully or optimally, that is, have healthy, autonomous, engaged and happy lives. As I have argued elsewhere, it is more enjoyable, easier and less costly to grow old in some places than in others (Golant, 1984, 2015a).

However, the decisions of older people to age in place are also of concern because of their relevance to other societal stakeholders. Large residential concentrations of older persons often result from their collective decisions to age in place in the same building or neighbourhood (Golant, 1975). Often their dwellings are older and are at greater risk of being in disrepair or in need of physical upgrading. Researchers link these indicators of physical obsolescence to the depressed housing values in their neighbourhoods and in turn smaller local government property tax revenues (Golant, 2008b).

Because the older residents who age in place are predominantly homeowners, the experts also worry that when they eventually decide to sell their dwellings, it will result in a market glut – and an oversupply of larger and older houses not easily absorbed by younger age groups – that will "upset the historic balance of buyers and sellers" (Myers and Ryu, 2008: 30). The result may be a large number of neighbourhoods with high vacancy rates and lower property values.

On the other hand, research also shows that these residential enclaves of older adults may offer their occupants more opportunities for supportive social relationships because of their shared lifestyles and values (Golant, 2015b). Furthermore, businesses and providers in both the private and public sectors have stronger organisational and financial incentives to target these older resident concentrations with their products and services (Golant, 2008c).

When physically vulnerable older people age in place instead of transitioning to long-term care facilities, they dramatically change who is responsible for their health and care. The caregiving burden falls mostly on family members, especially spouses and daughters (Tang and Lee, 2011). These responsibilities, however, come with costs. Family members often find that caring for their loved ones is a physically and emotionally draining experience. Moreover, despite good intentions, they often provide only poor-quality care. In the United States, a vast industry of home care providers and home remodelers has emerged to serve these ageing in place populations when family care is not enough (Milligan, 2015). Consequently, the livelihoods of a large number of workers and businesses depend on where older people opt to receive their services and assistance (Stone and Harahan, 2010).

Theoretical understanding of the ageing in place behaviours of older people

Researchers have relied on two theoretical models from geography and gerontology to explain why older people move from their dwellings or age in place: Wiseman's (1980) behavioural model and Litwak and Longino's (1987) developmental model (Perry, Anderson and Kaplan, 2014; Smetcoren *et al.*, 2015). Wiseman argues that residential relocation occurs when older people are dissatisfied with their current residences. He proposes that older individuals re-evaluate the desirability of where they live because of push and pull "triggering mechanisms" (Wiseman, 1980: 146). "Push" motivations arise because older persons experience disruptive life events such as widowhood or health declines or unfavorable changes in their current residential environments. "Pull" motivations result from older people believing that they can find more attractive recreational, social, economic and health opportunities in other residential locations. Wiseman treats the personal resources of the older person, such as income or health, as factors that facilitate or impede these moving decisions.

The developmental perspective model of Litwak and Longino (1987) proposes that three types of life events motivate the moves of older people. The first move is in response to retirement when older persons seek locations with more recreation or leisure opportunities or that are closer to friends and family. The second move occurs when chronic disabilities make it difficult for them to perform household tasks such as shopping, cooking, cleaning, emergency first aid and protecting themselves from crime. They initiate a third move to find nursing home care that can accommodate the onset of severe physical or cognitive impairments.

Problems faced by an ageing in place older population

Both these theories contribute to our understanding of the relocation reasoning (Granbom *et al.*, 2014) of older people. Certainly, major life events such as retirement, losing a spouse, financial difficulties, housing challenges and health issues and impairment increase the likelihood of moving (Bonnet, Gobillon and Laferrère, 2010; Hillcoat-Nalletamby and Ogg, 2014; Sharma, 2013, Stoeckel and Porell, 2010; Wilmoth, 2010). However, these influences are far from determinative, and there is not a simple linear cause–effect relationship (Hayward, 2004; Shen and Perry, 2014). Furthermore, what the Wiseman model characterizes as "environmental push motivations" fail to account for why the majority of older people age in place, even as significant proportions occupy dwellings and neighbourhoods with problems (Golant, 2015a). Studies throughout the world have identified these residential downsides (Cao *et al.*, 2014; Coleman, Kearns and Wiles, 2016; Golant, 2008c, 2015a; Golant and LaGreca, 1994; Hillcoat-Nalletamby and Ogg, 2014; Sixsmith and Sixsmith, 2008; Van Dijk *et al.*, 2015; Vasunilashorn *et al.*, 2012). The most important are summarized below:

- Ageing in place older adults may live in unaffordable dwellings and pay an excessive amount of their monthly incomes on their housing expenses. In the United States, those most at risk include the poor, homeowners still paying off their mortgages, African Americans and Hispanics, unmarried women and the less educated.
- When owned homes are older, they are more likely to have less energy efficient cooling and heating equipment, out-of-date interior designs and physical deficiencies (e.g., leaky windows and roofs, plumbing and electrical problems) and, thus, require costly dwelling repairs and infrastructure upgrades. Furthermore, these structures have a higher risk of having environmental hazards including dust, mold, dampness, air pollution, and termites, rodents and roaches.
- Because these older dwellings were usually designed with younger consumers in mind, physically impaired older persons often confront dwelling usability challenges including front door or second-storey stairs, high shelves in cupboards, fall risks because of slippery floors, poor lighting or the absence of grab bars.
- Because their spouses have recently died or good friends have left their communities, older persons may feel lonely or socially isolated. They also may be upset with the changed social composition of their neighbourhoods because occupants are now younger or identify with different ethnic, racial or religious groups.
- Older persons who age in place often are located in lower-density suburban or rural counties with no nearby transit. Consequently, when declines in their physical health and cognitive abilities make driving unsafe or impossible, they have difficulties accessing goods, services, people and activities.

This is especially true when they cannot count on others to take them as passengers. They may also have restricted walking opportunities, if their neighbourhoods have poor street lighting, inadequate walking paths, unsafe traffic conditions or higher levels of crime.

- They may have difficulty maintaining their independent households because of health problems and mobility limitations. Once taken-for-granted household tasks, such as preparing meals, doing housework, and managing financial affairs, become difficult. More seriously, they also may have trouble performing self-care tasks such as bathing, toileting, dressing and managing their complex medication regimens.
- Because they are sometimes separated by large geographic distances from their adult children, older persons with unmet health and long-term care needs cannot depend on them as caregivers.
- When older persons who age in place do receive *in situ* help with their health and long-term care needs, their residential environments sometimes become difficult to distinguish from care settings because they are frequented by informal (e.g., family members) and formal (paid staff) caregivers and filled with medical equipment and assistive devices, such as canes, walkers, scooters and wheelchairs (Golant, 2015a). In the extreme, these residential settings undesirably look and function like nursing homes.

A new residential decision-making framework

The residential normalcy theory outlined in the rest of this chapter explains why it is possible for older people to experience individual crises and residential problems and still opt to age in place. They may have offsetting positive feelings, understate or rationalize the seriousness of their problems, or lack efficacious or viable coping solutions (Golant, 2011, 2012, 2015c, 2015d). Unlike Wiseman's reliance on a unidimensional indicator of congruence – residential satisfaction – it proposes the multidimensional construct of *residential normalcy* as an emotion-based indicator of individual-environment fit. Unlike Litwak and Longino's model, it includes environmental factors as possible antecedents of older people's residential moves.

Achieving residential normalcy

The theory argues that the evaluation or *primary appraisal* (Lazarus, 1966) by older individuals of their residential settings depends on two independent categories of their emotional experiences. The first category, *residential comfort experiences*, captures whether they feel that their residential settings are pleasurable, appealing, comfortable and enjoyable places with minimal everyday hassles (Golant, 2012). For example, some older people will enjoy places that are warm year-round, while others will enjoy the change of seasons. Some will enjoy taking care of their lawns, but others will feel it is a nuisance chore. Some will enjoy urban settings with lots of restaurants, nightlife and shops, while others will enjoy the simplicity and tranquility of a rural setting. Some will thrive when they live in communities that offer them many opportunities to be active, such as volunteering or involvement in their religious congregations, while others will find a sedentary home-centered lifestyle more appealing. While some will feel emotionally attached to their personal possessions, others will view these material collections as junk. Some older persons will enjoy being alone, while others will feel fulfilled when they are interacting with others. When the experiences in this category judged as more salient are, on balance, positive, older people are theorized to be feeling in their *residential comfort zones*.

The second category, *residential mastery experiences*, captures whether older people feel competent, empowered and in control of their lives and environment (Schulz and Heckhausen, 1996). These types of emotional experiences are often elicited as a result of older people's everyday transactions with their dwelling and neighbourhood environments (Golant, 2012). For example, whether a frail older widow with arthritic knees feels vulnerable may depend on whether she occupies a one-storey dwelling or must continually negotiate stairs in a multi-storied house. Whether she feels confident about completing her shopping tasks will depend on whether stores are within walking distance or she can rely on others to drive her. Whether or not the older person who can no longer drive can still access his doctors will influence whether he feels independent. When the experiences in this category judged as more salient are, on balance positive, older people will feel in their *residential mastery zones*.

When both these categories of feelings are altogether positive, older people achieve *residential normalcy* and the theory predicts that they will stay put or age in place. Conversely, older people are more likely to contemplate moving when they are out of both their residential comfort and mastery zones. More frequently, older people find themselves in emotionally conflicted environments (Löfqvist *et al.*, 2013). Because of "the theorized orthogonality or independence of the residential comfort and residential mastery sets of emotional experiences, congruence is often not an all or nothing affair" (Golant, 2015d: 1553).

Consequently, even though older people have difficulty maintaining their independent households (*out of their residential mastery zones*), they can remain emotionally attached to their familiar homes, neighbourhoods and communities and can enjoy their amenities, friendships and religious affiliations (*in their residential comfort zones*). Alternatively, they can appraise their cold climates and socially inhospitable neighbourhoods as unenjoyable or unappealing (*out of their residential comfort zones*) but nonetheless feel secure, competent and empowered because they live close to an adult daughter who caters to their everyday needs (*in their residential mastery zones*). Such experiential contradictions blunt older people's overall motivation to move and help explain their residential inertia even as they report negative assessments.

Strategies to cope with adversity

It is not sufficient, however, to focus only on how older people subjectively assess or experience their residential environments. Human development theorists tell us that ageing successfully also is determined by how effectively older people cope with adversity (Brandtstadter and Greve, 1994). Consequently, when they occupy incongruent residential environments, it depends on what proactive steps older persons take "to engage and influence their environments in order to change their undesirable circumstances" (Golant, 2015c: 71).

Older people can initiate two very different types of coping strategies (Brandtstadter and Greve, 1994). First, they can rely on *accommodative* or mind-adaptive responses by which they rationalise or deny their incongruent residential arrangements. They convince themselves that their expectations of a satisfactory or "normal" place to live are too high or that their adverse circumstances are unimportant for their self-esteem or happiness. Alternatively, they minimise their problems by comparing themselves favorably to others. They also may ignore or deny their difficulties outright. When older people cope this way, it helps explain why even stressful life events or strong push–pull environmental factors identified by the experts do not result in their residential relocations.

Second, older people can rely on *assimilative* coping strategies whereby they initiate concrete corrective actions to solve their problems, such as changing the offensive aspects of where they live. If they fear falling, they install grab bars, or if they have difficulty getting to places, they hire a student to get their groceries. If they feel alone after the death of a spouse, they can look to social supports from their religious congregations. If they have trouble paying their property taxes, they can rent out their basement to a boarder for added income or can cut back on their heating to save on their utility bills. Consequently, the theory argues that even when older individuals experience disruptive life events or "environmental push" motivation factors, they can successfully cope with the incongruent aspects of their residential settings even as they continue to age in place. Moving to a different place of residence is but one of many possible actions by which older people can achieve residential normalcy.

Coping opportunities or repertoires

The Residential Normalcy theory further argues that the selection of these accommodative or assimilative strategies by older persons depends on how they subjectively evaluate their *coping opportunities or repertoires*, that is, their *secondary appraisal processes* (Lazarus, 1966). By distinguishing secondary from primary appraisal efforts, the theory emphasises that whether or not older people age in place (or move) may have less to do with their incongruent residential environments than with their motivations and abilities to cope concretely with their residential problems. The theory proposes another conceptual layer of factors to explain why some older people have better options than others, that is, have more enriched coping opportunities or repertoires. These older persons are more *aware* of alternative adaptive strategies and assess them as *efficacious* (leading to successful outcomes) and *viable*; that is, doable or implementable (Golant, 2015c).

How *aware* older people are of their coping opportunities will initially frame their ageing in place or moving actions. Obtaining complete and accurate information about "what solutions are out there" is often one of the most demanding coping tasks because older persons typically have little experience dealing with the challenges of getting old and are unfamiliar with their options. Whether they act on their acquired information will further depend on their perceptions of the credibility and trustworthiness of their communicators. Once aware of potential solutions, they will judge some alternatives as more *efficacious*, that is, enabling them to manage or eliminate their incongruent residential arrangements. They must decide whether their adaptive efforts will result in more comfortable or appealing dwellings and neighbourhoods where they feel more competent and in control. If their solutions include the possibility of moving to another place of residence, "they must feel they can rejuvenate the magnetic forces elsewhere that for so long nurtured their current residential inertia" (Golant, 2015a: 100).

Once judging their coping strategies as efficacious, they must also assess if they are *viable*, that is, doable or implementable. Renting a basement out to a boarder to obtain another source of income may alleviate their financial difficulties, but older people must be willing to sacrifice their privacy or ignore their fears of having a stranger in their houses. Alternatively, they can cope with their unaffordable home by moving to less-expensive accommodations elsewhere but only if there are potential buyers of their current homes who offer them an acceptable financial return. Similarly, even though some suburban older occupants seek to return to city environments that support their more active lifestyle preferences, unaffordable housing opportunities in their chosen destinations may stymie such moves.

Additionally, however positively older people judge other places to live, they must also appraise the act of moving itself as viable or doable, that is, as not requiring too much physical or psychological energy (Lieberman, 1991). "Moving day" stresses or the undesirable prospects of confronting environmental change cannot dissuade them. They must not be overly concerned about finding new doctors and friends in their new locale or never socially "belonging" in their new community (Golant, 2015a; Lieberman, 1991; Wahl, Iwarsson and Oswald, 2012). They must be willing to "disband" a lifetime's accumulation of personal possessions if they plan to downsize their residences (Ekerdt *et al.*, 2004).

Determinants of coping repertoires

To understand the coping repertoires of older persons, individual differences are key, not to predict who is more susceptible to negative environmental press or adversity (Lawton and Nahemow, 1973) but, rather, to predict who is likely to cope more successfully. Here, we are informed by a large literature arguing that older people with stronger adaptive abilities are more resilient (Aldwin and Igarashi, 2012; Skodol, 2010; Wiles, 2012). That is, they are less impaired, in better health, more highly educated and can draw on greater economic resources. They have stronger problem-solving skills and generally deal better with adversity. They have personalities that motivate and enable them to find creative and constructive solutions. They are more optimistic, more extroverted, have greater confidence in their ability to effectuate change and control their fates, and are more trusting and open to the solutions offered by others (Golant, 2015c).

The action strategies of older adults will depend not just on their personal resilience but also on the *resilience* of their current (or alternative) places of residence; that is, whether places offer effective and viable solutions to address their current housing problems. Even the most motivated and capable older persons may find that ageing in place is not possible if they live in a rural community and lack transportation to reach needed adult day-care centers or health care clinics. In contrast, other older persons will benefit from more attractive community resources – socially engaging activities and events for those living alone, more affordable home care for the economically disadvantaged, and Uber or paratransit transportation alternatives for non-drivers. Some places will have more capable, motivated, and caring leaders who recognize the importance of creating *age-friendly communities* offering products, services, and activities that enable their older constituents to age actively, safely and securely in their own dwellings (Golant, 2014; Scharlach and Lehning, 2016; World Health Organization, 2007).

The *influencers* of older people's residential deliberations matter (Golant, 2015c). The aforementioned migration models unrealistically assume a single and autonomous decision-maker. In practice, married older adults with potentially different and conflicting retirement and care plans are both appraising their ageing in place or moving options. Similarly, frail older people often seek the counsel of family members when they engage in their ageing in place deliberations. Alternatively, adult sons and daughters may impose their own coping solutions on an older parent (Morgan and Brazda, 2013). For example, the singular act by an adult son, whereby he takes away the car keys from his older father because of driving mishaps, may be the catalyst for his father's moving to a place more accessible to everyday goods and services. Complicating the process, such triggering events may not influence the moving behaviour of older people but, rather, the residential decisions of their influencers or significant others. Consequently, it is often the adult children who move closer to their ageing in place mothers and fathers when they want to help them cope with their debilitating health events, such as a stroke (Choi *et al.*, 2014; Zhang, Engelman and Agree, 2013).

Past migration models have also failed to conceptualize how *period* (or historical) effects can greatly increase the likelihood that older persons will cope with their problems by ageing in place (Golant, 2003). During the Great

Recession in the United States (roughly 2007–2010), moving was off the table for older homeowners even if they had multiple dwelling problems. Many had to postpone their retirement for financial reasons, and this kept them tied to their workplace locations. There were few potential buyers for their houses, and when they did receive offers, they were unrealistically low. In contrast, when the US housing market later recovered, relocations by the old became more feasible (Golant, 2015a).

Conclusion

Even as older adults throughout the world experience highly disruptive senti-nel events in their lives, such as retirement, widowhood, impaired bodies and declines in their income, and as they occupy residential settings no longer con-gruent with their needs and goals, they often opt not to change their individual and environmental circumstances by moving or migrating elsewhere. Rather, they age in place or stay put in their current residences as long as possible. Our current migration models fail to adequately explain why older people make these ageing in place decisions. This chapter outlined a new theoreti-cal model – the Residential Normalcy theory – that attempts to advance our understanding in two important ways. First, in recognition that older people often have conflicting positive and negative feelings about where they live, it conceptualizes individual–environment congruence as a resolution of two sets of relatively independent emotional experiences: residential comfort and resi-dential mastery. Second, it argues that the ageing in place or moving decisions of older people depend not just on how they appraise their current residential settings but also on how they evaluate their coping opportunities or reper-toires. Consequently, they may stay put if they are able to adapt their current environments to fit their changed lifestyles or capabilities. Alternatively, their residential inertia may be showcasing an inability to cope with their incongru-ent situations by moving because they believe efficacious or viable residential alternatives are unavailable. They feel trapped in place. Yet another group of older persons do not cope with their problems by taking concrete actions, such as moving, but, rather, justify their ageing in place decisions by relying on accommodative or mind strategies – they deny having any problems or ration-alize that they are not very serious.

References

Aldwin, C. and Igarashi, H. (2012). An ecological model of resilience in late life, in B. Hay-slip and G. C. Smith (eds), *Annual Review of Gerontology and Geriatrics*, v32(1), 115–130. New York: Springer Publishing Company. http://dx.doi.org/10.1891/0198-8794.32.115

Barrett, P., Hale, B. and Gauld, R. (2012). Social inclusion through ageing-in-place with care? *Ageing and Society*, 32(3), 361–378.

Bonnet, C., Gobillon, L. and Laferrère, A. (2010). The effect of widowhood on housing and location choices. *Journal of Housing Economics*, 19(2), 94–108. http://dx.doi.org/10.1016/j.jhe.2010.04.003

Brandtstadter, J. and Greve, W. (1994). The aging self: Stabilizing and protective processes. *Developmental Review*, 14(1), 52–80. http://dx.doi.org/10.1006/drev.1994.1003

Cao, M. J., Guo, X. L., Yu, H., Chen, L. Y. and McDonald, T. A. (2014). Chinese community-dwelling elders' needs: Promoting ageing-in-place. *International Nursing Review*, 61(3), 327–335.

Choi, H., Schoeni, R. F., Langa, K. M. and Heisler, M. M. (2014). Older adults' residential proximity to their children: Changes after cardiovascular events. *The Journals of Gerontology Series B: Psychological Sciences and Social Sciences*, 70(6), 995–1004. http://dx.doi.org/10.1093/geronb/gbu076

Coleman, T., Kearns, R. A. and Wiles, J. (2016). Older adults' experiences of home maintenance issues and opportunities to maintain ageing-in-place. *Housing Studies*, 31(8), 964–983. http://dx.doi.org/10.1080/02673037.2016.1164834

Ekerdt, D. J., Sergeant, J. F., Dingel, M. and Bowen, M. E. (2004). Household disbandment in later life. *Journal of Gerontology: Social Sciences*, 59(5), S265–S273. http://dx.doi.org/10.1093/geronb/59.5.S265

Golant, S. M. (1975). Residential concentrations of the future elderly. *The Gerontologist*, 15(1 Pt. 2), 16–23. http://dx.doi.org/10.1093/geront/15.1_Part_2.16

Golant, S. M. (1984). *A place to grow old: The meaning of environment in old age*. New York: Columbia University Press.

Golant, S. M. (2003). Conceptualizing time and space in environmental gerontology: A pair of old issues deserving new thought. *The Gerontologist*, 43(5), 638–648. http://dx.doi.org/10.1093/geront/43.5.638

Golant, S. M. (2008a). Low-income elderly homeowners in very old dwellings: The need for public policy debate. *Journal of Aging and Social Policy*, 20, 1–28. http://dx.doi.org/10.1300/J031v20n01_01

Golant, S. M. (2008b). Affordable clustered housing-care: A category of long-term care options for the elderly poor. *Journal of Housing for the Elderly*, 22(1–2), 3–44. http://dx.doi.org/10.1080/02763890802096906

Golant, S. M. (2008c). Commentary: Irrational exuberance for the aging in place of vulnerable low-income older homeowners. *Journal of Aging and Social Policy*, 20(4), 379–397. http://dx.doi.org/10.1080/08959420802131437

Golant, S. M. (2011). The quest for residential normalcy by older adults: Relocation but one pathway. *Journal of Aging Studies*, 25(3), 193–205. http://dx.doi.org/10.1016/j.jaging.2011.03.003

Golant, S. M. (2012). Out of their residential comfort and mastery zones: Toward a more relevant environmental gerontology. *Journal of Housing for the Elderly*, 26(1–3), 26–43. http://dx.doi.org/10.1080/02463893.2012.655654

Golant, S. M. (2014). *Age-friendly communities: Are we expecting too much?* Montreal: Institute for Research on Public Policy.

Golant, S. M. (2015a). *Aging in the right place*. Baltimore, MD: Health Professions Press.

Golant, S. M. (2015b). Should older Americans live in places segregated from the young. *The Conversation*. Available from: https://theconversation.com/should-older-americans-live-in-places-segregated-from-the-young-47340 [accessed 29 September 2015]

Golant, S. M. (2015c). Residential normalcy and the enriched coping repertoires of successfully aging older adults. *The Gerontologist*, 55(1), 70–82. http://dx.doi.org/10.1093/geront/gnu036

Golant, S. M. (2015d). Residential normalcy and the aging in place behaviors of older Americans. *Progress in Geography*, 34(12), 1535–1557. Available from: www.researchgate.

net/publication/290392455_Residential_Normalcy_and_the_Aging_in_Place_Behaviors_of_Older_Americans

Golant, S. M. and LaGreca, A. J. (1994). Housing quality of US elderly households: Does aging in place matter? *The Gerontologist*, 34(6), 803–814. http://dx.doi.org/10.1093/geront/34.6.803

Granbom, M., Himmelsbach, I., Haak, M., Löfqvist, C., Oswald, F. and Iwarsson, S. (2014). Residential normalcy and environmental experiences of very old people: Changes in residential reasoning over time. *Journal of Aging Studies*, 29, 9–19. http://dx.doi.org/10.1016/j.jaging.2013.12.005

Hayward, L. M. (2004). Mid-life patterns and the residential mobility of older men. *Canadian Journal on Aging*, 23(1), 73–89. http://dx.doi.org/10.1353/cja.2004.0007

Heatwole Shank, K. S. and Cutchin, M. P. (2016). Processes of developing 'community livability' in older age. *Journal of Aging Studies*, 39, 66–72. http://dx.doi.org/10.1016-j.jaging.2016.11.001

Hillcoat-Nalletamby, S. and Ogg, J. (2014). Moving beyond 'ageing-in-place': Older people's dislikes about their home and neighbourhood environments as a motive for wishing to move. *Ageing and Society*, 34(10), 1771–1796. http://dx.doi.org/10.1017/S0144686X13000482

Koss, C. and Ekerdt, D. J. (2017). Residential reasoning and the tug of the fourth age. *The Gerontologist*, 57(5), 921–929. Online first: 13 February 2016. http://dx.doi.org/10.1093/geront/gnw010

Lawton, M. P. and Nahemow, L. (1973). Ecology and the aging process, in C. Eisdorfer and M. P. Lawton (eds), *The psychology of adult development and aging* (pp. 619–674). Washington, DC: American Psychological Association.

Lazarus, R. S. (1966). *Psychological stress and the coping process*. New York: McGraw-Hill.

Lieberman, M. A. (1991). Relocation of the frail elderly, in J. E. Birrren, J. E. Lubben, J. C. Rowe and D. E. Deutchman (eds), *The concept and measurement of quality of life in the frail elderly* (pp. 120–141). New York: Academic Press.

Litwak, E. and Longino, C. F. (1987). Migration patterns among the elderly: A developmental perspective. *The Gerontologist*, 27(3), 266–272. http://dx.doi.org/10.1093/geront/27.3.266

Löfqvist, C., Granbom, M., Himmelsbach, I., Iwarsson, S., Oswald, F. and Haak, M. (2013). Voices on relocation and aging in place in very old age – a complex and ambivalent matter. *The Gerontologist*, 53(6), 919–927. http://dx.doi.org/10.1093/geront/gnt034

Milligan, C. (2015). Place and informal care in an ageing society: Reviewing the state of the art in geographical gerontology. *Progress in Geography*, 34(12), 1565–1576. http://dx.doi.org/10.18306/dlkxjz.2015.12.005

Morgan, L. A. and Brazda, M. A. (2013). Transferring control to others: Process and meaning for older adults in assisted living. *Journal of Applied Gerontology*, 32(6), 651–668. http://dx.doi.org/10.1177/0733464813494568

Myers, D. and Ryu, S. (2008). Aging baby boomers and the generational housing bubble. *Journal of the American Planning Association*, 74(1), 17–33. http://dx.doi.org/10.1080/0194436070182006

Perry, T. E., Andersen, T. C. and Kaplan, D. B. (2014). Relocation remembered: Perspectives on senior transitions in the living environment. *The Gerontologist*, 54(1), 75–81. http://dx.doi.org/10.1093/geront/gnt070

Pinquart, M. and Burmedi, D. (2003). Correlates of residential satisfaction in adulthood and old age: A meta-analysis, in H-W. Wahl, R. J. Scheidt and P. G. Windley (eds), *Annual review of gerontology and geriatrics: Focus on aging in context, socio-physical environments*, v23(1), 195–222. New York: Springer Publishing Co.

Rowles, G. D. and Ravdal, H. (2002). Aging, place and meaning in the face of changing circumstances, in R. S. Weiss and S. A. Bass (eds), *Challenges of the third age: Meaning and purpose in later life* (pp. 81–114). New York: Oxford University Press.

Rubenstein, R. L. (2003). Ecology and the aging self, in H-W. Wahl, R. J. Scheidt and P. G. Windley (eds), *Annual review of gerontology and geriatrics*, v23(1), *Aging in context: Sociophysical environments* (pp. 59–84). New York: Springer.

Scharlach, A. E. and Lehning, A. J. (2016). *Creating aging-friendly communities*. New York: Oxford University Press.

Schulz, R. and Heckhausen, J. (1996). A life span model of successful aging. *American Psychologist*, 51(7), 702–714. http://dx.doi.org/10.1037//0003-066X.51.7.702

Sharma, A. (2013). The chain is only as strong as the weakest link: Older adult migration and the first move. *Research on Aging*, 35(5), 507–532. http://dx.doi.org/10.1177/0164027512446836

Shen, H-W. and Perry, T. E. (2014). Giving back and staying put: Volunteering as a stabilizing force in relocation. *Journal of Housing for the Elderly*, 28(3), 310–328. http://dx.doi.org/10.1080/02763893.2014.930368

Sixsmith, A. and Sixsmith, J. (2008). Ageing-in-place in the United Kingdom. *Ageing International*, 32(3), 219–235. http://dx.doi.org/10.1007/s12126-008-9019-y

Skinner, M. W. and Power, A. (2011). Voluntarism, health and place: Bringing an emerging field into focus. *Health & Place*, 17(1), 1–6. http://dx.doi.org/10.1016/j.healthplace.2010.09.001

Skodol, A. (2010). The resilient personality, in J. W. Reich, A. J. Zautra and J. S. Hall(eds), *Handbook of adult resilience* (pp. 112–125). New York: Guilford Press.

Smetcoren, A-S., De Donder, L., Dury, S., De Witte, N., Kardol, T. and Verté, D. (2015). Refining the push and pull framework: Identifying inequalities in residential relocation among older adults. *Ageing and Society*, 37(1), 90–112. Online first. http://dx.doi.org/10.1017/S0144686X15001026

Stoeckel, K. J. and Porell, F. (2010). Do older adults anticipate relocating? The relationship between housing relocation expectations and falls. *Journal of Applied Gerontology*, 29(2), 231–250. http://dx.doi.org/10.1177/0733464809335595

Stone, R. and Harahan, M. F. (2010). Improving the long-term care workforce serving older adults. *Health Affairs*, 29(1), 109–115. http://dx.doi.org/10.1377/hlthaff.2009

Tang, F. and Lee, Y. (2011). Social support networks and expectations for aging in place and moving. *Research on Aging*, 33(4), 444–464. http://dx.doi.org/10.1177/0164027511400631

Van Dijk, H. M., Cramm, J. M., Van Exel, J. and Nieboer, A. P. (2015). The ideal neighbourhood for ageing in place as perceived by frail and non-frail community-dwelling older people. *Ageing and Society*, 35(8), 1771–1795. http://dx.doi.org/10.1017/S0144686X14000622

Vasunilashorn, S., Steinman, B. A., Liebig, P. S. and Pynoos, J. (2012). Aging in place: Evolution of a research topic whose time has come. *Journal of Aging Research*, 37(1), 90–112. http://dx.doi.org/10.1155/2012/120952

Wahl, H-W., Iwarsson, S. and Oswald, F. (2012). Aging well and the environment: Toward an integrative model and research agenda for the future. *The Gerontologist*, 52(3), 306–316. http://dx.doi.org/10.1093/geront/gnr154

Wiles, J. L., Leibing, A., Guberman, N., Reeve, J. and Allen, R.E.S. (2011). The meaning of "aging in place" to older people. *The Gerontologist*, 52(3), 357–366. http://dx.doi.org/10.1093/geront/gnr098

Wiles, J. L., Wild, K., Kerse, N. and Allen, R.E.S. (2012). Resilience from the point of view of older people: 'There's still life beyond a funny knee'. *Social Science & Medicine*, 74(3), 416–424. http://dx.doi.org/j.socscimed.2011.11.005

Wilmoth, J. M. (2010). Health trajectories among older movers. *Journal of Aging and Health*, 22(7), 862–881. http://dx.doi.org/10.1177/0898264310375985

Wiseman, R. F. (1980). Why older people move. *Research on Aging*, 2(2), 242–254. http://dx.doi.org/10.1177/016402758022003

World Health Organization (WHO). (2007). *Global age-friendly cities: A guide* Geneva, Switzerland: WHO. Available from: http://apps.who.int/iris/bitstream/10665/43755/1/9789241547307_eng.pdf

Zhang, Y., Engelman, M. and Agree, E. M. (2013). Moving considerations: A longitudinal analysis of parent – child residential proximity for older Americans. *Research on Aging*, 35(6), 663–687. http://dx.doi.org/10.1177/0164027512457787

16 Being in place

Identity and place attachment in late life

Graham D. Rowles

Introduction

A sense of *being in place* is the essence of well-being in late life. We are all products of where we have been, when we were there, and how this experience shapes our understanding of present circumstances. In this chapter, I focus on interpreting the changing relationship between identity and place over the life course with particular attention to later life. A primary objective is to illustrate how capacity for *personal place identification* (a trait of each person) and the *identity of places* (a characteristic of places) are uniquely blended through expression in *attachment to place* in a manner that, for each person, facilitates their degree of *being in place* (or its opposite, being out of place) that is intimately tied to well-being and quality of life, and especially so in old age.[1]

Being in place: interweaving self and place

The human quest for belonging is universal. There is an instinctive motivation toward attaining a sense of being in place or at one with one's environmental context (Rowles, 1991). This finds expression in many ways. We share with much of the animal kingdom a need for territory – a place of ownership and control (Ardrey, 1966; Porteous, 1976). Building on this ethological imperative, a large literature has explored the related expression of this concept in probing the psychologically, socially and culturally nuanced concept of home. Beyond a place of centering (territory), home embraces notions of permanence, ownership, responsibility, privacy, self-expression, refuge, belonging, comfort, emotional affiliation and – most important in this context – identity (Cooper Marcus, 1995; Rowles and Chaudhury, 2005).

Being in place finds expression in patterns of use of space that exhibit the rhythm and routines of daily life, selective cognitive orientation within the environments of our life, emotional affiliation with significant personal and shared places and vicarious immersion in settings displaced in space and/or time (Rowles, 1978, 1991, 2008). Against this backdrop, two additional constructs facilitate unpacking the essential characteristics of being in place.

Personal place identification

Every human being's response to places they encounter is shaped by previous life experience. A person raised in the wide open spaces of Wyoming
will experience a new environment very differently from an individual whose
childhood and adolescence was lived amidst the skyscrapers of downtown New
York. Central Manhattan may seem claustrophobic to the visitor from Wyoming. And Wyoming may generate a sense of exposure and vulnerability for
the New Yorker. Each person develops a personal mode of place identification – patterns of visceral reaction, dispositions and preferences that predispose
them to a distinctive manner of reacting to place. This *personal place identification* profile is shaped by personality. But there is more to it than the legacy
of genes. Rather, the way we experience and react to places is shaped by the
layered accumulation of life experiences in environments over our life span that
develops within us a place consciousness and sensitivity. Our way of reacting
to place, personal place identification, becomes a mind-set, an attribute of who
we are.

The identity of places

The second construct is the *identity of places*. Spaces have no meaning apart
from that which is imbued through habitation – meaning that transforms them
from spaces into places (Tuan, 1977). Beyond individual experience, meaning
is transferred over time through cultural legacy. When, in the summer of 2016,
I walked into England's Winchester Cathedral (first established in AD 645), on
one level it was merely a space; there was nothing to prevent me from shouting
or breaking into song. But like most people, I did not violate the sacredness
of a place created centuries ago. Hundreds of years later, I could share in the
intentionality of a medieval stone mason who assisted in constructing the nave,
leaving a legacy for me to appreciate with awe.

Understanding the identity of places is a complex process. It involves selective synthesis of myriad elements that shape a location. These include components of its natural environment, the ecological context. Thus, underlying each
place are fundamental elements of physical geography – geology, soils, climate
and weather, natural vegetation, flora and fauna. This provides the canvas on
which human habitation is painted through construction of the built environment, the cultural creation of social spaces and *genres de vie* (Buttimer, 1969),
and the complex interweaving of individual biographies over generations. Cultures, social groups and individuals play a part in shaping the identity of places
both directly and indirectly.

In his autobiography, *Born to Run*, Bruce Springsteen describes the identity
of his New Jersey hometown (Springsteen, 2016). He vividly recounts a visit
to the neighbourhood of his childhood during one November evening. The
church he had attended had not changed, although all was quiet and there was
no activity that evening. As he drove through the familiar block where he had

lived, he was at first shocked by the absence of a giant copper beech tree that had dominated the immediate landscape. Only telltale remnants of the roots remained as testimony to what once was here. But for him, the tree was still alive, still part of his intimate association with this place. So, too, were the people who had lived here and whose lives had created the identity of this place, "my clan, my blood, my place, my people" (Springsteen, 2016: 504). Springsteen's writing captures his identification with a place I have never physically experienced. Simply by reading his eloquent words, this location assumes an identity for me. Should I ever visit this site, its identity as a place would be intimately associated with my awareness of its meaning to Springsteen. His image becomes a part of the identity of this place, not just for him, but also for me and for all who read his words. And so it is with all places. Each is both a physical manifestation of what remains visible and a locus of expression – through stories (both personal and shared) of all that transpired in this place.

Only a fraction of the identity of each place is revealed because we have only partial and limited access to its story. This access is shaped by and shapes those who selectively characterize and portray the place. Over many centuries, Winchester Cathedral has been constructed, modified, renovated, maintained and revered as a consecrated place by generations who forged its place identity as a site of worship, religious administrative centre, tourist destination and cultural and historical symbol. Similarly, the identity of Bruce Springsteen's neighbourhood is a composite expression of all who have lived in this community. Inasmuch as there arises a consensus in framing the essence of a place, it assumes an inherent identity independent of any individual or group.

Scale: from body to world

Personal place identification begins at the body and at this level evolves over the life course. During infancy, it involves few and generally spatially limited places – being in place at mother's breast or within the bodily range of the protective confines of the crib. In childhood we begin to differentiate place around the body within a series of zones of personal space, security and comfort, extending away from the body (Hall, 1966; Sommer, 1969). On this level of proxemics, place is mobile, providing a template for engaging with others. We sense a being in place when people do not get too close without invitation. Our conceptions of personal space and personal place identification evolve with age and, especially in advanced old age, may differ from those of caregivers (leading to potential for conflict and misinterpretation; DeLong, 1970).

On the scale of our residence, as we grow up, we increasingly have the capability to sense being in place and "at home" through the patterned rhythm and routine of daily use of the dwelling and accumulation of meanings through the events that transpire in this location. As our independence expands, we exert increasing agency as we reinforce personal place identification by surrounding ourselves with artefacts and treasured possessions reflecting our persona and projecting our identity (Csikszentmihalyi and Rochberg-Halton, 1981; Jacobs and Malpas, 2013; Sherman and Dacher, 2005). We seek, and have more ability to intentionally achieve, consonance between our personal place identification and the identity of the place where we dwell.

This relationship becomes more complex beyond the threshold. Neighbourhood and community spaces are shared with others, each with their own identification with this space. Personal place identification becomes embedded and subsumed within shared identification with "our neighbourhood" that reflects group values, norms and expectations. The resulting social order and identity has been illustrated in literature on neighbourhoods spanning more than half a century, including classic work documenting how different ethnic populations imbue shared space with distinctive identities that enable them to live in the same neighbourhood in relative harmony (Suttles, 1968, 1972).

On the even larger scale of region and nation, evidence of the strength of personal place identification is provided by the fervor with which we support our regional sports franchise or in our willingness to make the ultimate sacrifice of our life for our country. A propensity for territorial identification may lead to forms of personal place identification that result in acts of altruism or aggression that transcend reason.

Acknowledging personal place identification as an inherently human trait leads us to a discussion of place attachment and the way in which this links to the identity of places (Altman and Low, 1992; Manzo and Devine-Wright, 2014). The two constructs are different but interrelated (Cross, 2015; Hernandez *et al.*, 2007). Personal place identification is a component of an individual's persona and way of relating to the world; it reflects the processes through which people come to identify themselves as belonging to, or part of, a place. Place attachment refers to the intensity of that belonging. It is "a positive affective bond people form with particular places where they feel comfortable and safe and desire to maintain their connection" (Cross, 2015: 494).

Place attachment

The literature on place attachment and its relationship to personal identity spans multiple disciplines, including architecture and housing studies (Jacobs and Malpas, 2013), geography (Hay, 1998), gerontology (Rowles, 1991), psychology (Anton and Lawrence, 2014), sociology (Cross, 2015) and social work (Jack, 2010). Indeed, Cross (2015: 515) has argued that "[t]he literature on place attachment has suffered from interdisciplinary diversity and conceptual complexity." Much of the literature has demonstrated associations between place attachment, quality of life and well-being (Gilleard, Hyde and Higgs, 2007; Jack, 2010; Tartaglia, 2013). The literature has also explored the obverse of attachment, social exclusion, alienation from place and its negative consequences (Buffel, Phillipson and Scharf, 2013; Manzo, 2014). Research has also explored the mechanisms that individuals and social groups employ to compensate for separation (Fullilove, 2014; Lewin, 2005). Finally, work in this domain has revealed the paradox of situations where attachment to place may, over time, become maladaptive (Fried, 2000), and especially so for older adults (Twigger-Ross and Uzzell, 1996). For example, older adults' resolute insistence on remaining in a familiar dwelling may become hazardous to their health and well-being.

Increasingly, the focus of inquiry is investigation of the processes through which place attachment evolves and is maintained. A variety of typologies have been developed (Low, 1992; Seamon, 2014). One of the most useful is Cross's empirically grounded identification of seven overlapping interactional processes: sensory, narrative, historical, spiritual, ideological, commodifying and material dependence (Cross, 2015). *Sensory attachment* reflects visceral response to an environment through our five senses, and perhaps even a sixth sense that arises from body subject or body awareness (Seamon, 1980, 2014). *Narrative attachment* is the selective outcome of interpersonal storytelling and cultural stories of place that shape place affiliation. The temporal accumulation and interweaving of personal life experiences, family history and cultural history nurtures *historical attachment. Spiritual attachment* reflects belonging, a deeply seated sense of sacred "at oneness" with place. Moral, ethical and legal commitments to place result in *ideological attachment*. On a more pragmatic level, *commodifying attachment* may result from the alignment of the desirable traits of a place with a person's immediate needs. Finally, processes of place attachment may reflect *material dependence* on social resources and physical attributes of a place.

Interaction among elements of these processes shapes place attachment for any person at any time in their life. Each element evolves over time in association with changing needs and preferences. Some, such as the commodifying attachment that results from finding a place that meets immediate needs, are ephemeral and give way to deeper levels of attachment. Others are highly ingrained and express the core of identity. Indeed, some processes of place attachment, for example, the spiritual, may be so deeply ingrained, enduring and taken for granted that they are difficult to articulate. For each individual, a level of incorporation of each process of place attachment, to a greater or lesser degree, becomes expressed in an overall sense of place attachment. It expresses the outcome of a unique interaction between personal place identification and the identity of places at a particular time in history and in a particular location.

Melding personal place identification, the identity of places and place attachment

Consider the situations of 83-year-old Connie who lives in a third-floor apartment in a run-down inner-city neighbourhood and her 78-year-old neighbour Elsie, who lives down the hall. Over a lifetime of residence, Connie has developed a rhythm and routine in her life, a mode of *personal place identification* that frames her world. She values privacy, settings where she has something to grab on to should she feel unsteady, places that are warm (75–80 degrees Fahrenheit) and settings that are uncluttered and uncrowded: she shuns brightly lit and fast-paced environments, including her local shopping mall. She has lived in this neighbourhood for more than 60 years and is part of its history. The neighbourhood, especially the area most proximate to her apartment, has its own history and identity (*identity of places*) that has evolved through phases of demolition and construction over many more decades than Connie has been a resident. She has lived through the most recent changes and has seen the service resource landscape change with turnover of cafés, stores and sidewalk vendors. She has observed the changing demographic and ethnic composition of the community. She is known by many in the neighbourhood; indeed, her life is interwoven with the narrative of its social and cultural identity to which she has contributed – a narrative that embraces all of the people who have lived here and contributed to its personality. Through decades of residence, Connie has developed strong *place attachment* to this setting. She is emotionally tied to where most of the events of her life have transpired. Every nook and cranny evoke images and memories that have become part of her identity. As she sits in her chair by the window, next to the table made by her long-deceased husband, looking at the bird feeder her daughter placed outside and watching familiar people pass by, she is at one with her environment; she experiences a sense of *being in place* and a quality of life and well-being that result from being a part of this place.

Elsie is not so fortunate. At the behest of her children, she recently moved to this neighbourhood from a farm in rural Iowa that, when her husband died, she found she could no longer manage by herself. She is unfamiliar with the neighbourhood and its services, is unrecognised when she walks down the street, has made few friends and feels claustrophobic in an apartment where looking out of her window all she can see across the street is the back of an office building. She does not like *being out of place* and wishes she were at home.

The stark contrast between Connie and Elsie puts faces on the somewhat abstract constructs that are at the core of this chapter (Figure 16.1). But the themes I have considered have crucial human implications. Being in place is characteristically associated with well-being. In contrast, being out of place is generally accompanied by distress and low levels of well-being. Thus, it behooves us to support and reinforce interventions that sustain and enhance being in place both for individuals and populations. In concluding this chapter, I argue for a clinical geography that focuses on maximising the potential for being in place by recognising and reinforcing personal place identification, strengthening the supportive identity of places and enabling and celebrating diverse processes of place attachment.

Toward a clinical geography of being in place in old age

During each phase of life, different aspects of being in place assume different levels of importance. While the conceptualisation developed in this chapter can be applied to persons of all ages, the focus here is on older adults. Consider

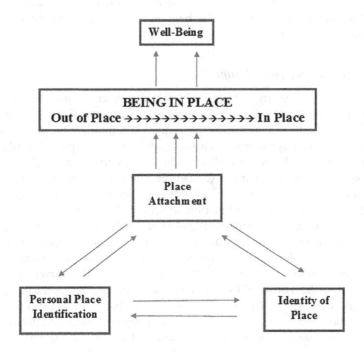

Figure 16.1 Toward a conceptualisation of being in place

how we might translate theoretical and empirical understanding into concrete action. At the core of my suggestions is the need for geographers to adopt an increasingly clinical focus. The need is to go beyond existing approaches (Andrews and Shaw, 2008), and employ awareness of being in place in interventions with older adults and in environmental planning.

At the individual level this involves recognising that individuals are constantly making and remaking place as they accommodate their modes of personal place identification to *in situ* changes in their environment or relocation to new environments. Over the life course, people develop greater or lesser degrees of experience and expertise in making or remaking place as they move from environment to environment, perpetually seeking a sense of being in place within the constraints of the place identity of each location they inhabit. With each environmental change, this involves processes of *transference, creation* and *abandonment* (Rowles and Bernard, 2013: 15 Figure 1.2; Rowles and Watkins, 2003). Elements of being in place are transferred from previous settings, created through embracing opportunities provided by the new setting and discarded when they are no longer viable. Within this context, our task is to support interventions on the individual level to enhance the abilities of older adults to achieve and sustain an optimal sense of being in place. At the community level, the need is to acknowledge that the identity of places can be deliberately and purposefully modified through perceptive environmental design and planning and conscious social action to make them more consonant with the needs and aspirations of older adults.

Reinforcing personal place identification

Regardless of our efforts and success in delaying onset, physiological and sensory impairments and cognitive decrements are inevitable features of growing old. This generates circumstances that require accommodation. A key need is empathy for the ways in which an older adult's personal place identification may be changing. Reduced ability to climb stairs, to reach a high shelf in a cupboard, to maintain a garden, to read the fine print in instructions, to hear the knock at the door, or to smell the odor of a pet who has "missed" the litter box gradually limit effective functioning within the immediate environment and lead to the need for behavioral and lifestyle adjustments. More time may be spent in the immediate environment of home. Environmental centralisation may occur as people limit themselves to a single room and surround themselves with needed items (Lawton, 1990; Rubinstein, 1989). The surveillance zone, space within the visual field of the residence, may assume increasing importance (Rowles, 1981). But there is more to it than this. Interventions and adjustments must recognise the diversity and depth of place attachment outlined earlier – sensory attachment and the benefits of a familiar physical space that compensate for cognitive deficits, the narrative and historical attachment that stems from perpetual re-storying of the space that forged its experienced identity, and the spiritual attachment that stems from at-oneness. The challenge becomes recognising and reinforcing, as much as is possible, the older adult's personal place identification. This is difficult to accomplish in a world where practicality, cost, efficiency and convenience in providing physical care tend to override maintaining meaning in people's lives.

The recent emergence of Certified Ageing in Place Specialists, a programme nurtured by the National Association of Home Builders (NAHB) in collaboration with Home Innovation Research Labs, NAHB 50+ Housing Council and the AARP, has provided both an opportunity and a mechanism offering clinical advice and service to older adults and their families in home remodeling, accommodating home environments to changing physical capabilities and maintaining a safe environment. Beyond such practical advice, sensitive counseling can help older adults bring to conscious awareness elements of their relationship to place that they generally take for granted. There is a need for "place therapists," people who can empathise, who understand the meaning of being in place for older adults and how it can be nurtured (Scheidt and Norris-Baker, 1999). Nurturing such coming-to-awareness makes it possible to counsel older adults in strategies that reinforce their being in place in their current setting. The potential contributions of place therapists are considerable. Such individuals might employ strategies including in-depth, life-history-oriented therapeutic interviewing or place-oriented reminiscence therapy to unveil dimensions of personal place identification that are generally implicit but when brought to consciousness may become highly therapeutic (Chaudhury, 2003). Pioneering exemplars of such an approach are provided by Godkin's (1980) work with persons grappling with alcoholism and by Chaudhury's (2008) innovative interventions with nursing home residents, including persons with dementia. In the latter case, the intervention reinforces and nurtures identity and sense of self in the alienating environment of a long-term care facility by building on their engagement with place.

When a relocation becomes inevitable, this mutual knowledge can be used in counseling to ease the transition process and assist people in remaking place in their new abode. Concrete actions might include assisting people to sort through possessions during inevitable downsizing to ensure that the items transferred are those that most strongly reinforce identity and facilitate transference of elements of being in place (Ekerdt, Luborsky and Lysack, 2012).

Closely related is the need for education to sensitize professionals who come into contact with older adults to recognise the nature of personal place identification and the importance of being in place. Occupational therapists, nurses, social workers, home care workers and others who visit older adults in their home can become informed through the routine inclusion of information and training on the importance of being in place as a component of their continuing education (Rowles, 1991, 2000). In addition, efforts to increase family awareness can be enhanced through public awareness initiatives.

Reframing the identity of places

Few would dispute that the identity of places is constantly changing. Indeed, there is a voluminous literature on the processes by which places evolve as settings for habitation (McClay and McAllister, 2014; Rowles and Bernard, 2013). A motel room presents a place identity for the traveler during an overnight stay. The identity of a home may evolve over decades and in some cases over multiple generations of a family's residence (Rowles and Chaudhury, 2005). Similarly, neighbourhoods develop an identity over time as they assume the personality of their residents and become the site of distinctive patterns of behavior, social mores and cultural conventions. Here the focus is on the manner in which intentional intervention can facilitate the development of places with identities that nurture being in place and well-being for their inhabitants.

Recent years have seen the emergence of a literature on the way in which natural and garden landscapes can become therapeutic (Cooper Marcus and Sachs, 2013; Finlay *et al.*, 2015; Gesler, 1992). It is but a short step from this perspective to the recognition that all landscapes can be designed to have therapeutic benefit.

On the micro scale, we are talking about elements of design and furnishing that nurture well-being, are consonant with personal place identification and facilitate being in place, either temporarily (as in the case of the design and accoutrements of a motel room), or more permanently through the design of living spaces.

Conscious manipulation of the identity of places in ways that are consonant with people's personal place identification is the essence of environmental design and over the past half century has spawned the field of environmental psychology. Human-factors-sensitive furniture design attuned to the physical needs and capabilities of older adults (Fisk *et al.*, 2009), smart homes and ambient assisted-living housing design (Wilson, Hargreaves and Hauxwell-Baldwin, 2015) and universal design (Sanford, 2012) have become central themes in a societal priority and preoccupation with ageing in place (Rowles, 1993).

On the larger scale this has translated into widespread concern with the negotiability and social supportiveness of behavior settings and neighbourhoods and the relationship of neighbourhood supportiveness and identity to well-being (von Wirth *et al.*, 2016). Finally, this has translated into nation-wide and even global concern with creating age-friendly communities, initially stimulated by a World Health Organization age friendly cities initiative in 2007 (Scharlach and Lehning, 2016). The quest for making communities age friendly has been reinforced through the priorities of the AARP and the desire of cities throughout the world to improve livability for all citizens. All these initiatives are targeted toward manipulating place identity in a manner that is consonant with personal place identification and facilitates being in place. Older adults are among those who are benefitting most from these trends.

As initiatives on every scale from the design of furniture and rooms to the creation of supportive communities flourish, we may well be on the threshold of a new era in which being in place becomes explicitly recognised as the essence of a good life and a universal mantra for guiding progress. This will be good news for us all.

Note

1 It is essential to acknowledge two features of this chapter. First, it is written almost entirely from the perspective of a Western developed society, although I contend that the underlying thesis represents a universal construct. Second, it is critical to point out that many people, especially those who are disadvantaged including the poor, refugees and others who are excluded in contemporary societies, do not have the opportunity to fully realise a confluence of personal place identification and the identity of the places in which they live.

References

Altman, I. and Low, S. M. (1992). *Place attachment*. New York: Plenum Press.
Andrews, G. J. and Shaw, D. (2008). Clinical geography: Nursing practice and the (re)making of institutional space. *Journal of Nursing Management*, 16(4), 463–473. http://dx.doi.org/10.1111/j.1365-2834.2008.00866.x
Anton, C. E. and Lawrence, C. (2014). Home is where the heart is: The effect of place of residence on place attachment and community participation. *Journal of Environmental Psychology*, 40, 451–461. http://dx.doi.org/10.1016/j.jenvp.2014.10.007
Ardrey, R. (1966). *The territorial imperative: A personal inquiry into the animal origins of property and nations*. New York: Atheneum.

Buffel, T., Phillipson, C. and Scharf, T. (2013). Experiences of neighbourhood exclusion and inclusion among older people living in deprived inner-city areas in Belgium and England. *Ageing and Society*, 33(1), 89–109. http://dx.doi.org/10.1017/S0144686X12000542

Buttimer, A. (1969). Social space in interdisciplinary perspective. *Geographical Review*, 59(3), 417–426. http://dx.doi.org/10.2307/213485

Chaudhury, H. (2003). Quality of life and place therapy. *Journal of Housing for the Elderly*, 17(1–2), 85–103. http://dx.doi.org/10.1300/J081v17n01_07

Chaudhury, H. (2008). *Rediscovering the self in dementia*. Baltimore, MD: Johns Hopkins University Press.

Cooper Marcus, C. (1995). *House as a mirror of self: Exploring the deeper meaning of home*. Berwick: Conari Press.

Cooper Marcus, C. and Sachs, N. A. (2013). *Therapeutic landscapes: An evidence based approach to designing healing gardens and restorative outdoor spaces*. New York: John Wiley and Sons.

Cross, J. E. (2015). Processes of place attachment: An Interactional framework. *Symbolic Interaction*, 38(40), 493–520. http://dx.doi.org/10.1002/symb.198

Csikszentmihalyi, M. and Rochberg-Halton, E. (1981). *The meaning of things: Domestic symbols and the self*. Cambridge: Cambridge University Press.

DeLong, A. J. (1970). The micro-spatial structure of the older person: Some implications of planning the social and spatial environment, in L. A. Pastalan and D. H. Carson (eds), *Spatial behavior of older people* (pp. 68–87). Ann Arbor, MI: The University of Michigan – Wayne State Institute of Gerontology.

Ekerdt, D. J., Luborsky, M. and Lysack, C. (2012). Safe passage of goods and self during relocation in later life. *Ageing and Society*, 32, 833–850. http://dx.doi.org/10.1017/So144686X11000705

Finlay, J., Franke, T., McKay, H. and Sims-Gould, J. (2015). Therapeutic landscapes and well-being in later life: Impacts of blue and green spaces for older adults. *Health and Place*, 34, 97–106. http://dx.doi.org/10.1016/j.healthplace.2015.05.001

Fisk, A. D., Rogers, W. A., Charness, N., Czaja, S. J. and Sharit, J. (2009). *Designing for older adults: Principles and creative human factors approaches*. Boca Raton: CRC Press.

Fried, M. (2000). Continuities and discontinuities of place. *Journal of Environmental Psychology*, 20(3), 193–205. http://dx.doi.org/10.1006/jevp.1999.0154

Fullilove, M. T. (2014). "The Frayed Knot" What happens to place attachment in the face of serial forced displacement? in L. C. Manzo and P. Devine-Wright (eds), *Place attachment: Advances in theory, methods and applications* (pp. 141–153). New York: Routledge.

Gesler, W. M. (1992). Therapeutic landscapes: Medical issues in light of the new cultural geography. *Social Science & Medicine*, 34(7), 735–746. http://dx.doi.org/10.1016/0277-9536(92)90360-3

Gilleard, C., Hyde, M. and Higgs, P. (2007). The impact of age, place, aging in place, and attachment to place on the well-being of the over 50s in England. *Research on Aging*, 29(6), 590–605. http://dx.doi.org/10.1177/0164027507305730

Godkin, M. A. (1980). Identity and place: Clinical applications based on notions of rootedness and uprootedness, in A. Buttimer and D. Seamon (eds), *The human experience of space and place* (pp. 73–85). New York: Routledge.

Hall, E. T. (1966). *The hidden dimension*. New York: Doubleday.

Hay, R. (1998). Sense of place in developmental context. *Journal of Environmental Psychology*, 18(1), 5–29. http://dx.doi.org/10.1006/jevp.1997.0060

Hernandez, B., Hidalgo, M. C., Salazar-Laplace, M. E. and Hess, S. (2007). Place attachment and place identity in natives and non-natives. *Journal of Environmental Psychology*, 27(4), 310–319. http://dx.doi.org10.1016/j.envp.2007.06.003

Jack, G. (2010). Place matters: The significance of place attachments for children's well-being. *British Journal of Social Work*, 40, 755–771. http://dx.doi.org/10.1093/bjsw/bcn142

Jacobs, K. and Malpas, J. (2013). Material objects, identity and the home: Towards a relational housing agenda. *Housing, Theory and Society*, 30(3), 281–292. http://dx.doi.org/10.1080/14036096.2013.767281

Lawton, M. P. (1990). Residential environment and self-directedness among older people. *American Psychologist*, 45(5), 638–640. http://dx.doi.org/10.1037/0003-066X.45.5.638

Lewin, F. A. (2005). Elderly migrants and the concept of home: A Swedish perspective, in G. D. Rowles and H. Chaudhury (eds), *Home and identity in late life* (pp. 143–170). New York: Springer Publishing Company.

Low, S. M. (1992). Symbolic ties that bind: Place attachment in the plaza, in I. Altman and S. Low (eds), *Place attachment* (pp. 165–186). New York: Plenum Press.

Manzo, L. C. (2014). Exploring the shadow side: Place attachment in the context of stigma, displacement and social housing, in L. C. Manzo and P. Devine-Wright, P. (eds), *Place attachment: Advances in theory, methods and application* (pp. 178–190). New York: Routledge.

Manzo, L. C. and Devine-Wright, P. (eds). (2014). *Place attachment: Advances in theory, methods and application*. New York: Routledge.

McClay, W. M. and McAllister, T. (eds). (2014). *Why place matters: Geography, identity, and civic life in modern America*. New York: Encounter Books.

Porteous, J. D. (1976). Home: The territorial core. *Geographical Review*, 66(4), 383–390. http://dx.doi.org/10.2307/213649

Rowles, G. D. (1978). *Prisoners of space? Exploring the geographical experience of older people*. Boulder, CO: Westview Press.

Rowles, G. D. (1981). The surveillance zone as meaningful space for the aged. *The Gerontologist*, 21(3), 304–311. http://dx.doi.org/10.1093/geront/21.3.304

Rowles, G. D. (1991). Beyond performance: Being in place as a component of occupational therapy. *American Journal of Occupational Therapy*, 45(3), 265–271. http://dx.doi.org/10.5014/ajot.45.3.265

Rowles, G. D. (1993). Evolving images of place in aging and "aging in place". *Generations*, 17(2), 65–70.

Rowles, G. D. (2000). Habituation and being in place: *Occupational Therapy Journal of Research*, 20(1 Supplement), 52S–67S. http://dx.doi.org/10.1177/15394492000200S105

Rowles, G. D. (2008). Place in occupational science: A life course perspective on the role of environmental context in the quest for meaning. *Journal of Occupational Science*, 15(3), 127–135. http://dx.doi.org/10.1080/14427591.2008.9686622

Rowles, G. D. and Bernard, M. (2013). The meaning and significance of place in old age, in G. D. Rowles and M. Bernard (eds), *Environmental gerontology: Making meaningful places in old age* (pp. 3–24). New York: Springer Publishing Company.

Rowles, G. D. and Chaudhury, H. (eds). (2005). *Home and identity in late life: International perspectives*. New York: Springer Publishing Company.

Rowles, G. D. and Watkins, J. F. (2003). History, habit, heart and hearth: On making spaces into places, in K. W. Schaie, H-W. Wahl, H. Mollenkopf and F. Oswald (eds), *Aging independently: Living arrangements and mobility* (pp. 76–96). New York: Springer Publishing Company.

Rubinstein, R. L. (1989). The home environments of older people: A description of psychosocial processes linking person to place. *Journal of Gerontology*, 44(2), S45–S53. http://dx.doi.org/10.1093/geronj/44.2.S45

Sanford, J. A. (2012). *Design for the ages: Universal design as a rehabilitation strategy*. New York: Springer Publishing Company.

Scharlach, A. E. and Lehning, A. J. (2016). *Creating age-friendly communities*. New York: Oxford University Press.

Scheidt, R. J. and Norris-Baker, C. (1999). Place therapies for older adults: Conceptual and interventive approaches. *International Journal of Aging and Human Development*, 48(1), 1–15. http://dx.doi.org/10.2190/KWBT-8PYT-LYA6-1WPT

Seamon, D. (1980). Body-subject, time-space routines, and place-ballets, in A. Buttimer and D. Seamon (eds), *The human experience of space and place* (pp. 148–165). New York: St. Martin's Press.

Seamon, D. (2014). Place attachment and phenomenology: The synergistic dynamism of place, in L. C. Manzo and P. Devine-Wright (eds), *Place attachment: Advances in theory, methods and applications* (pp. 11–22). New York: Routledge.

Sherman, E. and Dacher, J. (2005). Cherished objects and the home: Their meaning and roles in later life, in G. D. Rowles and H. Chaudury (eds), *Home and identity in late life* (pp. 63–79). New York: Springer Publishing Company.

Sommer, R. (1969). *Personal space: The behavioral basis of design*. Englewood Cliffs, NJ: Prentice-Hall.

Springsteen, B. (2016). *Born to run*. New York: Simon & Schuster.

Suttles, G. D. (1968). *The social order of the slum: Ethnicity and territory in the inner city*. Chicago: University of Chicago Press.

Suttles, G. D. (1972). *The social construction of communities*. Chicago: University of Chicago Press.

Tartaglia, S. (2013). Different predictors of quality of life in urban environment. *Social Indicators Research*, 113(3), 1045–1053. http://dx.doi.org/10.1007/s11205-012-0126-5

Tuan, Y. F. (1977). *Space and place: The perspective of experience*. Minneapolis, MN: University of Minnesota Press.

Twigger-Ross, C. and Uzzell, D. L. (1996). Place and identity processes. *Journal of Environmental Psychology*, 16(3), 205–220. http://dx.doi.org/10.1006/jevp.1996.0017

Von Wirth, T., Gret-Regamey, A., Moser, C. and Stauffacher, M. (2016). Exploring the influence of perceived urban change on residents' place attachment. *Journal of Environmental Psychology*, 46, 67–82. http://dx.doi.org/10.1016/j.jenvp.2016.03.001

Wilson, C., Hargreaves, T. and Hauxwell-Baldwin, R. (2015). Smart homes and their users: A systematic analysis and key challenges. *Personal and Ubiquitous Computing*, 19(2), 463–476. http://dx.doi.org/10.1007/s00779-014-0813-0

17 Active relationships of ageing people and places

Malcolm P. Cutchin

Introduction

Located in conceptual proximity to place attachment, ageing in place, embodiment and emotion, and therapeutic landscape dimensions of geographical gerontology is the active and ongoing engagement of older people and places in everyday life. We know too little about such active relationships, primarily because geographical gerontology has too often treated space and place as static or fixed aspects of ageing (Andrews, Evans and Wiles, 2013; Andrews *et al.*, 2009). The 'relational turn' in geography and geographical gerontology (Andrews *et al.*, 2013; Andrews, Cutchin and Skinner, Chapter 2; Milligan and Tarrant, Chapter 4; Skinner, Cloutier and Andrews, 2015; Wiles, Chapter 3) appears to have generated a greater focus on the active relationships of people and places in daily life. Some of the earliest geographical gerontology had, however, already opened up the issue of the interactions, or transactions, of place and person in later life. Indeed, over three decades ago the focus in environmental gerontology on Lawton and Nahemow's (1973) competence-press model was joined with behavioural geography (Golant, 1984). Golant's focus on environmental transactions of older people has been a fruitful effort that has evolved into more complex models of such transactions (Golant, 2011; see also Chapter 15), and environmental gerontology has become theoretically nuanced about place and activity (Diaz Moore, 2014; see also Chapter 7) Nonetheless, Golant's perspective on the relationship of older people and place resides squarely in the tradition of behavioural geography. His model of residential normalcy is heavily weighted toward certain cognitive and behavioural factors that can lead to better or worse outcomes for the older person living in an environment. Moreover, much of the work in this vein has focused on housing for older adults across the spectrum from single-family houses to nursing care.

At approximately the same time as the behavioural approach to person–place relationships in later life was developing, a more humanistic geography of ageing emerged with a focus on the intricate, and often sublime, dimensions inherent in the relations between ageing and place. Informed by phenomenological approaches to the human experience of place, Rowles's early work on older people's lives in both urban (Rowles, 1978) and rural places (Rowles,

1981; 1983a, 1983b) was seminal. Yet while Rowles's work became well known and cited for its contributions regarding the 'insideness' of older people whose identity had merged with place, and the attachment to place through memory and artefacts, his contributions on the active relationship between person and place and its ongoing reconstruction have often been overlooked. His studies of older people in 'Colton' showed that gathering with others at particular sites and surveilling the area around one's home are active forms of engagement and that those mundane activities are central to how the relationship with place is generated and maintained for older people. A key insight of which Rowles reminds us is that as ageing creates change for the person, so too does the changing circumstance of place – the combination of which creates a dynamic affair that challenges society and individuals, particularly in the ongoing search for meaning (Rowles and Ravdal, 2002; Rowles, 2008). His work continues in this vein with a more recent turn to ways in which such knowledge can inform interventions for older adults and places in need of help (Rowles and Bernard, 2013; Rowles, Chapter 16).

In this chapter, I set out to provide a somewhat selective discussion of the more recent literature that contributes to the understanding of such active relationships of older people and places in everyday life, and I pay special attention to work that deepens the theoretical and conceptual basis of our understanding of those relationships. The contribution by Andrews, Cutchin and Skinner (Chapter 2) and the chapters in Part II of this book lay out larger frames for thinking about relational perspectives on ageing and place, and other chapters contribute fine analyses of place attachment, ageing in place, embodiment and emotion, and therapeutic landscape dimensions of geographical gerontology. For that reason, I intentionally pay more close attention to processes or dynamics that that feed into those – and which can be somewhat overlooked because of their taken-for-granted everydayness. With geography's move to include philosophical positions beyond behavioural and phenomenological thought (see Andrews, Cutchin and Skinner, Chapter 2), geographical gerontology has taken some novel approaches to the subject. At the same time, more traditional views have evolved and continue to contribute important insights.

Relational dynamics: instabilities, transitions and negotiations

What is it about the place–person relationship that demands our attention; that is, why is a relational perspective considered helpful in geographical gerontology? This is a key question because, if not adequately explained, a relational focus could be viewed as academic posturing with the 'new'. Indeed, the best way to understand the utility of work that focuses on active relationships of older people and places is to organise our approach around the more common dynamics that a relational perspective addresses. Instabilities, transitions and negotiations can be viewed as primary causes underlying the dynamism of person–place relationships.

Instabilities

Wiles (2005) provided a powerful argument of the ways that place can and should be considered a *process* by geographical gerontologists. She pointed to the dynamic, negotiated, contested, contextual and complex processes of place that are relevant when thinking about gerontological issues. Wiles's argument was part of a movement within *geography* to recast theory and incorporate new theories from outside the field and a move in geographical *gerontology* to focus on the ongoing change at the heart of ageing in, and through, places. Part of the latter movement has been oriented toward aspects of the dynamism and change that I group under the term *instability*.

Although 'the relational' in ageing and place inquiry had been signaled by Rowles's pioneering studies, Cutchin (2001) suggested a need to move beyond the humanistic (phenomenological) theorisation of the relationship. The argument was to utilise John Dewey's pragmatism as an action-based perspective on place experience, and as such, the person and environment distinction is discounted in favor of the relational unity of person and place. An emphasis is put on change, a continual source of challenges to the integrity of the person–place whole. The concept of 'place integration' provides a way to underscore the ongoing transactions of older people and place to address instability and change through a process of coordination or re-integration. Instead of thinking about place attachment – an outcome of such processes – the focus is on the active process that connects person and place and continually transforms them and their relationship. Place (including social, symbolic, emotional, and aesthetic aspects) and person co-constitute each other in an ongoing way through constant change.

The movement signaled in part by Wiles's and Cutchin's theoretical arguments also was exemplified in empirical work on closely related themes. Some important studies established the 'blurring of boundaries' between home and institutional care for older people receiving care at home (Dyck *et al.*, 2005; Milligan, 2006). Others pointed to the 'ambiguity' of congregate care places and their meaning for older residents (Cutchin, 2003; Cutchin, 2007a). The 'liminality' of the relationship between older people and their new assisted living residence was denoted by Frank (2002) as the defining characteristic. 'Uncertainty' is another concept of relevance to the instability of the person–place relationship. Some older people (or family members) perceive assisted living (or other residential care contexts) as only a "way-station" on the route to a more institutional setting (Cutchin, 2004: 211). Such views make the formation of strong and positive relationships with place difficult, creating more uncertainty about the future.

Transitions

While instabilities of person–place relationships in later life are not always concurrent with significant transitions, they sometimes are. By transitions, I mean the phases of adjustment and change of the person–place relationship often following events such as retirement, spousal death, downsizing, health changes, neighbourhood decline and relocation. Indeed, much of the work on transitions focused on person–place relationships has used the event of relocation as the starting point.

Rowles and Watkins (2003) addressed the issue of place-making after a move by an older adult. They explained the process of transforming a generic space into a place that has meaning for the older person and develops the 'hearth' aspects of home. Rowles and Watkins (2003) added an active, relational dimension to the place-making argument, however, by suggesting the need to reshape the habits tied to the old home into ones that fit the new home. Habits are a complex and overlooked dimension of the daily relationship with place, and Cutchin (2007b, 2013) has drawn on pragmatism and the work of Bourdieu to stress the importance of habits and how they are integral to understanding relocation transitions and creative place-making. Empirical work on the patterns of engagement in daily activities found that they may change with the move into a retirement community, and such change is often a positive and desired part of the new person–place relationship (Cutchin, Marshall and Aldrich, 2010).

Although such perspectives tend to focus more on the person component of the relationship, others take a more place-focused approach. Byrnes (2011) studied a subsidised senior housing community in a blighted part of Detroit. She discovered a "city within a city", where residents had re-created spaces in the building into equivalents of a corner store, sidewalks, and a coffee shop. The combination of problematic aspects of place and socio-spatial practices generated a cultivation of place by residents to render the housing community better. Yet the constraints of the surrounding neighbourhood and the housing community limited how successful their efforts could be. Taking the view of the neighbourhood as a process in which relationships are constructed and transition, Gardner (2011) discovered natural neighbourhood networks of daily life that enhanced well-being. She proposed 'third places' (e.g., parks, cafés, barber shops), 'thresholds' (hybrid third places, e.g., driveways, back yards, balconies), and 'transitory zones' (e.g., streets and sidewalks, lines in the grocery store, seats on busses) are places in daily life that connect older people to neighbourhood networks. Those places function as nodes by enabling relationships of proximity, service, and chance for her older adult participants as they carried out life in their neighbourhood. The transitioning landscape of senior communities within larger urban contexts selling themselves as places of active leisure is another albeit different example of such processes (McHugh and Larson-Keagy, 2005). The dialectic of insider/outsider, as well as idyllic haven/fortress, mentality shows how places largely populated by older adults are in ongoing transition – and how that transition in place and through time shapes the experience of residents.

A less-studied transition relevant to the active person–place relationship is residential downsizing. Luborsky, Lysack and Van Nuil (2011) conducted an analysis of narratives of older people who had moved to smaller quarters and reduced possessions. Their stories provided entrée into lifelong cultural processes of development and change, and Luborsky *et al.* argued that "downsizing evokes and unsettles a powerful sense of place experience; there are places in multiple timelines as downsizing is a major transition that deeply engages in multiple streams of personal, family, social and cultural life" (Luborksy *et al.*, 2011: 245). The authors suggested older adults who downsize are changing place but simultaneously struggling with "enduring streams of life engagement in sets of social relationships, identities and experiences" (Luborsky *et al.*, 2011: 245). This conceptualisation, as well as others mentioned earlier, hint at ongoing person–place negotiations as part of the active person–place relationship.

Negotiations

All people encounter the world in a continual way through transactions with various types of places. The instabilities of person and place, and their transitions through time, necessitate a continual set of adjustments. Those adjustments are often considered negotiations in a relational view to account for holism and the co-constitution of place and person. Wiles and colleagues' (2009) concept of 'social space' is a useful example of how this process works for older adults. Their research demonstrates a 'complex engagement' with social space at various scales, from home to neighbourhood and beyond, and the engagement involves change and the negotiation of changing circumstances as well as the emotional dimensions thereof. Although this process of negotiation is not dependent on a significant vulnerability of the older person, Wiles (2011) also points out how vulnerability may be intricately linked to person–place negotiations:

> For older people experiencing vulnerability through living with chronic or long-term illnesses or injuries and receiving care, there is thus ongoing re-negotiation of sense of self, past and future experiences and relationships with others and with places, and of embodied subjectivity. . . . Though it may incorporate fragility or dependence, the process of recognising and accepting vulnerability may also be a potentially transformative process incorporating collective and individual journeys of change as well as more corporeal and material changes to people, places, and relationships.
>
> (Wiles, 2011: 582–583)

Dyck, Kontos, Angus and McKeever (2005) focused explicitly on such negotiations for people receiving home care in Canada. They found complex and intertwined negotiations of body, home, knowledge, and meanings and noted how the various landscapes of care at home are "constituted through specific social practices and micro-politics of power" (Dyck *et al.*, 2005: 177). The home as a site of care is particularly powerful in terms of negotiation because of the deep-rooted sense of self which motivates a person to grapple with the changing relationship. This is even true for those who are living at home with advanced cancer (Maersk, 2017). Similar types of contestations and negotiations occur in senior housing developments as older adults and management often clash on the meanings and uses of space and place (Cutchin, 2003; Kontos, 1998). Although care provision for the chronically ill is a common thread here, the provision of home modifications for those becoming frailer also is an issue of instability in, and negotiation of, the relationship with place. Using a place integration perspective, Johansson, Josephsson and Lilja (2009) studied Swedish older adults going through this process. They discovered that the problems and possibilities of the place-person relationship were continuously changing and necessitated complex transactions and creativity in how modifications were implemented, adjusted or foregone.

Other processes of negotiation occur for older adults in their relationship with places at larger geographic scales than the residence, such as the neighbourhood or community. Peace's (2013) study of older adults' use of spaces and places in an English town suggested that patterns of activity in the centre were governed by policy as well as socially developed norms. Although older adults could sometime use places without issue because of their space-time patterns of activity, the use of public places was sometimes contested or negotiated by different age groups. While that work begins to reveal the complexity in everyday activity in public spaces, there are other layers of complexity for some older adults. Those unfamiliar with specific urban environments because of redevelopment or dementia have been largely overlooked until recently. Phillips, Walford and Hockey (2011) argue that lack of history in, and unfamiliarity with, a place can be overcome by meaning and sense of place conveyed by community environments. Those authors argue that the keys to that process for those unfamiliar with a place are aesthetics, access and activities/interactions with place. Brorsson *et al.* (2011) carried out a detailed investigation of such issues with people who had dementia. They pointed out how older people with cognitive challenges face numerous types of problematic situations related to everyday technologies, crowded places with fast-paced activity and noise, and changing landmarks; such challenges decreased feelings of accessibility and increased difficulties in carrying out activities in public space.

What these more recent studies begin to point toward is a greater focus on time and daily activities in the negotiation of person–place relationships. Some of this orientation comes from the fact that occupational scientists – those who study everyday activities and their intersection with other significant dimensions of human life – have contributed to the advancement of geographical gerontology. Writing about ageing in place versus migration, Johansson *et al.* (2013) provide a case for why occupations are important in person–place relationships:

> Knowledge about how such relationships are negotiated and expressed among older persons' everyday lives through daily occupation is of relevance in understanding and supporting people who are growing older in a place other than a so-called place of origin. . . . Place making draws attention to how occupation can be drawn upon to negotiate relationships that connect people to different places around the world, how the negotiated

relations are embedded within the occupations that fill daily lives, and how this process is contextualized and enacted in relation to resources.

(Johansson *et al.*, 2013: 113, 116)

Focusing on how older adults in a community participate in everyday life, Heatwole Shank and Cutchin (2016) disclose how 'community livability' is the outcome of several person–place processes rather than the collection of community attributes. Older individuals plan and strategise daily life so that the process of participating is coordinated with issues of time and space as well as personal abilities, priorities and needs. That 'negotiating daily participation' enables older people to continuously develop social infrastructure and enact the culturally pervasive ideology of successful ageing.

In a similar vein, Lager, Van Hoven and Huigen (2016) discovered that rhythms of everyday life in a neighbourhood slowed down as people age. That temporal relationship with place also was at the root of discontinuities between generations as well as the way older residents tried to structure time and activities in place to increase meaning. Also focusing on the neighbourhood, Fritz and Cutchin (2017) studied older African-Americans' relationship to place during a period of great stress on that relationship. Living in neighbourhoods that had experienced significant changes because of economic turmoil, those older adults described changing relationships with their neighbourhoods through modifications of their activities. Becoming more vigilant about threats in their neighbourhoods, most older residents also began to restrict activities by space and time, as well as withdrawing from social participation in the neighbourhood. Others became involved in new activities in the attempt to maintain or improve neighbourhood conditions. On a longer timeframe and including larger geographic scales of influence, Andrews (2016) illustrates the complex and fluid relationality of older residents and Teignmouth, England. He discovered how the historical experiences with physical and social dimensions of place during World War II – and with wider spheres through those relationships – very much influenced and shaped the ongoing relationship with Teignmouth in the present.

Emerging methodological approaches

As the narrative so far suggests, the complex relational understanding of active person–place transactions has developed significantly during the last decade. A concomitant evolution of methodological approaches to understand and explain person-place processes has occurred. In line with transdisciplinary work on the conceptual development of the field, geographers, anthropologists, sociologists, occupational scientists and gerontologists have been developing or implementing novel methods – often multiple methods used in combination – to address many of the concerns discussed in the previous section. Such methodological creativity can propel further knowledge creation about active relationships of older people and their places. Therefore, those methodological developments merit discussion here.

Hand *et al.* (2017) produced the first scoping review of methodological approaches combining participatory geospatial and qualitative methods in gerontological research. Their rationale was consistent with the argument in this chapter:

> Health, well-being, participation and engagement emerge from transactions between elements of the environment and the persons within it. Locations such as neighbourhoods are thus conceptualized as dynamic places with attached meanings that, in turn, require qualitative, interpretive methods to fully explore.
>
> The interwoven nature of person and place also leads us to look to methods that can explore this relationship *in situ*. A grounding of persons in places can be provided through more in-depth application of geospatial methods to research. In particular, participatory geospatial methods, that involve the person in the research process, may be useful to this area of study, for example, conducting data collection with the person in their daily environments through observation or global positioning system (GPS) tracking. Research without this grounding is at risk of generating superficial understandings of the relationships between person and place and missing the many details about neighbourhoods that are not verbalized during interviews or otherwise captured
>
> (Hand *et al.*, 2017: 3).

The review process results included 15 studies from 1995 to 2015 that used the photovoice method, GPS tracking plus interview, or go-along interviews.

Hand *et al.'s* (2017) findings are insightful. Photovoice approaches (Chaudhury *et al.*, 2016; Lockett, Willis and Edwards, 2005; Mahmood *et al.*, 2012) were participatory and geospatial because they allowed participants to select subject matter and take photographs of neighbourhood features related to physical activity. Participants then documented and discussed the photos and their meanings. Global Positioning System (GPS) tracking plus interview approaches have been more common (Huang *et al.*, 2012; Rosenberg *et al.*, 2013; Vine *et al.*, 2012, 2014; Zeitler *et al.*, 2012; Zeitler and Buys, 2015). These approaches typically create maps of activity patterns from the GPS data and then use them to focus interviews, sometimes adding a diary component. The go-along interview approaches reflect the work of Gardner (2011, 2014) who used an ethnographic methodology in which photos and participant observation were included within the go-along interviews. She walked with older adults as they went about their daily life in the Toronto community to glean how they related to the various spaces, places and people there. Hand *et al.* (2017) note that the geospatial dimension of these approaches is under-utilised and should be strengthened to better study older person and place transactions. The authors also discuss the common issue of privacy and confidentiality when using the methods, and they note that adopting an explicit methodology (e.g., ethnography) is not common in the studies to date.

Because of the timing and orientation of their scoping review, Hand *et al.* missed some other examples of novel approaches. Finlay and Bowman (2016) reported on the utility of 'mobile interviews' with older adults, and they used those interviews on brief walks from participants' homes after a sit-down interview. Heatwole Shank and Cutchin (2016) employed a combination of GPS tracking, activity observation, and multiple interviews in their study of processes of community livability. Photovoice approaches are not the only photographic approaches used to date in order to gain access to older person–place relationships. Byrnes's (2011) study of a 'city within a city' was pioneering in its use of photo elicitation methods in gerontology. That approach uses participant-generated photos to serve as the basis for eliciting participant insights into their relationship with place during interviews. Fritz and Cutchin (2017) used this approach in their study of older African-Americans in Detroit neighbourhoods.

There are at least two other methods that have recently been documented as useful for studying the active relationships of older people and their places. Andrews' (2016) study of Teignmouth offers a strong argument for the power of oral history interviews to weave together modern history and place experience in a relational way. Fritz *et al.* (2017) used a smartphone-based EMA (ecological momentary assessment) approach with older adults. The EMA design allowed collection of GPS coordinates, activity reports, social settings and emotional responses in real time and *in situ*. Both of these very different approaches to data collection hold promise for future research on older people and their relationships with place.

Conclusion

In my narrative about geographical gerontology focused on the transactional (i.e., active and ongoing) relationships between older people and places, I focused on three dynamics that are at the heart of more recent scholarship: instabilities, transitions and negotiations. It is almost certainly the case that my approach to the subject matter is not exhaustive. Furthermore, I do not suggest that those three dimensions are the only ones capable of reflecting important dynamics of person–place relationships in later life. I also should note that the works cited in each area do not focus only on the dimension with which I have associated them. In some ways the framework presented is artificially constructed for the sake of clarity and organisation, yet in other ways the geographical gerontology on the transactional relations of people and place substantiate my approach.

I have noted – as Andrews *et al.* (2007) reported in their review 10 years ago – that geographers and gerontologists of other stripes are conducting geographical gerontology on active relationships of people and places. From my vantage point, however, the teams and the projects being carried out are becoming more transdisciplinary in character. Furthermore, both the conceptual and methodological approaches are becoming increasingly sophisticated. The result is a more nuanced understanding of what happens in the daily life of older people and places – and how we might intervene to improve those situations. For me, this is 'where the action is', and my appraisal gives me hope that the field of geographical gerontology has an important and sustainable role in gerontology.

References

Andrews, G. J. (2016). The relational making of people and place: The case of the Teignmouth World War II homefront. *Ageing & Society*, 37(4), 725–752. Online first. http://dx.doi.org/10.1017/S0144686X15001415

Andrews, G. J., Cutchin, M., McCracken, K., Phillips, D. R. and Wiles, J. (2007). Geographical gerontology: The constitution of a discipline. *Social Science & Medicine*, 65(1), 151–168. http://dx.doi.org/10.1016/j.socscimed.2007.02.047

Andrews, G. J., Evans, J. and Wiles, J. (2013). Re-spacing and re-placing gerontology: Relationality and affect. *Ageing and Society*, 33(8), 1339–1373. http://dx.doi.org/10.1017/SO144686X12000621

Andrews, G. J., Milligan, C., Phillips, D. R. and Skinner, M. W. (2009). Geographical gerontology: Mapping a disciplinary intersection. *Geography Compass*, 3(5), 1641–1659. http://dx.doi.org/10.1111/j.1749-8198.2009.00270.x

Brorsson, A., Öhman, A., Lundberg, S. and Nygård, L. (2011). Accessibility in public space as perceived by people with Alzheimer's disease. *Dementia*, 10(4), 587–602. http://dx.doi.org/10.1177/1471301211415314

Byrnes, M. E. (2011). A city within a city: A "snapshot" of aging in a HUD 202 in Detroit, Michigan. *Journal of Aging Studies*, 25(3), 253–262. http://dx.doi.org/10.1016/j.jaging2011.03.011

Chaudhury, H., Campo, M., Michael, Y. and Mahmood, A. (2016). Neighbourhood environment and physical activity in older adults. *Social Science & Medicine*, 149, 104–113. http://dx.doi.org/10.1016/j.socscimed.2015.12.011

Cutchin, M. (2001). Deweyan integration: Moving beyond place attachment in elderly migration theory. *The International Journal of Aging and Human Development*, 52(1), 29–44. http://dx.doi.org/10.2190/AF2D-A0T4-Q14C-1RTW

Cutchin, M. (2003). The process of mediated aging-in-place: A theoretically and empirically based model. *Social Science & Medicine*, 57(6), 1077–1090. http://dx.doi.org/10.1016/S0277-9536(02)00486-0

Cutchin, M. P. (2004). A Deweyan case for the study of uncertainty in health geography. *Health & Place*, 10(3), 203–21. http://dx.doi.org/10.1016/j.healthplace.2003.06.001

Cutchin, M. P. (2007a). Therapeutic landscapes for older people: Care with commodification, liminality, and ambiguity. In A. Williams (ed.), *Therapeutic landscapes: Advances and applications* (pp. 181–198). Aldershot: Ashgate.

Cutchin, M. P. (2007b). From society to self (and back) through place: Habit in transactional context. *OTJR: Occupation, Participation, and Health*, 27(Suppl), 50S–59S. http://dx.doi.org/10.1117/153944920702705107

Cutchin, M. P. (2013). The complex process of becoming at-home in assisted living. In G. D. Rowles and M. Bernard (eds), *From knowledge to practice in environmental gerontology: Making meaningful places in old age* (pp. 105–124). New York: Springer.

Cutchin, M. P., Marshall, V. W. and Aldrich, R. M. (2010). Moving to a continuing care retirement community: Occupations in the therapeutic landscape process. *Journal of Cross-Cultural Gerontology*, 25(2), 117–132. http://dx.doi.org/10.1007/s10823-010-9113-y

Diaz Moore, K. (2014). An ecological framework of place: Situating environmental ger-ontology within a life course perspective. *The International Journal of Aging and Human Development*, 79(3), 183–209. http://dx.doi.org/10.2190/AG.79.3.a

Dyck, I., Kontos, P., Angus, J. and McKeever, P. (2005). The home as a site for long-term care: Meanings and management of bodies and spaces. *Health & Place*, 11(2), 173–185. http://dx.doi.org/10.1016/j.healthplace.2004.06.001

Finlay, J. M. and Bowman, J. A. (2016). Geographies on the move: A practical and theoreti-cal approach to the mobile interview. *The Professional Geographer*, 69(2), 263–274. http://dx.doi.org/10.1080/00330124.2016.1229623

Frank, J. B. (2002). *The paradox of aging in place in assisted living*. Westport, CT: Greenwood Publishing Group.

Fritz, H. and Cutchin, M. P. (2017). Changing neighborhoods and occupations: Experiences of older African-Americans in Detroit. *Journal of Occupational Science*, 24(2), 140–151. http://dx.doi.org/10.1080/14427591.2016.1269296

Fritz, H., Tarraf, W., Saleh, D. J. and Cutchin, M. P. (2017). Using smartphone-based ecolog-ical momentary assessment protocol with community dwelling older African Americans. *The Journals of Gerontology: Social Sciences*, 72(5), 876–887. http://dx.doi.org/10.1093/geronb/gbw166

Gardner, P. J. (2011). Natural neighborhood networks – important social networks in the lives of older adults aging in place. *Journal of Aging Studies*, 25(3), 263–271. http://dx.doi.org/10.1016/j.jaging.2011.03.007

Gardner, P. J. (2014). The role of social engagement and identity in community mobility among older adults aging in place. *Disability and Rehabilitation*, 36, 1249–1257. http://dx.doi.org/10.3109/09638288.2013.837970

Golant, S. M. (1984). *A place to grow old: The meaning of environment in old age*. New York: Columbia University Press.

Golant, S. M. (2011). The quest for residential normalcy by older adults: Relocation but one pathway. *Journal of Aging Studies*, 25(3), 193–205. http://dx.doi.org/10.1016j.jaging.2011.03.003

Hand, C., Huot, S., Rudman, D. L. and Wijekoon, S. (2017). Qualitative-geospatial methods of exploring person-place transactions in aging adults: A scoping review. *The Gerontologist*, 57(3), e47–e61. http://dx.doi.org/10.1093/geront/gnw130

Heatwole Shank, K. and Cutchin, M. P. (2016). Processes of developing 'community liv-ability' in older age. *Journal of Aging Studies*, 39, 66–72. http://dx.doi.org/10.1016/j.jaging.2016.11.001

Huang, D. L., Rosenberg, D. E., Simonovich, S. D. and Belza, B. (2012). Food access patterns and barriers among midlife and older adults with mobility disabilities. *Journal of Aging Research*, Article ID 231489, 8 pp. http://dx.doi.org/10.1155/2012/231489

Johansson, K., Josephsson, S. and Lilja, M. (2009). Creating possibilities for action in the pres-ence of environmental barriers in the process of 'ageing in place'. *Ageing & Society*, 29(1), 49–70. http://dx.doi.org/10.1017/So144686X08007538

Johansson, K., Rudman, D. L., Mondaca, M., Park, M., Luborsky, M., Josephsson, S. and Asaba, E. (2013). Moving beyond 'Aging in place' to understand migration and aging: Place making and the centrality of occupation. *Journal of Occupational Science*, 20(2), 108–119. http://dx.doi.org/10.1080/14427591.2012.735613

Kontos, P. C. (1998). Resisting institutionalization: Constructing old age and nego-tiating home. *Journal of Aging Studies*, 12(2), 167–184. http://dx.doi.org/10.1016/S0890-4065(98)90013-5

Lager, D., Van Hoven, B. and Huigen, P.P.P. (2016). Rhythms, ageing and neighbour-hoods. *Environment and Planning A*, 48(8), 1565–1580. http://dx.doi.org/10.1177/0308518X16643962

Lawton, M. P. and Nahemow, L. (1973). Ecology and the aging process, in C. Eisdorfer and M. P. Lawton (eds), *The psychology of adult development and aging* (pp. 619–674). Washington, DC: American Psychological Association.

Lockett, D., Willis, A. and Edwards, N. (2005). Through seniors' eyes: An exploratory qualitative study to identify environmental barriers to and facilitators of walking. *Canadian Journal of Nursing Research*, 37(3), 48–65. Available from: www.research.ed.ac.uk/portal/files/8279147/s4.pdf

Luborsky, M. R., Lysack, C. L. and Van Nuil, J. (2011). Refashioning one's place in time: Stories of household downsizing in later life. *Journal of Aging Studies*, 25(3), 243–252. http://dx.doi.org/10.1016/j.jaging.2011.03.009

Maersk, J. (2017). *Forming the self in daily life with advanced cancer*. Ph.D. diss., Southern Denmark University, Odense.

Mahmood, A., Chaudhury, H., Michael, Y. L., Camp, M., Hay, K. and Sarte, A. (2012). A photovoice documentation of the role of neighborhood physical and social environments in older adults' physical activity in two metropolitan areas in North America. *Social Science Medicine*, 74(8), 1180–1192. http://dx.doi.org/10.1016/j.socscimed.2011.12.039

McHugh, K. E. and Larson-Keagy, E. M. (2005). These white walls: The dialectic of retirement communities. *Journal of Aging Studies*, 19(2), 241–256. http://dx.doi.org/10.1016/j.jaging.2004.07.004

Milligan, C. (2006). Caring for older people in the 21st century: 'Notes from a small island'. *Health & Place*, 12(3), 320–331. http://dx.doi.org/10.1016/j.healthplace.2004.12.002

Peace, S. (2013). Social interactions in public spaces and places: A conceptual overview, in G. D. Rowles and M. Bernard (eds), *Environmental gerontology: Making meaningful places in old age* (pp. 25–52). New York: Springer.

Phillips, J., Walford, N. and Hockey, A. (2011). How do unfamiliar environments convey meaning to older people? Urban dimensions of placelessness and attachment. *International Journal of Ageing and Later Life*, 6(2), 73–102. http://dx.doi.org/10.3384/ijal.1652-8670.116273

Rosenberg, D. E., Huang, D. L., Simonovich, S. D. and Belza, B. (2013). Outdoor built environment barriers and facilitators to activity among midlife and older adults with mobility disabilities. *The Gerontologist*, 53(2), 268–279. http://dx.doi.org/10.1093/geront/gns119

Rowles, G. (1978). *Prisoners of space? Exploring the geographical experience of older people*. Boulder, CO: Westview Press.

Rowles, G. D. (1981). The surveillance zone as meaningful space for the aged. *The Gerontologist*, 21(3), 304–311. http://dx.doi.org/10.1093/geront/21.3.304

Rowles, G. D. (1983a). Geographical dimensions of social support in rural Appalachia. In G. D. Rowles and R. Ohta (eds), *Aging and milieu: Environmental perspectives on growing old* (pp. 111–130). New York: Academic Press.

Rowles, G. D. (1983b). Place and personal identity in old age: Observations from Appalachia. *Journal of Environmental Psychology*, 3(4), 299–313. http://dx.doi.org/10.1016/S0272-4944(83)80033-4

Rowles, G. D. (2008). Place in occupational science: A life course perspective on the role of environmental context in the quest for meaning. *Journal of Occupational Science*, 15(3), 127–135. http://dx.doi.org/10.1080/14427591.2008.9686622

Rowles, G. D. and Bernard, M. (2013). *Environmental gerontology: Making meaningful places in old age.* New York: Springer Publishing.

Rowles, G. D. and Ravdal, H. (2002). Aging, place and meaning in the face of changing circumstances, in R. S. Weiss and S. A. Bass (eds), *Challenges of the third age: Meaning and purpose in later life* (pp. 81–114). New York: Oxford University Press.

Rowles, G. D. and Watkins, J. F. (2003). History, habit, heart and hearth: On making spaces into places, in K. W. Schaie, H.-W. Wahl, H. Mollenkopf and F. Oswald (eds), *Aging independently: Living arrangements and mobility* (pp. 77–96). New York: Springer.

Skinner, M., Cloutier, D. and Andrews, G. (2015). Geographies of ageing: Progress and possibilities after two decades of change. *Progress in Human Geography*, 39(6), 776–799. http://dx.doi.org/10.1177/0309132514558444

Vine, D., Buys, L. and Aird, R. (2012). The use of amenities in high density neighbourhoods by older urban Australian residents. *Landscape and Urban Planning*, 107(2), 159–171. http://dx.doi.orb=g/10.1016/j.landurbplan.2012.04.013

Vine, D., Buys, L. and Aird, R. (2014). Conceptions of 'community' among older adults living in high-density urban areas: An Australian case study. *Australasian Journal on Ageing*, 33(2), E1–E6. http://dx.doi.org/10.1111/ajag.12115

Wiles, J. (2005). Conceptualizing place in the care of older people: The contributions of geographical gerontology. *Journal of Clinical Nursing*, 14(S2), 100–108. http://dx.doi.org/10.1111/j.1365-2702.2005.01281.x

Wiles, J. L. (2011). Reflections on being a recipient of care: The vexing concept of vulnerability. *Social and Cultural Geography*, 12(6), 573–588. http://dx.doi.org/10.1080/14649365.2011.601237

Wiles, J. L., Allen, R.E.S., Palmer, A. J., Hayman, K. J., Keeling, S. and Kerse, N. (2009). Older people and their social spaces: A study of well-being and attachment to place in Aotearoa New Zealand. *Social Science and Medicine*, 68(4), 664–671. http://dx.doi.org/10.1016.j.socscimed.2008.11.030

Zeitler, E. and Buys, L. (2015). Mobility and out-of-home activities of older people living in suburban environments: 'Because I am a driver, I don't have a problem'. *Ageing & Society*, 34(4), 785–808. http://dx.doi.org/10.1017/S0144686X13001086

Zeitler, E., Buys, L., Aird, R. and Miller, E. (2012). Mobility and active ageing in suburban environments: Findings from in-depth interviews and person-based GPS tracking. *Current Gerontology and Geriatrics Research*, 2012, Article ID 257186, 10pp. http://dx.doi.org/10.1155/2012/257186

18 Older persons, place and health care accessibility

Neil Hanlon

Introduction

One of the central tenets of geographical gerontology is that older individuals and the spaces and places they occupy are deeply interconnected. The person–place nexus is produced and reinforced in myriad life events, but the interweaving of networks of care and support in everyday social spaces constitutes a particularly important instance of what Malpas (1999) terms the 'emplacement' of personal experience. People typically require more health care and social support as they age, and the anticipation of this takes on greater weight over time as people think about, and act on, their relationships to home, neighbourhood, community, region, and, indeed, wider concepts of society and humanity. Living in a place that enhances access to a wide range of health services and social support is especially important to one's sense of security and belonging, especially in later stages of life (Menec *et al.*, 2011; World Health Organization (WHO), 2007). The emplacement of health care accessibility has far-reaching implications for present and future life trajectories, including one's quality of interpersonal relations, degree of social connectivity, well-being, feelings of self-worth and identity. The different aspects of health care accessibility (e.g., availability, reliability, suitability) are, therefore, among the most prominent concerns of geographical gerontology.

Health care accessibility is a multidimensional concept that describes the relative ease of entry into a health care system (Andersen, 1995), and the degree of fit between people and the health services available to them (Penchansky and Thomas, 1981). The processes by which 'entry' and 'fit' are achieved encompass a range of factors, including geographical proximity, temporal availability, affordability, acceptability, awareness, and the norms and procedures governing the use of a particular service (Gulliford *et al.*, 2002). These factors may operate independently and/or interactively depending on contextual circumstances (Field and Briggs, 2001). Making use of health care, or even making considerations to do so, raises issues of location, distance and the effort needed to connect people and resources. Not surprisingly, then, health geographers have a long-standing interest and engagement in accessibility research (Andrews *et al.*, 2007; Joseph and Phillips, 1984; McLafferty, 2003).

And yet it has been suggested that geographical approaches to health care accessibility have not had much impact outside of the discipline (Powell, 1995). If so, one explanation for this is that the spatial experience of health care accessibility is far more complex than is traditionally considered in models of accessibility and that physical distance, by itself, is not a particularly powerful determinant of utilization (Andrews, Evans and Wiles, 2013). At the same time, the way in which health care accessibility has been framed over the past several decades is coming under increasing scrutiny, with researchers calling into question the emphasis on atomistic (i.e., individual-level, single episode) measures of health care utilization behaviour as the basis for understanding complex and multidimensional facets of accessibility (Gulliford *et al.*, 2002; Penchansky and Thomas, 1981; Rosenberg and Hanlon, 1996). New ways of framing socio-spatial behaviour (Wiles, Chapter 3) and the experience of health care may offer geographers an opportunity to take a more prominent role in this area of research.

I begin this chapter with an overview of more conventional ways of thinking about accessibility as a product of the health care utilization behaviour of older persons. I then offer some thoughts about approaches to accessibility that begin to come to terms with the central challenges, and potential, of geographical gerontology. That is, I consider accessibility issues as an integral feature of the way in which older persons experience health and social care and how these experiences are necessarily bound up with processes of emplacement. I conclude with a call for a greater engagement with the contextual and institutional aspects of health care accessibility as a means to realize the full potential of geographical gerontology.

Accessibility as behaviour

The literature on health care access and utilization has been dominated for nearly five decades by a "behavioural" framework first introduced by Ronald Andersen (1968), and subsequently fine-tuned with the help of colleagues John Newman and Lu Ann Aday (Andersen and Newman, 1973; Aday and Andersen, 1974). The original model, hereafter referred to as the behavioral framework, considered the family to be the basic unit of analysis. This proved difficult to operationalize when analysing population-based health survey questionnaire data, which has been the preferred study design of health care accessibility research over this period (Andersen, 2008). Later iterations of the framework settled on factors influencing health care utilization at the level of the individual consumer, with family, household and other local contextual considerations regarded as variables or factors acting on individual health care utilization behaviour. According to this behavioural framework, health care use, or what the authors term *realized access*, is determined by the interaction of broader societal factors (e.g., social norms, technology), the particular institutional arrangements by which health care is organized and delivered, and a combination of individual characteristics (i.e., predisposing, enabling and illness level) that influence the type and quantity of health services consumed, as well as the level of satisfaction with care consumed. While the behavioural framework purports to take a broader social systems approach, its application has emphasized the individual characteristics of health care consumers, with scant attention to wider social, environmental and health care institutional considerations (Phillips *et al.*, 1998). The remainder of this section offers a brief outline of the individual characteristics in the Andersen framework, particularly as they pertain to older health care consumers.

Need and predisposing factors

The behavioural framework posits that an ideal health care system is organized principally to respond to the level of illness present in the population it is intended to serve. This level of illness, or need, is determined by a combination of professional assessment and consumer self-determination. Beyond these normative factors, the framework suggests that certain characteristics of individuals will predispose them, more so than others, to consume more health care. These predisposing features (i.e., age, sex, race, ethnicity and beliefs) are considered in the following paragraphs.

Age is the most prominent and influential of these individual predispositions. Both ends of the age spectrum exhibit a greater propensity to consume health care, but older adults as a group are by far the most frequent and intense users of health care (WHO, 2002). In Canada, for instance, the population aged 65 years and older (i.e., considered the 'seniors population' according to Statistics Canada) account for 4.5 times greater per capita public expenditures on health care than the population aged 20 to 64 years, and this predisposition to consume health care is shown to intensify with each decade of life lived (Canadian Institute for Health Information (CIHI), 2011). The magnitude of these rates of consumption is not entirely driven by health needs; indeed, frail and socially isolated elders are at heightened risk of unnecessary and inappropriate levels of prescription medication and hospitalization (Cloutier-Fisher and Kobayashi, 2009). But improvements in the way health care is delivered to older populations will, at best, merely delay the onset of frailty (i.e., achieve further compression of morbidity and mortality) and its corresponding needs for health care (Breyer, Costa-Font and Felder, 2010). In other words, age as a predisposition can be better moderated but not avoided.

The other demographic characteristics identified as predisposing factors are more difficult to generalize. The category of 'sex' is a good example. While aggregate measures of health care utilization suggest that women consume a disproportionate share of health care, these differentials disappear after taking into account services related to fertility and childbirth and the fact that women tend to outlive men in many parts of the world (Mustard *et al.*, 1998). Many studies suggest that older men seek and use health care at rates comparable to women of the same age, with the exception of services related to emotional and mental health issues (Koopmans and Lamers, 2007; MacIntyre, Hunt and

Sweeting, 1996; WHO, 2007). The emerging consensus is that considerations of 'gender' as a *social construct* are far more important in explaining differences in the health care seeking and utilization behavior of older women and men than considerations of sexual anatomy (see Joseph and Joseph, Chapter 21).

'Race' and 'ethnicity' are also commonly included in accessibility studies, but these require conceptual nuance and cautious interpretation. In many multicultural contexts, for instance, it has been observed that older members of visible minority groups use health care less often than their demographic contemporaries in the general population, in spite of exhibiting comparable levels of health care need (Gee, Kobayashi and Prus, 2004). Recent work employing a critical interpretive lens on access has shed light on the ways in which older members of 'visible minority' groups experience barriers to health care entry based on beliefs and assumptions held by dominant groups, such as the healthy immigrant effect or the assumption that elderly minorities prefer to be cared for informally by members of their own ethnic group (Dixon-Woods *et al.*, 2006; Koehn, 2009). The key point here is that the "predisposition" of racial and ethnic groups to seek and use health care is best understood as a product of the prevailing relations between dominant and minority groups rather than any presumed innate behaviour of minority groups themselves.

Andersen and colleagues also suggest that those holding certain 'beliefs' (e.g., values, attitudes, understandings) in the efficacy of health care are more inclined to seek and use these services. Many studies employ proxy variables such as level of education attained, occupation, and the size and strength of social networks to stand in for health care beliefs. The expectation is that persons with higher levels of formal education, wider social networks and in more 'skilled' types of employment are more aware of health care and more confident about obtaining it (Aday and Andersen, 1974). Critics of the behavioural framework have long questioned whether this aspect of the behavioural framework has been adequately conceptualized (Penchansky and Thomas, 1981). In particular, the psycho-social processes by which older people form attitudes and opinions about health and health care are not well understood, nor is the role of place and space as context for the formation of health care beliefs at different stages of the life course (Heenan, 2006).

Enabling factors

The behavioural framework also considers a set of household and community level characteristics that are thought to help enable health care utilization. At the scale of households, higher levels of savings and income, and the presence of health insurance, are factors considered the most critical in gaining entry to health care. Using health care services incurs a variety of costs (e.g., transportation, professional fees, absence from work) that typically must be borne by consumers. Health insurance is the principal means by which the high cost and unpredictability of health care consumption is managed. Older persons living in societies with more generous welfare programmes, including universal entitlement to health care insurance, or older individuals in households able to afford comprehensive private insurance, are expected to be better enabled to take advantage of available health care.

Yet, health care insurance plans are not necessarily well suited to older consumers, regardless of their affordability or universality. In Canada, for instance, health insurance schemes run by provincial governments tend to cover only services deemed 'medically necessary,' which in practice are those services offered by a physician or in hospital. Many services and resources (e.g., prescription drugs for those not in acute or continuing care, nursing care provided outside of institutional settings) particularly germane to the care of older people with chronic and degenerative conditions are covered only partially, or not at all, by provincial health insurance schemes (Chappell and Hollander, 2013). Thus, while health insurance 'enables' access to services by making care more affordable through risk pooling, the particular ways in which services are bundled together has the potential to constrain the optimal needs-based use of health care by older clients.

The behavioural framework also considers particular aspects of geography as enabling health care utilization. That is, certain places are expected to offer more variety and/or opportunity for residents to gain entry to the health system by virtue of their attractiveness to health care providers, such as population (market) size and amenities. Andersen and colleagues discussed this 'community' dimension in fairly broad geographic terms (e.g., neighbourhood, district, region) and categories (e.g., rural, urban, inner city, suburban). The expectation is that places with larger populations are more attractive to health care providers because their market size is capable of supporting a greater variety and number of health professionals, facilities, and services. At the same time, Andersen and associates recognized the importance of considering the population pressure on available resources, or what they referred to as a community's level of potential accessibility. For instance, a community with a lower ratio of people per general practitioner is said to have a higher level of potential accessibility to these health care resources. Notwithstanding the noted shortcomings of the measures of potential accessibility (e.g., cross-boundary spillover, modifiable area unit problem), such indicators of community enablement have enjoyed a modest degree of impact in accessibility research and policy (McLafferty, 2003).

Making sense of the spatial behaviour of older persons

Andersen and associates had very little to say about how older people moved about in space to obtain health care, and consideration of this was left to others, including health geographers, who typically regarded distance and travel time as enabling factors. This is premised on the notion that rational individuals will choose the nearest location capable of satisfying a demand for health care (Joseph and Phillips, 1984). The corollary of this is that the likelihood of using a service diminishes with increased distance to obtain the service, or what geographers have termed the 'friction of distance'. The extent to which a population bypasses the nearest care delivery site signals a degree of irrationality in the system, or else a level of consumer dissatisfaction or lack of confidence in local services (e.g., Hanlon, 2003; Hanlon and Skedgel, 2006).

Friction of distance is potentially an important barrier to access for certain groups of elders, namely those facing economic hardship, physical, and cognitive challenges, and social isolation (Fortney *et al.*, 2002). For many older persons, however, physical distance has not been found to be an especially onerous constraint (Alsnih and Hensher, 2003; Nemet and Bailey, 2000). Gains in years of disability-free life expectancy mean that a majority of older people continue to maintain a level of spatial mobility comparable to what they experienced in earlier phases of their adult lives (Webber, Porter and Menec, 2010). In addition, technological developments and improved standards of material well-being have substantially reduced the cost of mobility for all but the frailest or most economically disadvantaged. Older people living in rural areas, for example, now have more service options available to them (e.g., telehealth, traveling clinics, home care, transportation services) that mitigate, if not eliminate, geographic barriers (Cutchin, 2002; Ryser and Halseth, 2012).

Older people today likewise benefit from widespread social and technological developments that have reduced the effort needed to traverse space and, hence, the personal costs of travel (Reimer and Bollman, 2010). Technological change coupled with rising living standards over the past century have resulted in steadily rising rates of private automobile ownership among the elderly. While a gap between personal transportation still exists for older men and women, these gaps appear to be narrowing. Thus, while older women remain at greater risk of social isolation than older men (Webber *et al.*, 2010), they are also now more likely to hold a driver's license, own their own vehicle, and have access to employment and pension income to secure transportation (Alsnih and Hensher, 2003).

These changing social and technological conditions signal the need for better models of spatial behaviour. That is, the ways in which older people move about in space are influenced and constrained by a wide range of personal and social considerations and not simply by travel time and distance. For instance, in their study of the health care utilization by older rural residents, Nemet and Bailey (2000) report that the daily 'activity spaces' are a more powerful predictor of health care utilization than spatial proximity to services. Similarly, temporal dimensions of accessibility should not be reduced strictly to the time it takes for someone to travel to a clinic or health facility. Delays in waiting rooms may be sufficiently long to prompt health care users to travel longer distances to obtain care. Likewise, limitations imposed by practitioners on the amount of time per clinical appointment may prove unacceptable to older clients, especially those dealing with chronic disease, degenerative health issues and co-morbidity (Field and Briggs, 2001; Konrad *et al.*, 2010). These and other instances of service acceptability may exert a greater influence on health care seeking and utilization behaviour than a desire to minimize costs of travel, even for routine check-ups. Environmental (e.g., physical design of facilities) and institutional (e.g., provider behaviour) features of health care determine questions of suitability and acceptability that, in turn, shape the spaces of daily activity, or what geographical gerontologists refer to as the *social space* of older persons (Andrews *et al.*, 2009; Andrews *et al.*, 2013; Wiles *et al.*, 2009; see also Andrews, Cutchin, and Skinner, Chapter 2).

Emplacement: the lived experience of health care accessibility

The gerontology literature on health services accessibility is dominated by analyses of health survey questionnaires, many of which continue to employ the Anderson behavioural framework. These approaches to health care accessibility for older populations continue to offer valuable insights about factors associated with accessibility behaviour. At the same time, much of the contextual complexity of health care accessibility is necessarily overlooked in these approaches, or else reduced to 'indicators' or a very small number of analytic categories, in order to allow for more efficient measurement of, and generalization to, a much wider population.

A more in-depth and experiential account of health care accessibility for older people should take account of place as a central frame of reference (Wiles, 2005, Chapter 3). Place is a particularly powerful concept that embraces questions of emotions, instincts, values, intentions, opportunity, behaviour and interaction as an orientation for experience and being-in-the-world (Casey, 1993; Malpas, 1999). More than a mere spatial container to be filled with health care activity, place plays an active role in shaping health care exchanges. That is, place is the resource through which networks of health care production and consumption intersect, and the situated structure of opportunity in and through which older persons negotiate entry to health care and social support (Hanlon, 2009).

Calls to take place seriously are not entirely without precedent in the gerontology literature. The subfield of environmental gerontology has, for several decades, been a home for geographers and others to promote more place-centred ways of thinking. The theory of environmental press suggests that the fit, or equilibrium balance, between a person's needs (e.g., for health care and social support) and the features and conditions of their lived everyday environments (e.g., housing, neighbourhood location) undergo periodic challenges over the life course that require adaptations on the part of the person and/ or environment (Byrnes, Lichtenberg, and Lysack, 2006; Cutchin, 2003). The capacity for adaptation, however, becomes increasingly strained with age and the onset of declining health and mobility, eventually undermining one's environmental equilibrium. The longer one lives in a state of dis-equilibrium, the greater the state of environmental press, whereby the needs of older persons and the capacities of their living arrangements are out of sync. This framework has been widely used to identify a range of environmental factors that impede or otherwise do not support active living and participation for older persons, including unsafe neighbourhoods, unsuitable housing, inaccessible or age-inappropriate amenities, social and physical barriers to social interaction and, most important, for our purposes here, inadequate access to health care (e.g., Hodge, 2008). The model itself is the conceptual foundation for international efforts to encourage 'age-friendly' community development (Kahana *et al.*, 2003; Menec *et al.*, 2011; WHO, 2007).

Taking place seriously entails recognizing the ways in which the prevailing resources and social opportunities of place condition the understandings, expectations, and behaviour of older residents with respect to accessing health services and social support. Care and social support are negotiated through place-embedded negotiations among family, friends, volunteers, health professionals, employers and other social agents (Bowlby, 2011; Hanlon *et al.*, 2007; Skinner and Joseph, 2011). In this sense, health and social care present themselves through place, a process that Malpas (1999) refers to as 'emplacement'.

An older person's entry into and fit with the health care system are only possible through the process of emplacement. Ageing in place typically entails poignant changes in one's connection and relationship to place; principal among these changes are the physical and emotional challenges associated with emerging frailty and declining well-being. Malcolm Cutchin's (2003, 2016) concept of place integration is especially helpful here in thinking about these person–place interrelationships, more generally, and the experiential aspects of health care accessibility, more specifically. Cutchin's (2003) study of the experiences of older individuals in adult day centres and assisted living residences is illustrative of the value of such an approach in revealing how consuming these forms of 'eldercare' brings about a profound transformation in the meaning of home, community, and place. Framing experiential studies of health care accessibility and utilization in this way draws attention away from the declining state of older bodies and casts light instead on the ways in which dominant social relations of place-making contribute to the disablement and marginalization of elders (Blewett and Hanlon, 2016; Macpherson, 2009, 2010; see also Walsh, Chapter 20). More in-depth consideration of elder care emplacement helps reveal the ways in which socio-spatial processes play a critical part in life course transitions (Horschelmann, 2011; Wiles *et al.*, 2009; see also Cutchin, Chapter 17).

Concluding remarks

Space and place remain central concerns of geographic gerontology. At the same time, the ways in which these key concepts are conceptualized and operationalized to look at issues of health care accessibility are in need of updating. In particular, geographical gerontology should promote more relational and experiential understandings of place that have pervaded the wider discipline over the past several decades. This entails less an emphasis on the influence of geography on questions of health care accessibility and utilization and more on how the negotiation of health care accessibility necessarily alters the relationship between older people and their lived daily environments. Moving such a project forward will require experimentation with novel research designs and broad empirical testing of emplacement theory.

The accessibility of health care and social support are critical features of person–place interrelationships. Curiously, the call to take place more seriously is not so much an outright rejection of the prevailing behavioural approach to accessibility as it is a plea for a more meaningful engagement with the long-overlooked environmental and contextual dimensions of the Andersen framework. Much of the social gerontology literature on health care accessibility, and indeed the traditional health geography literature, is fixated on health care behaviour at the level of the individual 'consumer'. Questions of geography have been relegated to discussions of enabling factors (e.g., urban versus rural location, travel times and distance) as they relate to the availability and use of a narrow range of health services. This has come at the expense of considering the emplaced experience of health care and social support. A re-invigoration of accessibility theory and research promises to herald new opportunities for geographical gerontology to make deeper and more lasting contributions.

References

Aday, L. A. and Andersen, R. M. (1974). A framework for the study of access to medical care. *Health Services Research*, 9, 208–220.

Alsnih, R. and Hensher, D. A. (2003). The mobility and accessibility expectations of seniors in an aging population. *Transportation Research Part A*, 37(10), 903–916. http://dx.doi.org/10.1016/S0965-8564(03)00073-9

Andersen, R. (1968). *A behavioral model of families' use of health services*. Chicago: Center for Health Administration Studies, University of Chicago.

Andersen, R. M. (1995). Revisiting the behavioral model and access to medical care: Does it matter? *Journal of Health and Social Behavior*, 36(1), 1–10.

Andersen, R. M. (2008). National health surveys and the behavioral model of health service use. *Medical Care*, 46(7), 647–653. http://dx.doi.org/10.1097/MLR.0b013e31817a835d

Andersen, R. M. and Newman, J. F. (1973). Societal and individual determinants of medical care utilization in the United States. *Milbank Memorial Fund Quarterly*, 51(1), 95–124. http://dx.doi.org/10.1111/j.1468-0009.20005.00428.x

Andrews, G. J., Cutchin, M., McCracken, K., Philips, D. R. and Wiles, J. (2007). Geographical gerontology: The constitution of a discipline. *Social Science and Medicine*, 65(1), 151–168. http://dx.doi.org/10.1016/j.socscimed.2007.02.047

Andrews, G. J., Evans, J. and Wiles, J. L. (2013). Re-spacing and re-placing gerontology: Relationality and affect. *Ageing and Society*, 33(8), 1339–1373. http://dx.doi.org/10.1017/S0144686X12000621

Andrews, G. J., Milligan, C., Phillips, D. R. and Skinner, M. W. (2009). Geographical gerontology: Mapping a disciplinary intersection. *Geography Compass*, 3(5), 1641–1659. http://dx.doi.org/10.1111/j.1749-8198.2009.00270.x

Blewett, J. and Hanlon, N. (2016). Disablement as inveterate condition: Living with habitual ableism in Prince George, British Columbia. *The Canadian Geographer*, 60(1), 46–55. http://dx.doi.org/10.1111/cag.12254

Bowlby, S. (2011). Friendship, co-presence and care: Neglected spaces. *Social and Cultural Geography*, 12(6), 605–622. http://dx.doi.org/10.1080/14649365.2011.601264

Breyer, F., Costa-Font, J. and Felder, S. (2010). Ageing, health, and health care. *Oxford Review of Economic Policy*, 26(4), 674–690. http://dx.doi.org/10.1093/oxrep/grq032

Byrnes, M., Lichtenberg, P. A. and Lysack, C. (2006). Environmental press, aging in place, and residential satisfaction of urban older adults. *Journal of Applied Science*, 23(2), 50–77. http://dx.doi.org/10.1177/19367244062300204

Canadian Institute for Health Information. (2011). *Health care in Canada, 2011 – a focus on seniors and aging*. Ottawa: The Author.

Casey, E. (1993). *Getting back into place: Toward a renewed understanding of the place-world*. Bloomington, IN: Indiana University Press.

Chappell, N. and Hollander, M. J. (2013). *Aging in Canada*. Don Mills, Ontario: Oxford University Press.

Cloutier-Fisher, D. and Kobayashi, K. (2009). Examining social isolation by gender and geography: Conceptual and operational challenges using population health data in Canada. *Gender, Place, and Culture*, 16(2), 181–199. http://dx.doi.org/10.1080/09663690902795787

Cutchin, M. (2002). Virtual medical geographies: Conceptualizing telemedicine and regionalization. *Progress in Human Geography*, 26(1), 19–39. http://dx.doi.org/10.1191/0309132502ph352ra

Cutchin, M. (2003). The process of mediated aging-in-place: A theoretically and empirically based model. *Social Science and Medicine*, 57(6), 1077–1090. http://dx.doi.org/10.1016/S0277-9536(02)00486-0

Cutchin, M. (2016). Place integration: Notes on a Deweyan framework for community inquiry. In: M. Skinner and N. Hanlon (eds), *Ageing resource communities: New frontiers of rural population change, community development and voluntarism* (pp. 24–37). London and New York: Routledge.

Dixon-Woods, M., Cavers, D., Agarwal, S., Annandale, E., Arthur, A., Harvey, J., Hsu, R., Katbamna, S., Olsen, R., Smith, L., Riley, R. and Sutton, A. J. (2006). Conducting a critical interpretive synthesis of the literature on access to healthcare by vulnerable groups. *BMC Medical Research Methodology*, 6(35), 13pp. http://dx.doi.org/10.1186/1471-2288-6-35

Field, K. S. and Briggs, D. J. (2001). Socio-economic and locational determinants of accessibility and utilization of primary health care. *Health and Social Care in the Community*, 9(5), 294–308. http://dx.doi.org/10.1046/j.0966-0410.2001.00303.x

Fortney, J., Chumbler, N., Cody, M. and Beck, C. (2002). Geographic access and service use in a community-based sample of cognitively impaired elders. *Journal of Applied Gerontology*, 21(3), 352–367. http://dx.doi.org/10.1177/073346480202100305

Gee, E. M., Kobayashi, K. M. and Prus, S. G. (2004). Examining the healthy immigrant effect in mid- to later life: Findings from the Canadian Community Health Survey. *Canadian Journal on Aging*, 23(Suppl 1), S56–S63.

Gulliford, M., Figueroa-Munoz, J., Morgan, M., Hughes, D., Gibson, B., Beech, R. and Hudson, M. (2002). What does 'access to health care' mean? *Journal of Health Services Research and Policy*, 7(3), 186–188. http://dx.doi.org/10.1258/135581902760082517

Hanlon, N. (2003). Measuring aspects of devolved health authority performance: Nova Scotia patients who travel further than necessary to obtain hospital care. *Healthcare Management Forum*, 16(2), 8–13. http://dx.doi.org/10.1016/S0840-4704(10)60213-X

Hanlon, N. (2009). Access and utilization reconsidered: Towards a broader understanding of the spatial ordering of primary health care, in V. Crooks and G. Andrews (eds), *Primary health care: People, practice, place* (pp. 43–56). Farnham and Burlington: Ashgate.

Hanlon, N., Halseth, G., Clasby, R. and Pow, V. (2007). The place embededness of social care: Restructuring work and welfare in Mackenzie, B.C. *Health and Place*, 13(2), 466–481. http://dx.doi.org/10.1016/j.healthplace.2006.05.006

Hanlon, N. and Skedgel, C. (2006). Cross-district utilization of general hospital care in Nova Scotia: Policy and service delivery implications for rural districts. *Social Science and Medicine*, 62(1), 145–156. http://dx.doi.org/10-1016/j.socscimed.2005.05.018

Heenan, D. (2006). The factors influencing access to health and social care in the farming communities of County Down, Northern Ireland. *Ageing and Society*, 26(3), 373–391. http://dx.doi.org/10.1017/S0144686X06004697

Hodge, G. (2008). *The geography of aging: Preparing communities for the surge in seniors*. Montreal and Kingston: McGill-Queen's University Press.

Horschelmann, K. (2011). Theorising life transitions: Geographical perspectives. *Area*, 43(4), 378–383. http://dx.doi.org/10.1111/j.1475-4762.2011.01056.x

Joseph, A. and Phillips, D. (1984). *Accessibility and utilization: Geographical perspectives on health care delivery*. New York: Harper and Row.

Kahana, E., Lovegreen, L., Kahana, B. and Kahana, M. (2003). Person, environment, and person-environment fit as influences on residential satisfaction of elders. *Environment and Behavior*, 35(3), 434–453. http://dx.doi.org/10.1177/0013916503035003007

Koehn, S. (2009). Negotiating candidacy: Ethnic minority seniors' access to care. *Ageing Society*, 29(4), 585–608. http://dx.doi.org/10.1017/S0144686X08007952

Konrad, T. R., Link, C. L., Shackelton, R. J., Marceau, L. D., von dem Knesebeck, O., Siegrist, J., Arber, S., Adams, A. and McKinlay, J. B. (2010). It's about time: Physicians' perceptions of time constraints in primary care medical practice in three national health care systems. *Medical Care*, 48(2), 95–100. http://dx.doi.org/10.1097/MLR.0b013e3181c12e6a

Koopmans, G. T. and Lamers, L. M. (2007). Gender and health care utilization: The role of mental distress and help-seeking propensity. *Social Science and Medicine*, 64(6), 1216–1230. http://dx.doi.org/10.1016/j.socscimed.2006.11.018

MacIntyre, S., Hunt, K. and Sweeting, H. (1996). Gender differences in health: Are things really as simple as they seem? *Social Science and Medicine*, 42(4), 617–624. http://dx.doi.org/10.1016/0277-9536(95)00335-5

Macpherson, H. (2009). The incorporeal emergence of landscape: Negotiating sight, blindness and ideas of landscape in the British countryside. *Environment and Planning A*, 41(5), 1042–1054. http://dx.doi.org/10.1068/a40365

Macpherson, H. (2010). Non-representational approaches to body-landscape relations. *Geography Compass*, 4(1), 1–13. http://dx.doi.org/10.1111/j.1749-8198.2009.00276.x

Malpas, J. (1999). *Place and experience: A philosophical topology*. Cambridge and New York: Cambridge University Press.

McLafferty, S. L. (2003). GIS and health care. *Annual Review of Public Health*, 24, 25–42. http://dx.doi.org/10.1146/annurev.publhealth.24.012902.141012

Menec, V. H., Means, R., Keating, N., Parkhurst, G. and Eales, J. (2011). Conceptualizing age-friendly communities. *Canadian Journal on Aging*, 30(3), 479–493. http://dx.doi.org/10.1017/S0714980811000237

Mustard, C., Kaufert, P., Kozyrskyj, A. and Mayer, T. (1998). Sex differences in the use of health care services. *New England Journal of Medicine*, 338(23), 1678–1683. http://dx.doi.org/10.1056/NEJM199806043382307

Nemet, G. and Bailey, A. (2000). Distance and health care utilization among the rural elderly. *Social Science and Medicine*, 50(9), 1197–1208. http://dx.doi.org/10.1016/S0277-9536(99)00365-2

Penchansky, R. and Thomas, J. W. (1981). The concept of access: Definition and relationship to consumer satisfaction. *Medical Care*, 19(2), 127–140.

Phillips, K. A., Morrison, K. R., Andersen, R. and Aday, L. U. (1998). Understanding the context of healthcare utilization: Assessing environmental and provider-related variables in the behavioral model of utilization. *Health Services Research*, 33(3), 571–596.

Powell, M. (1995). On the outside looking in: Medical geography, medical geographers and access to health care. *Health and Place*, 1(1), 41–50. http://dx.doi.org/10.1016/1353-8292(95)00005-7

Reimer, B. and Bollman, R. D. (2010). Understanding rural Canada: Implications for rural development policy and rural planning policy, in D.J.A. Douglas (ed.), *Rural planning and development in Canada* (pp. 10–52). Toronto: Nelson Education.

Rosenberg, M. W. and Hanlon, N. T. (1996). Access and utilization: A continuum of health service environments. *Social Science and Medicine*, 43(6), 975–983. http://dx.doi.org/10.1016/0277-9536(96)00007-X

Ryser, L. and Halseth, G. (2012). Resolving mobility constraints impeding rural seniors' access to regionalized services. *Journal of Aging and Social Policy*, 24(3), 328–344. http://dex.doi.org/10.1080/08959420.2012.683329

Skinner, M. W. and Joseph, A. E. (2011). Placing voluntarism within evolving spaces of care in ageing rural communities. *GeoJournal*, 76, 151–162. http://dx.doi.org/10.1007/s10708-009-9238-8

Webber, S. C., Porter, M. M. and Menec, V. H. (2010). Mobility in older adults: A comprehensive framework. *The Gerontologist*, 50(4), 443–450. http://dx.doi.org/10.1093/geront/gnq013

Wiles, J. (2005). Conceptualizing place in the care of older people: The contribution of geographical gerontology. *Journal of Clinical Nursing*, 14(2), 100–108. http://dx.doi.org/10.1111/j.1365-2702.2005.01281.x

Wiles, J., Allen, R. E., Palmer, A. J., Hayman, K. J., Keeling, S. and Kerse, N. (2009). Older people and their social spaces: A study of well-being and attachment to place in Aotearoa New Zealand. *Social Science and Medicine*, 68, 664–671. http://dx.doi.org/10.1016/j.socscimed.2008.11.030

World Health Organization (WHO). (2002). *Active aging: A policy framework*. Madrid: United Nations World Assembly on Ageing. Available from: http://apps.who.int/iris/bitstream/10665/67215/1/WHO_NMH_NPH_02.8.pdf

World Health Organization (WHO). (2007). *Global age-friendly cities: A guide*. Geneva: WHO. Available from: www.who.int/ageing/publications/Global_age_friendly_cities_Guide_English.pdf?ua=1

19 Mobilities and ageing

"We're quite outgoing people"

Anthony C. Gatrell

Introduction

Social gerontologists, social geographers and transport researchers have frequently studied how older people move across space using different modes of transport that take them to transient or more permanent destinations (O'Hern and Oxley, 2015). They have also examined the constraints placed upon older people by the built environment or their own physical abilities (Chudyk *et al.*, 2015). Recent overviews of geographical gerontology have also touched on movement, illustrating this in terms of retirement migration or the extent to which older people are separated spatially from their younger family members (see, for example, Andrews *et al.*, 2007, 2009). Thanks to research by geographers and other social scientists we know a good deal about older adults' access to public services, including health care (see Hanlon, Chapter 18, for an overview), while much research has also been conducted into transport in both urban areas (Buffle and Phillipson, Chapter 10) and rural areas (Skinner and Winterton, Chapter 11). Researchers have also considered how local environmental context shapes movement, with numerous studies published on the walkability of urban environments for older adults (see, for example, Mitra, Siva and Kehler, 2015).

Here, I want to draw on a different body of literature: that from the 'mobilities turn' in social science, a perspective that has emerged over the last 10 to 15 years to shake up our understanding of movement but one that has, as yet, had disappointingly little to say about older adult lives. I seek to convey both what is new about mobilities research and to suggest ways in which mobilities researchers and social gerontologists can engage in mutually beneficial intellectual encounters: how they might 'move closer together', as it were.

Mobilities

One of the leading social scientists in mobilities research, Mimi Sheller, suggests that it is a multidisciplinary field of inquiry dealing with "the spatial mobility of humans, non-humans and objects; the circulation of information, images and capital, including critical theories of the affective and psycho-social implications of such mobility; as well as the study of the physical means for movement, such as infrastructures, vehicles, and software systems that enable travel and communication to take place" (Sheller, 2015: 46–47). It is immediately clear that mobility is not simply the act of moving from one location to another. Rather, it is movement charged with meaning, emotions and (crucially) relations, whether with other people or the technologies that enable or impede movement. Cresswell draws an analogy with location and place. Places are "locations imbued with meaning and power" (1996: 3), and mobility is the dynamic equivalent; more than simply movement from *a* to *b*, it, too, has meaning and operates within a context of relations between the moving object or person and other actors and agents, whether human or non-human.

Mobility studies, therefore, take us well beyond traditional (though important) concerns with migration patterns of older people or the changing possibilities in accessing goods and services or meeting other people. Instead, mobilities literature draws heavily on post-positivist theory, not least the kind of non-representational (or 'more-than-representational') approaches promulgated by Nigel Thrift (2008) and others (for example, Lorimer, 2005), where performance and affect are key (see Skinner, Cloutier and Andrews, 2015, for helpful insights into how non-representational theory can illuminate our understanding of ageing). In relation to social gerontology, a concern with mobilities means, as Mansvelt (2015: 398) puts it, going beyond a "focus on the physical and functional aspects of older people's bodies and their capacity to displace space". However, the landmark texts on mobilities (Adey, 2010; Cresswell, 2006; Merriman, 2012a; Urry, 2007) make little if any reference to age and ageing, and the same is true of published research in the eponymous journal, *Mobilities* (an exception is Stjernborg, Wrestrand and Tesfahuney, 2015). In contrast, some of the contributions in a special issue of *Ageing & Society* (see Schwanen and Ziegler, 2011, for the introduction) have drawn on the mobilities turn. Whether discussing fast mobilities (by air), or the slower mobilities of walking, the contemporary mobilities literature tends to consider the relatively youthful mobile subject.

A key concept in mobilities research is that of *motility*, introduced by Kaufmann (2002) in order to convey the capacity or potential to be mobile. Since motility "captures the unequal distribution of power over mobility, mobility resources and access to space" (Stjernborg *et al.*, 2015: 384), it is particularly relevant to the study of older people who may have limited access to different modes of transport and may be constrained by physical capabilities. Another key concept is *mobility capital*, the resource that can be put to use to move about in a relatively unconstrained way. Such capital is, clearly, in part a function of *economic* capital, but there are mutually reinforcing relations with other forms of capital. For example, motility can create social and cultural capital (access to friends and places of interest), while having stocks of such capital motivates the actor to travel. Last, it is important to signal the fact that mobility can refer to engagement in virtual or cyberspaces, not merely physical space. This is of obvious interest in social gerontology, where connections in virtual space can perhaps substitute for those in physical space. Having introduced the main elements of mobilities research and bemoaned its neglect of older adults, I want now to examine some of the possible connections. I begin with slow mobilities.

Walks and dances

The mobilities literature has given considerable attention to the experience of walking, but as with so much else, these experiences are of the younger mobile. In contrast, literature on transport and social gerontology does discuss walking, often in terms of the constraints placed by physical deterioration or by the environment. For the older adult, as for those at other ages, walks may be utilitarian (for example, to shop or to attend a health care appointment; Hanlon, Chapter 18), or discretionary activities (for example, to visit friends and family, or for exercise; see Siren, Hjorthol and Levin, 2015).

Context is everything during the walk. In urban settings seats, benches, hand rails and door frames all provide the material resources to enable rest. Slow mobilities may be forced on the older person because of infirmity but the flipside of this is the ability (not disability) to stop and enjoy surroundings; to be more aware of what is going on. Yet the environment within which the older adult is walking may be littered with obstacles and debris, and this contrasts dramatically with the walk through an attractive environment. Finlay and her colleagues (2015) illustrate this in their study of the experience of green and blue spaces (parks and gardens, lakes and the ocean) in Vancouver. These motivate the walker. As one of their respondents says, "I just know I have to go out and do walking because if I don't I'm going to get old too fast" (Finlay *et al.*, 2015: 101). But their paper shows how the walker is not the only mobile actor in this network of affect. Branches rustle, birds fly and people appreciate and draw stimulation from the waves on the shore, the running water of the stream or even the courtyard fountain.

The environmental context within which walking takes place has, in recent years, spawned a substantial literature on 'walkability': the extent to which the built environment is conducive to urban walking. Walkability may be characterised by mixed land uses and a connected street network, although attempts to measure it using objective databases do not seem to include features (such as steepness of terrain) that might discourage walking among older adults. King and her colleagues (2011) studied over 700 older walkers in Seattle and Baltimore and found that those in more walkable neighbourhoods were much more likely to walk, and to have lower body mass index, than those in less walkable parts of the cities. A longitudinal study of older adults in two German cities confirms that various aspects of the environment can play a crucial positive or negative role (Mollenkopf, Hieber and Wahl, 2011). Of course, the natural environment can be important as well, not least if pavements (sidewalks) are ice-covered and slippery in winter months.

One of the key texts on mobilities (Cresswell, 2006) has an entire chapter on dance, though the emphasis is on the politics and regulation of the activity, not how it is performed in contemporary society, still less by older adults. But organisations such as Age UK have promoted dance as a means of maintaining good health and fitness, as well as strengthening social relationships. Evidence from a longitudinal study of 469 adults, aged more than 75 years (at baseline), suggests that regular dancing is significantly associated with a reduced risk of dementia, so, too, is playing a musical instrument involving manual and digital dexterity (Verghese *et al.*, 2003). From a non-representational theory perspective, Merriman points to "all of those forms of personal, informal and, indeed, 'not-quite' dance practices that we perform more-or-less regularly – from dancing in nightclubs to shimmying down the street or in one's house" (Merriman, 2012b: 22), though whether this speaks meaningfully to older adults is a moot point. Nonetheless, for many older people, dancing is a recreation. I next consider further examples of recreational activities that help to shape the mobile subject.

Recreations

As people transition out of paid work into later life, the opportunities for leisure and recreation may increase. These activities help maintain physical fitness, cognitive abilities and social connections.

The front page of my local newspaper recently carried an article headlined 'Mo Zimmer'. It featured a keen runner in his 80s, Brian Forster, and contrasted this with a photograph of him in his youth, riding a motorcycle. The name 'Mo Zimmer' referenced both the British long-distance world champion Mo Farah and the walking frame that some older people use. There is condescension in the headline, which implies that he is subverting our expectations by being able to run at all. His behaviour is *transgressive*, or 'out of place', not in the sense of fixed places or as an intentional resistance (as considered by Cresswell, 1996), but in terms of being unexpected, surprising and countering expected norms of behaviour.

Whitaker (2005) conducted ethnographic research with a group of older Italian men ranging in age from 49 to 89 years, who ride for several hours each day (and up to 12,000 km a year). While many of her participants allude to the health benefits (and physical challenge) of regular cycling, for most there are more general benefits for psychosocial well-being. One 83-year-old claimed never to have had his blood pressure measured and says that "the only pressure he cares about is that of his bicycle's tires" (Whitaker, 2005: 16). Federico, aged 59, refers to "intimate voyages", and Whitaker asserts that "the aesthetic is perceived through the material, which, in turn, is in movement through rhythmic, physically taxing activity" (Whitaker, 2005: 25). Furthermore, this is deeply social mobility: membership in cycling clubs strengthens social connections, a point reinforced in a more recent study of older cyclists in Vancouver, Canada (Winters *et al.*, 2015).

The stereotypical idea of recreation among older men is golf. That caricature ignores the proportion of older women who play and enjoy golf, though for anyone the sport demands a degree of economic, as well as mobility, capital to play, since equipment and club membership can both be expensive. Nonetheless, it provides exercise for those walking the course, while those unable to walk the course can rent 'buggies' to get around. It is itself a slow mobility, but since it is an intensely social sport it sets the scene for considering the sociality of the mobile older adult.

Connections

The subtitle of this chapter, taken from a respondent in the important study undertaken by Ziegler and Schwanen (2011), suggests that while mobility may take the individual on a trip, it also offers the prospect of meeting others. Ziegler and Schwanen stress the importance of the older individual's determination and motivation to engage with the world. They argue that "the mobility of the self and positive attitudes *can to some extent* replace movement in physical space as a basis for the older person's engagement with the world" (my italics; Ziegler and Schwanen, 2011: 770). But while social and psychological engagement matters, many older adults will be impacted by the loss of physical capabilities that enable movement in physical space. Positive attitudes are all well and good, but the necessary visit to (for example) a hospital or health centre where both the journey and perhaps car parking are very difficult make for a potentially emotionally draining experience. Again, such experiences tend to be neglected in the mobilities literature.

Mollenkopf *et al.* (2011) confirm the importance of out-of-home mobility. As one of their participants says, "Really, it's getting out what makes up life, isn't it? When you stay at home you can watch TV, but that's not life, that's dying slowly" (Mollenkopf *et al.*, 2011: 789). Getting out of the house is still important and has *meaning*. Of course, the house is not the same as the home. Wiles and her colleagues (2009) make the point that home is a fluid space that extends beyond the household; it embraces nearby places for meetings and other social interactions. Furthermore, while older adults may be restricted in terms of physical movement, they can connect in other spaces using traditional modes of communication (letters and landline calls), or more modern social media: as one of Wiles's respondents put it neatly, "[We] do our visiting on the phone" (Wiles *et al.*, 2009: 669). Ziegler and Schwanen (2011) further note that mobility can also be imaginary, with memories of perhaps distant (spatially and temporally) places and people being recalled in a way that substitutes for physical mobility. Rowles (1978) made the same point almost 40 years ago. However, while technologies such as Skype and the exchange of photographs online can bring older people together with friends and family members, for other people the receipt of mobile physical objects – a handwritten letter, postcard, or printed photograph – may be more appreciated.

Connections are not just with other people. Pets offer companionship, and since dogs, in particular, need to be walked, this offers both potential health benefits and opportunities for social interaction. Empirical work with focus groups in Hampshire, the United Kingdom, conveys the importance of these mobilities. As one older adult expressed it, "I have met so many people just through dogs. . . . If you just walked past a person without a dog, they prob-ably wouldn't talk to you. . . . you still feel like you have friends" (Knight and Edwards, 2008: 447). Degeling and Rock (2012) conducted a study with 10 older people in Calgary, Canada, looking at the spaces (mostly streets) used for walking. Their dogs helped motivate physical activity – "get you going in the morning", as one respondent put it (p. 401). These mobilities are pro-foundly relational, tying older adults, dogs, encounters with others and the places walked, into an ecology of movement.

Recognition of the importance of maintaining connections and avoiding social isolation has triggered policy interventions directed at older adults. The 'Men in Sheds' initiative, originating in Australia, brings older men together in workshops to engage in purposeful social activity that also engen-ders bodily mobility in the sense of making things and thereby using motor skills. Milligan and her colleagues have evaluated this intervention in the United Kingdom, and their interviews are revealing. One man reports, "My wife died and I used to sit at home and look at the wallpaper . . . And she [the social worker] told me about Men in Sheds . . . And it got me out of the house. It got me to meet other people" (Milligan, 2015: 134). The sheds provide a kind of therapeutic landscape outside the home, in much the same way the community garden (allotment) is a space of social and physical activ-ity that involves physical mobility and social engagement (Milligan, Gatrell and Bingley, 2004). Gardening (including mowing the lawn) in the domestic setting is another example of a slow, but productive, mobility. Volunteer time given by older adults also provides sets of possible connections that can, to some extent, replace those that may disappear when full-time employment ceases.

Enablements

As noted in the introduction, a key feature of the 'mobilities turn' is how the mobility of human subjects interacts with objects, including infrastructure, devices, vehicles, and software systems. John Urry speaks of the "utter relation-ality of humans with objects and machines" (Adey and Bissell, 2010: 7), while Sheller puts it in terms of actor-network theory: "mobilities theorists pay close attention to the more-than-human infrastructures, technical objects, prostheses and practices that assist (or disable) mobility" (Sheller, 2015: 50). Much of this concern with the relations of older adults to technologies is not new to the social gerontologist or transport researcher interested in how older people get about (Musselwhite, Holland and Walker, 2015). Some research maintains a deficit-based perspective, focusing on how older adults lose the ability to drive a car safely, or at all. However, it is clearly a mistake to consider that all drivers above some arbitrary age are cognitively deficient, since many will be as compe-tent as any younger driver. Moreover, as many older people continue employ-ment into later life they continue to rely on private transport, though the advent of driverless vehicles in years to come should be of considerable benefit.

But the objects and machines mentioned by Urry extend well beyond the car. They include other forms of transport (such the bus and, as noted earlier, the bicycle), computer and mobile (cell) phones (Hardill and Olphert, 2012), walking frames and sticks, as well as the objects that have to be negotiated as movement takes place (the pavements, street furniture, steps, pedestrian crossings), not to mention other mobile actors (cyclists, drivers, pedestrians and so on). New technologies may be used to track movement. Shoval and others (2011) used GPS to track older people (both cognitively impaired and cognitively intact) in Tel-Aviv for 28 days in order to look at their mobilities. Those with mild cognitive impairment had a more restricted spatial range of movement than those without. But studies such as these raise serious ethical issues, as well as potential bias inasmuch as mobility range could be affected by the wearable technologies.

Restricted mobility is what I refer to as 'bio-geographical disruption', and I turn finally to other forms of disruption to movement.

Disruptions

Not all mobilities are welcomed. Older adults may have moved several, or many, times, and while earlier in life these residential moves may have presented as opportunities, relocations in later life will be influenced by a range of factors and may be emotionally disruptive. Peace, Holland and Kellaher (2011) invited their 54 respondents, drawn from different places in the United Kingdom, to talk about the places they had lived. Their biographies cannot be separated from the mobilities they have enjoyed (perhaps endured) before and during later life. "For *most* [original emphasis] people in later life . . . the prospect of moving becomes especially difficult when they live alone or are in poor health, because the disruption, organisation and associated costs are serious concerns" (Peace *et al.*, 2011: 743). And for some, relocating was not something they wished to contemplate, preferring to remain in their present home as long as possible.

Falls are an unwanted, disruptive mobility, whether occurring in the home or outside. While older adults with previous experience of falling may find a variety of assistive technologies useful, it is important not to reduce people to "just a faller" (Bailey *et al.*, 2011: 838). Again, context matters, even in the home, since rugs, stairs, furniture and doors are all material things that can create risks even for the mobile elder.

Getting lost is an unwelcome mobility, in both the literal sense of losing one's way and metaphorically – becoming 'lost' in the sense of getting disconnected from, or losing, friends. There are periodic news reports of older adults getting lost in car journeys. In December 2011, Londoner Dennis Leighton set out to drive 55 miles to visit his daughter but drove for many hours around the M25 London motorway before being found two days later by police, who excused this behaviour with the comment that "he is after all 82-years-old" (Orr, 2011: 4) as if his chronological age explained everything. Getting lost is a possible hazard for the older adult with dementia. But, as Graham (2015) puts it, "[w] hen a person without dementia goes for a walk, it is called going for a stroll, getting some fresh air, or exercising. When a person with dementia goes for a walk beyond prescribed parameters, it is typically called wandering" (Graham, 2015: 1). Graham resists pathologising this kind of behaviour and prefers the phrase 'people who like to walk' or the term 'wayfaring', suggesting it is purposeful to the individual, not aimless. This behaviour can be facilitated by environmental and design changes such as provision of safe walking areas: again, context matters. Other, quite different, research explores technologies that can be used to 'monitor' such behaviour. An example is iWander (Sposaro, Danielson and Tyson, 2010), technology that collects data on movement, assesses the likelihood of 'wandering' and is capable of providing directions home as well as notifying the caregiver of the location of the walker.

Natural hazards are well known to disrupt lives, and the impacts can be severe for older people with physical or cognitive limitations. A good example of this is the forced relocation of people following Hurricane Katrina in 2005. Interviews with older people in Mississippi (Langan and Palmer, 2012) showed how mobilities impacted behaviour and well-being. Things that usually flow (power and water) were no longer mobile. Some people preferred to remain in situ (for example, to protect property from theft or vandalism) rather than evacuating. For those who did move there were associated challenges: accessing medications, getting adequate shelter, negotiating evacuation routes clogged with traffic and staying connected with phone services disabled. In this case, the material things that should have *enabled* mobilities of people, services, or information, were themselves *dis*abled.

It is important not to construct all older people affected by such disasters as helpless victims; many will show resilience. Nonetheless, the mobilities that may be triggered by such disasters are involuntary and most likely intersect with race and poverty. Mobility capital (in terms of 'resources' to move) is unevenly distributed.

Conclusion

I have tried to convey here that, as Ziegler and Schwanen put it: "mobility in later life is more than a means of accessing people and places in distant space, and more than a mere function of or resource for undertaking the activities of daily living" (Ziegler and Schwanen, 2011: 774). The mobilities literature has much to offer the social gerontologist, and I have only scratched the surface; there is a rich vein of material to be tapped. For those wishing to pursue a non- (or more-than-)representational theory perspective, there are gains to be had. Lorimer suggests that such an approach focuses on "how life takes shape and gains expression in shared experiences, everyday routines, fleeting encounters, embodied movements, precognitive triggers, practical skills, affective intensities, enduring urges, unexceptional interactions and sensuous dispositions" (Lorimer, 2005: 84). These everyday moments are all mobilities of one kind or another, and excavating the way they unfold among older adults can be illuminating. But we should not neglect the ways in which mobilities continue to be constrained by context, physical abilities and the material resources available to individual actors.

References

Adey, P. (2010). *Mobility*. London: Routledge.

Adey, P. and Bissell, D. (2010). Mobilities, meetings, and futures: An interview with John Urry. *Environment and Planning D*, 28(1), 1–16. http://dx.doi.org/10.1068/d3709

Andrews, G. J., Cutchin, M., McCracken, K., Phillips, D. R. and Wiles, J. (2007). Geographical gerontology: The constitution of a discipline, *Social Science & Medicine*, 65(1), 151–168. http://dx.doi.org/10.1016/j.socscimed.2007.02.047

Andrews, G. J., Milligan, C., Phillips, D. R. and Skinner, M. W. (2009). Geographical gerontology: Mapping a disciplinary intersection. *Geography Compass*, 3(5), 1641–1659. http://dx.doi.org/10.1111/j.1749-8198.2009.00270.x

Bailey, C, Foran, T. G., Ni Scanaill, C. and Dromey, B. (2011). Older adults, falls and technologies for independent living: A life space approach. *Ageing & Society*, 31(5), 829–848. http://dx.doi.org/10.1017/S0144686X10001170

Chudyk, A. M., Winters, M., Moniruzzaman, M., Ashe, M. C., Gould, J. S. and McKay, H. (2015). Destinations matter: The association between where older adults live and their travel behaviour. *Journal of Transport & Health*, 2(1), 50–57. http://dx.doi.org/10.1016/j.jth.2014.09.008

Cresswell, T. (1996). *In place/out of place: Geography, ideology, and transgression*. Minneapolis, MN: University of Minnesota Press.

Cresswell, T. (2006). *On the move: Mobility in the modern Western World*. London: Routledge.

Degeling, C. and Rock, M. (2012). 'It was not just a walking experience': Reflections on the role of care in dog-walking. *Health Promotion International*, 28(3), 397–406. http://dx.doi.org/10.1093/heapro/das024

Finlay, J., Franke, T., McKay, H. and Sims-Gould, J. (2015). Therapeutic landscapes and wellbeing in later life: Impacts of blue and green spaces for older adults. *Health & Place*, 34, 97–106. http://dx.doi.org/10.1016/j.healthplace.2015.05.001

Graham, M. (2015). From wandering to wayfaring: Reconsidering movement in people with dementia in long-term care. *Dementia*, 16(6), 732–749. Online first: 29 October. http://dx.doi.org/10.1177/1471301215614572

Hardill, I. and Olphert, C. W. (2012). Staying connected: Exploring mobile phone use amongst older adults in the UK. *Geoforum*, 43(6), 1306–1312. http://dx.doi.org/10.1016/j.geoforum.2012.03.016

Kaufmann, V. (2002). *Re-thinking mobility*. Farnham: Ashgate.

King, A. C., Sallis, J. F., Frank, L. D., Saelens, B. E., Cain, K., Conway, T. L., Chapman, J. E., Ahn, D. K. and Kerr, J. (2011). Aging in neighborhoods in walkability and income: Associations with physical activity and obesity in older adults. *Social Science & Medicine*, 73(10), 1525–1533. http://dx.doi.org/10.1016/j.socscimed.2011.08.032

Knight, S. and Edwards, V. (2008). In the company of wolves: The physical, social and psychological benefits of dog ownership. *Journal of Aging and Health*, 20(4), 437–455. http://dx.doi.org/10.1177/0898264308315875

Langan, J. C. and Palmer, J. L. (2012). Listening to and learning from older adult Hurricane Katrina survivors. *Public Health Nursing*, 29(2), 126–135. http://dx.doi.org/10.1111/j.1525-1446.2011.00996.x

Lorimer, H. (2005). Cultural geography: The busyness of being 'more-than-representa tional'. *Progress in Human Geography*, 23(1), 83–94. http://dx.doi.org/10.1191/0309132 505ph531pr

Mansvelt, J. (2015). Elders, in P. Adey, D. Bissell, Hannam, P. Merriman and M. Sheller (eds), *The Routledge handbook of mobilities* (pp. 398–408). London: Routledge.

Merriman, P. (2012a). *Mobility, space and culture*. London: Routledge.

Merriman, P. (2012b). Human geography without time-space. *Transactions of the Institute of British Geographers*, 37(1), 13–27. http://dx.doi.org/10.1111/j.1475-5561.2011.00455.x

Milligan, C., Gatrell, A. and Bingley, A. (2004). Cultivating health: Therapeutic landscapes and older people in northern England. *Social Science & Medicine*, 58(9), 1781–1793. http://dx.doi.org/10.1016/S0277-9536(03)00397-6

Milligan, C., Payne, S., Bingley, A. and Cockshott, Z. (2015). Place and wellbeing: Shedding light on activity interventions for older men. *Ageing & Society*, 35(1), 124–149. http://dx.doi.org/10.1017/S0144686X13000494

Mitra, R., Siva, H. and Kehler, M. (2015). Walk-friendly suburbs for older adults? Exploring the enablers and barriers to walking in a large suburban municipality in Canada. *Journal of Ageing Studies*, 35, 10–19. http://dx.doi:10.1016/j.jaging.2015.07.002

Mollenkopf, H., Hieber, A. and Wahl, H-W. (2011). Continuity and change in older adults' perceptions of out-of-home mobility over ten years: A qualitative-quantitative approach. *Ageing & Society*, 31(5), 782–802. http://dx.doi.org/10.1017/S0144686X10000644

Musselwhite, C., Holland, C. and Walker, I. (2015). The role of transport and mobility in the health of older people. *Journal of Transport and Health*, 2(1), 1–4. http://dx.doi. org/10.1016/j.jth.2015.02.001

O'Hern, S. and Oxley, J. (2015). Understanding travel patterns to support safe active transport for older adults. *Journal of Transport & Health*, 2(1), 79–85. http://dx.doi.org/10.1016/j. jth.2014.09.016

Orr, J. (2011). Lost pensioner Dennis Leighton spends 30 hours driving around M25. *The Telegraph*, 14 December. Available from: www.telegraph.co.uk/motoring/news/8956844/ Lost-pensioner-Dennis-Leighton-spends-30-hours-driving-around-M25.html

Peace, S., Holland, C. and Kellaher, L. (2011). 'Option recognition' in later life: Varia tions in ageing in place. *Ageing & Society*, 31(5), 734–757. http://dx.doi.org/10.1017/ S0144686X10001157

Rowles, G. D. (1978). *Prisoners of space? Exploring the geographical experiences of older people*. Boulder, CO: Westview Press.

Schwanen, T. and Ziegler, F. (2011). Wellbeing, independence and mobility: An introduc tion. *Ageing & Society*, 31(5), 719–773. http://dx.doi.org/10.1017/S0144686X10001467

Sheller, M. (2015). Sociology after the mobilities turn, in P. Adey, D. Bissell, Hannam, P. Merriman and M. Sheller (eds), *The Routledge handbook of mobilities* (pp. 45–54). London: Routledge.

Shoval, N., Wahl, H-W., Auslander, G., Isaacson, M., Oswald, F., Edry, T., Landau, R. and Heinik, J. (2011). Use of the global positioning system to measure the out-of-home mobility of older adults with differing cognitive functioning. *Ageing & Society*, 31(5), 848– 869. http://dx.doi.org/10.1017/S0144686X10001455

Siren, A., Hjorthol, R. and Levin, L. (2015). Different types of out-of-home activities and well-being amongst urban residing old persons with mobility impediments. *Journal of Transport & Health*, 2(1), 14–21. http://dx.doi.org/10.1016/j.jth.2014.11.004

Skinner, M. W., Cloutier, D. and Andrews, G. J. (2015). Geographies of ageing: Progress and possibilities after two decades of change. *Progress in Human Geography*, 39(6), 776–799. http://dx.doi.org/10.1177/0309132514558444

Sposaro, F., Danielson, J. and Tyson, G. (2010). *iWander: An Android application for dementia patients*, in 32nd Annual Conference International Conference of the IEEE EMBS Buenos Aires, Argentina, 31 August–4 September, pp. 3875–3878. http://dx.doi.org/10.1109/IEMBS.2010.5627669

Stjernborg, V., Wretstrand, A. and Tesfahuney, M. (2015). Everyday life mobilities of older persons – a case study of ageing in a suburban landscape in Sweden. *Mobilities*, 10(3), 383–401. http://dx.doi.org/10.1080/17450101.2013.874836

Thrift, N. (2008). *Non-representational theory: Space, politics, affect*, Abingdon: Routledge.

Urry, J. (2007). *Mobilities*. London: Sage.

Verghese, J., Lipton, R. B., Katz, M. J., Hall, C. B., Derby, C. A., Kuslansky, G., Ambrose, A. F., Sliwinski, M. and Buschke, H. (2003). Leisure activities and the risk of dementia in the elderly. *New England Journal of Medicine*, 348(25), 2508–2516. http://dx.doi.org/10.1056/NEJMoa022252

Whitaker, E. D. (2005). The bicycle makes the eyes smile: Exercise, aging, and psychophysical well-being in older Italian cyclists. *Medical Anthropology*, 24(1), 1–43. http://dx.doi.org/10.1080/01459740590905633

Wiles, J. L, Allen, R.E.S., Palmer, A. J., Hayman, S., Keeling, S. and Kerse, N. (2009). Older people and their social spaces: A study of well-being and attachment to place in Aotearoa New Zealand. *Social Science & Medicine*, 68(4), 664–671. http://dx.doi.org/10.1016/j.socscimed.2008.11.030

Winters, M., Sims-Gould, J., Franke, T. and McKay, H. (2015). "I grew up on a bike": Cycling and older adults. *Journal of Transport and Health*, 2(1), 58–67. http://dx.doi.org/10.1016/j.jth.2014.06.001

Ziegler, F. and Schwanen, T. (2011). 'I like to go out to be energised by different people': An exploratory analysis of mobility and wellbeing in later life. *Ageing & Society*, 31(5), 758–781. http://dx.doi.org/10.1017/S0144686X10000498

20 Constructions of old-age social exclusion

In place and shaped by place

Kieran Walsh

Introduction

As testified by a number of the contributing chapters to this text, the intertwinement of the older adult life course and the geographic concept of place is multifaceted and multilevel. Multifaceted, in terms of the variety of different ways that such connections occur over the lifetime of an older person. Multilevel, in terms of the scalar contexts, such as a person's home, local places, and national and global settings that it can implicate. It is this intertwinement that renders questions around what makes a place exclusionary or indeed inclusionary[1] in later life so difficult to answer.

In this chapter I explore the role of place in shaping the social exclusion of older people I draw on the findings of a scoping review of the international literature on old-age social exclusion (Walsh, Scharf and Keating, 2017) as a foundation for the analysis. The chapter is situated amidst broader knowledge deficits concerning the exclusion of older adults and well-established conceptual debates on the ambiguous and sometimes problematic nature of the social exclusion concept itself. The analysis presented here builds on traditional concerns with disadvantage in social gerontology and the long-standing focus on geographies of inclusion and exclusion in human and health geography (e.g., Cameron, 2006; Wood, 2015).

Old-age exclusion and understandings of place

In a practical way, the importance of place in old-age exclusion is reinforced by the behavioural pattern of older people spending more time in their immediate environment (Buffel, 2012; Rowles, 1978), and a capability trajectory where age-related declines can mean older people are more susceptible to environmental forces (Golant, 2011; Wahl, Iwarsson and Oswald, 2012). To help position the concept of place within current knowledge, we can describe old-age exclusion as involving

> interchanges between multi-level risk factors, processes and outcomes. Varying in form and degree across the older adult life course, its complexity,

impact and prevalence is amplified by accumulated group-related disadvantage, old-age vulnerabilities, and constrained opportunities to ameliorate exclusion. Old-age exclusion leads to inequities in choice and control, resources and relationships, and power and rights in key domains of: neighbourhood and community; services, amenities and mobility; material and financial resources; social relations; socio-cultural aspects of society; and civic participation. Old-age exclusion implicates states, societies, communities and individuals.

(Walsh, Scharf and Keating, 2017: 93)

Place is, as such, operationalized differently within many contemporary discourses – in this instance being linked to both inequities with respect to neighbourhood and community (lines 5–7) and the process of exclusion itself (line 9). To some degree this reflects the various disciplinary treatments of place within ageing research (Skinner, Cloutier and Andrews, 2015).

Just as Andrews, Evans and Wiles (2013) note for gerontology, in general, place within the old-age exclusion literature is recognised as possessing basic agency, which either assists older people in their lives or serves to produce challenges to overcome. Agency itself is a central tenet in exclusion, symbolising the key act of exclusion (Atkinson, 1998). It is just a question of the extent to which the agency of place is conceptually detailed within exclusion discourses. This serves as a critical nuance when considering place in the ageing and exclusion literature.

Neighbourhood context studies represent an extensive body of work that contributes to the broad evidence base on old-age exclusion and place. However, in such literature, place is viewed mainly as a contextual construct. Research, primarily quantitative, focuses on the consequences of different place characteristics for health and well-being outcomes (Buys *et al.*, 2015; Marquet and Miralles-Guasch, 2015). Therefore, while incorporating understandings akin to environmental gerontology, of how place can influence older adult lives, these studies do not feature extensive interpretations of exclusionary relationships (Walsh, Scharf and Keating, 2017). Nor do they always explore in detail how the agency of place reflects wider societal and contextual factors.

Where its role in exclusion is considered in more depth, the focus is typically on place in multidimensional terms. Work in this area conceives place as encapsulating the following: spatial characteristics, such as location (Dwyer and Hardill, 2011); physical and built environment elements, such as housing and urban development (Keene and Ruel, 2013); localised service infrastructure, such as health and transport services (Ryvicker, Gallo and Fahs, 2012); social and cultural aspects, such as relational communities (Vitman, Iecovich and Alfasi, 2014; Walsh and Gannon, 2011); and crime (De Donder, Verte and Messelis, 2005). Within this literature, place is usually categorised across the urban–rural continuum with studies on old-age exclusion dominated by urban perspectives (Grant *et al.*, 2010) but with growing consideration given to rural areas (Milbourne and Doheny, 2012). Research has considered localised settings, such as neighbourhoods and communities; broader place manifestations, such as cities and rural regions; and the intertwinement of both with macroenvironments.

Cross-cutting these dimensions, a number of studies consider place as involving more intimate processes (Andrews *et al.*, 2013). This research charts a reciprocal, mutually reinforcing existence between place and older people, acknowledging the agency of both. It explores how exclusion impacts and is shaped by the subjective and symbolic meaning wrapped up in place-based roles, attachments, cultural narratives and identity (Walsh, 2016; Ziegler, 2012). Pointing to more fluid aspects, work on exclusion and ageing has also conceived place as a set of dynamic relationships, highlighting linkages to external processes, such as globalisation, and other scalar levels of place and power (Phillipson, 2007). Thus, place, and its role in exclusion, is not isolated from broader macro socio-economic factors, nor is it always treated as a bounded concentration of risk and responsibility (Cameron, 2006).

These various dimensions of place are also reflected in the formal conceptual frameworks on old-age exclusion, which detail the interconnected domains of life where older people experience exclusion (e.g., Barnes *et al.*, 2006/Kneale, 2012[2]; Feng, 2012; Guberman and Lavoie, 2004; Jehoel-Gijsbers and Vrooman, 2008; Scharf and Bartlam, 2008; Scharf, Phillipson and Smith, 2005; Walsh, O'Shea and Scharf, 2012; Walsh, Scharf and Keating, 2017). Generally, the degree of conceptual elaboration on place is still relatively limited. Yet, several frameworks are place- and context-specific: Scharf *et al.* (2005) focus on deprived urban areas in England, Scharf and Bartlam (2008) on rural areas in England, Feng (2012) on urban and rural regions in China and Walsh, O'Shea and Scharf (2012) on different kinds of rural communities in Ireland and Northern Ireland.

While the vast majority of conceptualisations recognise place as being involved in the exclusion of older people,[3] the extent of the role that place fulfils is conceived differently. In the main, existing conceptual frameworks represent place as a domain of exclusion (e.g., Barnes *et al.*, 2006; Kneale, 2012; Scharf *et al.*, 2005), that is, *place-based exclusion*. This, however, involves more than just recognising place as a location of disadvantage. As with the earlier definition, place is acknowledged as one of the core domains (like social relations or material and financial resources) of an older person's life within which he or she can experience exclusion. A small number of frameworks go further, expounding on the capacity of place to mediate multiple forms of exclusion, that is, *place-mediated exclusion*. Scharf and Bartlam (2008), for instance, emphasise changes within rural communities as a significant pathway to exclusion for older rural residents across different areas of life. Walsh O'Shea and Scharf (2012) demonstrate that geographic and natural elements, local population processes and cohesion, and life-course feelings of attachment and belonging can, in combination with other factors, mediate age-related exclusion in rural areas. These characteristics can protect against or intensify exclusionary experiences in different life domains.

Thus, there is a subtle but crucial distinction in how place is conceptualised in formal frameworks. As a domain, the characteristics of place, and the factors that influence those characteristics, shape exclusionary experiences of place. However, as a mediator of exclusion, place functions as a fundamental determinant of exclusionary experiences in old age, not just in relation to place but across multiple areas of life.

The interconnections between different domains, and the acknowledgement that an outcome of one exclusionary process can function as a part of other processes, means that many conceptual frameworks implicitly recognise how one form of exclusion contributes to other forms. Nevertheless, this distinction between domains and mediators is critical. It is this differential positioning that is explored in subsequent sections.

Place as a domain of exclusion

The place exclusion domain reflects an operationalisation of the different dimensions of place within the exclusion process. As such, place-based exclusion can, in itself, be broken down into a series of dimensions, with many interrelated.

First, *embedded services, amenities and the built environment* encompass exclusion from or as a result of services located in and delivered into place (e.g., health clinics; transport) and the physical infrastructure of a place (e.g., housing; footpaths). Exclusion within this dimension can be intertwined with geographic characteristics. For instance, spatial isolation can cement the importance of embedded services – as demonstrated by Dwyer and Hardill (2011) for rural older dwellers. Geographic elements can also combine with differential resourcing across different places to construct spatial disparities in provision – such as that illustrated in relation to home care services (Grenier and Guberman, 2009). Central to many analyses, are concerns with transport. Deficient access to public transport systems has, in some cases, been found to exacerbate placed-based service issues and, in others, to be the root cause (Shergold and Parkhurst, 2012). Public service re-structuring, in the form of service retrenchment, is another central theme and is highlighted as a pervasive policy underlying exclusion in these aspects of place (Walsh, Scharf and Shucksmith, 2014). However, exclusion in this dimension can occur in more subtle ways. Grenier and Guberman (2009) describe how home care policies and practices transformed the homes of older people into professional care environments, disembedding them from their private contexts.

Second, *place socio-economic aspects* describe the susceptibility of older residents to exclusion when living in spatial concentrations of structural disadvantage. Focusing primarily on measures of material and social deprivation in local contexts, research demonstrates the intersection of this dimension of place-based exclusion with a range of other forms of disadvantage. Older people living in deprived urban settings have, for instance, been found to experience multiple risk factors of exclusion (Scharf *et al.*, 2002; Scharf *et al.*, 2005). Similarly, concentrations of lower-income older people in rural communities are considered more vulnerable to disadvantage in general (Moffat and Glasgow, 2009) and more likely to experience place-based exclusion (Milbourne and Doheny, 2012). Spatial concentrations of poverty have also been linked to barriers in health care utilisation and negative health outcomes (Hossen and Westhues, 2010).

Third, *social and relational aspects* of place-based exclusion centres on exclusion arising from relational communities-in-place. Encompassing social and cultural participation, social capital, social cohesion and feelings of safety, a number of studies describe the changing nature of place and its impacts on the local relational environment. Walker *et al.* (2013) and Walsh *et al.* (2012) show how transformative shifts in rural areas (e.g., population shifts; changes in socialisation) can, in part, threaten older people's capacity to engage socially. Walker and Hiller (2007) noted concerns for diminishing relationships with neighbours and their impact on neighbourhood trust and reciprocity. Other work raises associations between deprived neighbourhood contexts and exclusion from sociocultural aspects. Krause (2006) found that living in run-down areas was linked to more negative social interactions, whereas Stoeckel and Litwin (2015) identified how a sense of connectedness with neighbours was positively associated with life satisfaction for older residents in deprived settings. A number of studies also tap into the psycho-social perspectives within this dimension, factoring in older people's subjective perceptions of place. This includes the work of Buffel *et al.* (2013a), who highlight the role of neighbourhood perceptions in overall social participations levels; Shiovitz-Ezra (2015), who links reduced loneliness to feeling a part of the local area; and De Donder, Verté and Messelis (2005), who show increased neighbourhood involvement is associated with decreased fear of crime.

Fourth, *socio-political power structures* within older people's neighbourhoods and communities point to how places can be politically performed as exclusionary instruments. Work on this dimension is dominated by studies on the alienation of older people from local-level decision-making processes. This includes research in relation to the execution of environmental rights in Scottish neighbourhoods (Day, 2010), the disempowering impact of water management tariffs on Australian older rural dwellers (Rogers *et al.*, 2015) and the lack of opportunities for older people to influence service provision in deprived English neighbourhoods (Abbott and Sapsford, 2005). The international literature also includes papers that highlight the marginalisation of local places or regions from the distribution of power. Grant *et al.* (2010) found differences in the capacity to influence walkability between higher- and lower-socio-economic neighbourhoods. Whereas, Giarchi (2006) highlighted differential levels of power in the interplay between urban and rural contexts in Europe.

Fifth, *place-based policy* can generate exclusion that impacts on how the lives of older people in place are positioned and recognised. In some cases, this relates to older people being considered marginal to policy development and implementation, as noted by Milne *et al.* (2007) for rural dwelling older adults. In other cases, it relates to the tendency of local policy, or broader regional approaches, to categorise and frame particular contexts as productive or non-productive places. In recent years, this aspect of research has been dominated by critical perspectives on age-friendly city and community programmes. Scharlach and Lehning (2013) note that while age-friendly initiatives have the potential to foster the social inclusion of older people, they can also reflect a residual social welfare approach, exacerbating existing disparities. Others have highlighted how age-friendly policy can fail to address the needs of older adults in changing environments (Keating, Eales and Phillips, 2013) or to support adaptive practices in places that enhance the lives of older people (Walsh *et al.*, 2014). Research in this area also demonstrates how many manifestations of age-friendly programmes appear to ignore the influence of macro forces, such as urban and rural transformation and global economic conditions (Buffel and Phillipson, 2016; Walsh, 2015).

Place as a mediator of exclusion

There are a number of mechanisms that transform place into a key mediator of exclusionary experiences in later life. These mechanisms can emerge from place or stem from older people's relationship with place and can both serve to protect individuals from disadvantage or intensify exclusion. Many of the mechanisms are embedded within the descriptions contained in the previous section. The focus here, however, is on how they impact overall multidimensional disadvantage and life domains outside of immediate place contexts. This chapter will focus on two mechanisms in particular.

First, neighbourhood and community change is an integral part of a place functioning as a mediator of old-age exclusion. Interconnected sets of forces operate at different scalar levels to produce changes in local settings and older people's position within these settings (Buffel, Phillipson and Scharf, 2013b; Walsh, 2015). Place-based factors can underlie such changes, but are often intertwined with broader macroeconomic and social trends related to globalisation. These include increased mobility, the growth of city economies, and the influence of international policy institutions, privatisation and risk individualisation (Phillipson, 2015; Warburton, Scharf and Walsh, 2016). Phillipson (2007) notes the potential for globalisation to fragment the experience of place for older people, disembedding individuals from their own localities. The recent global economic crisis, and the implementation of austerity programmes, has intensified these trends further, having a direct impact on the social and economic structures of older adults' neighbourhoods and communities (Walsh, 2015).

Research has identified how these processes can dilute connections with place for older people and even contribute to a sense of dislocation. In a fundamental way, population change, in the form of demographic ageing, general turnover and more intense population churn, can result in compositional shifts in places (Burns, Lavoie and Rose, 2012; Keating *et al.*, 2013). For rural areas, this can, for example, be driven by lower employment and higher outward migration as a result of the contraction of the agricultural sector and other resource-based industries, such as mining and forestry (Skinner and Hanlon, 2016; Walsh, O'Shea and Scharf, 2016). It can also involve counterurbanisation flows, such as commuter populations and retirement immigration. Walsh *et al.* (2012) highlighted how, in combination with demographic change, profound economic, social and cultural transformations can contribute to reduced health and social care services, the decline in the availability of support networks and downward pressure on reciprocity in some rural settings.

In contrast, much of the research on old-age exclusion in urban areas shows how urban development and urban renewal processes can create new demographic, socio-economic and cultural landscapes in older people's locales (Buffel and Phillipson, 2016). In an analysis of ageing in working-class Montreal neighbourhoods, Burns *et al.* (2012) noted the role of gentrification in three different forms of exclusion. Exclusion from services arose from the redevelopment of commercial shops and the loss of key institutions, such as churches. Sociopolitical exclusion was evident in terms of increased marginalisation from local decision making and social relations exclusion arose from the displacement of social spaces. Similarly, Lager, Van Hoven and Huigen (2013) found that older people in lower-socio-economic areas in the Netherlands did not experience the expected benefits from urban renewal in middle-class areas. Temelová and Slezáková (2014) note that other forms of cultural transformation, namely the transition from a post-socialist society in the Czech Republic, produced risks related to facilities decline and a breakdown of communities for older people living in social housing.

Sometimes, of course, these changes, can help sustain older people's places and contribute to new levels of vibrancy within their localities. Galčanová and Sýkorová (2015) show how reconstruction and renovation in neighbourhoods in three cities in the Czech Republic have produced enabling outcomes, creating more livable environments for senior citizens. Places, and older residents, can also function as interpreters and manipulators of change and macro processes. A growing body of international research on ageing resource-dependent communities illustrates how local contexts can unearth new pathways of diversification and voluntary innovation in response to contracting local industries in order to support healthy ageing (Gjertsen, Ryser and Haleth, 2016). In an examination of the 'age-friendly community' during austerity conditions in Ireland, Walsh (2015) notes how the global age-friendly movement also appears to have been co-opted by local communities to strengthen the culturally valued link between people and places.

Second, older people's life-course relationship with place is increasingly recognised as determining how place can influence trajectories of exclusion in later life and is considered to be laden with subjective and symbolic meaning (Wiles *et al.*, 2012). First, and reflecting associations between the life-course and disadvantage, the past life events of older residents are thought to influence the level of exclusion-in-place (Scharf and Bartlam, 2008). Often these events are grounded in a local context or become interwoven with experiences of place to shape lives in neighbourhoods and communities. For example, Scharf *et al.*

(2005) note how individuals' low socio-economic status preceding retirement, the continuation of long-standing difficult social relationships and particular life experiences or age-related losses conspired to initiate exclusion for many older people in disadvantaged neighbourhoods. Similarly, Walsh, O'Shea and Scharf (2012) highlight how some rural dwelling individuals have simply experienced more fragmented lives than others; with transitions around ill health and dependency, bereavement and loss and ageing itself, acting in singular or cumulative terms to construct risks of exclusion. Ziegler (2012), in an intersectional life-course analysis of older women in disadvantaged urban areas, found that lifelong patterns of, and opportunities for, social interaction shape older people's relational practices in later life.

Second, the extent to which older people feel a sense of belonging in place can inform perceived levels of exclusion. Wrapped up in this relationship are ideas of how place-based identity frame a relative prioritisation. Scharf *et al.* (2003) found that despite living in some of England's most-deprived urban neighbourhoods, older people demonstrated a long-standing commitment to their communities. The authors note that, in part, this was likely to reflect the desire of some older participants to maintain a sense of identity in a changing urban environment. Russell, Hill and Basser (1998) suggest that local places can provide a meaningful context for the everyday lives of older residents, even in the face of potential environmental exclusions. Older urban dwellers were found to bolster their identity as autonomous individuals through interactions with challenging aspects of inner-city environments (traffic issues, pedestrian safety). Walsh, O'Shea and Scharf (2012) observed how elements of the aesthetic and workable landscape, being native to an area, and efforts invested in place could create multiple kinds of attachments and, ultimately, a sense of home, in rural communities. This led some older people to place less significance on the impact of material and structural exclusion. For other individuals – typically more recent arrivals – feelings of dislocation, and a sense that home was elsewhere, functioned to exacerbate exclusionary experiences.

The broader literature on ageing in place and 'home' that is typically regarded outside of the scholarly discourse on old-age exclusion is also relevant here. For instance Rowles (1983), and as expanded by him in this volume (Chapter 16), suggests that older, long-term residents may experience a physical, social and autobiographical 'insideness' in place, fostering close links between person and place over the life course. The notion of 'place integration' (e.g., Cutchin, 2003) indicates a continuous set of processes involved in this relationship, highlighting how personal and place changes can either lead to relocation or to the remaking of meaning and identity in place. The residential life-course trajectory (Rowles, 2008), and the personal narrative of place that is constructed across time and across multiple residential locations, is of central importance here. The work of Cutchin, for example, illustrates how place-making (an active process of integration and creation of meaning in place) during residential transitions can help older people enhance personal well-being (see Chapter 17). This reinforces the notion of older people as active agents in place-making and their relationship with place (Buffel *et al.*, 2013b). Tentatively, then, older people are also likely to be active in determining how place can mediate disadvantage and circumvent risks of exclusion. This might involve their conscious efforts, their subconscious actions, or their accumulated expectations and sense of place and home developed over the life course.

Intertwined trajectories: future directions

In this chapter, I set out to explore the role of place in shaping exclusionary experiences in later life. In doing so, I hoped to establish the significance of old-age exclusion as a topic of focus for empirical and conceptual study within geographical gerontology. I also hoped to illuminate how place is operationalised within existing scholarly work on old-age exclusion, and push beyond any notions of place as a simple location of disadvantage.

Whether presented within the general ageing and exclusion literature, or within formal conceptual frameworks, the depth of understanding around the intersection of place, ageing and exclusion is still underdeveloped. The limitations in how we conceive place in processes of late-life disadvantage are matched and compounded by the limitations in our understanding of the old-age exclusion construct. The analysis offered here must be framed in this broader set of scholarly circumstances. It must also be framed within the context of not having sufficient scope to explore, in depth, work from human geography that, while may not concentrate on gerontological questions, offers significant insight into geographies of exclusion.

This chapter has shown how the associations between old-age exclusion and place can be considered both in terms of place-based exclusion and place-mediated exclusion. Thinking about place as a domain where an older person can be disadvantaged helps illustrate the weight of local settings in older people's everyday lives. Thinking about place as a mediator of exclusion demonstrates how place can function as a fundamental force influencing multidimensional experiences of old-age social exclusion.

We need, however, to also consider how the older adult place relationship – particularly in terms of place-mediated exclusion – is a dominant influence in the intersection of place, ageing and disadvantage. This link is implicit within much of the material presented in this analysis, and explicit at a number of points throughout the chapter. It can certainly be argued that the older adult place relationship, in itself, shapes the level of exclusion experienced by an older person in place, and the degree to which place protects against or intensifies disadvantage. Thus, experiences of exclusion involving place are not just a sole function of place and external forces of change, and their agency. Instead, and reflecting work not specific to exclusion (e.g., Rowles, Chapter 16, and Cutchin, Chapter 17, in this volume), they are likely to also be influenced by how people interact with, and exercise their agency in relation to, the physical and infrastructural, the social and cultural and the experiential aspects of place. Interrogating the construction of old-age exclusion in this way may help to provide a more appropriate platform, from which to disentangle how the intertwinement of the older adult life course and place functions to inform exclusionary experiences in later life.

Intriguing associations are only now beginning to emerge from the inter-relationship of place, exclusion and critical life-course rupture points in old age. Whether these ruptures concern transitions such as those arising from bereavement, the onset of chronic ill health or, indeed, forced migration, they illustrate the spatial as well as temporal significance of life-course and exclusion issues. We can then think about the ways in which place might be involved in exclusionary processes arising from such critical events in later life. Figure 20.1 presents the potential associations between place, exclusion and such ruptures. Here, place is considered both as one of a number of exclusion domains and

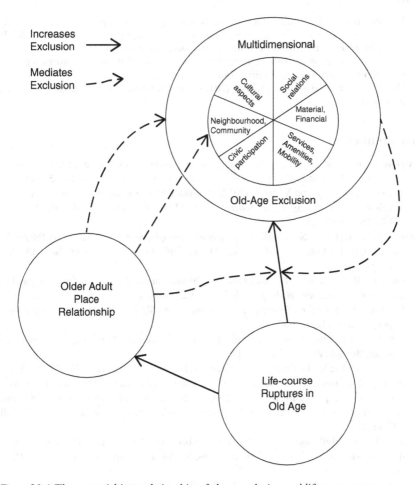

Figure 20.1 The potential interrelationship of place, exclusion and life-course ruptures

as an exclusion mediator represented within the older adult place relationship. Life-course ruptures may, for example, not only give rise to multidimensional forms of old-age exclusion but also may disrupt an older person's relationship with place. Conversely, the older adult relationship with place may not only mediate multidimensional exclusion, in general, but also that arising specifically from these ruptures.

Ultimately, further empirical and conceptual research on place and old-age exclusion is critically needed. Without it, we will always run the risk of simplifying the role of place in the exclusion and inclusion of older adults. We will also run the risk of failing to fully understand how older people's life-course interchanges with place can influence trajectories of social exclusion amongst older people.

Notes

1 Despite notable efforts, and ethical scholarly reasoning, to conceptualise social inclusion (e.g Cameron, 2006), it remains an under-theorised concept (Scharf and Keating, 2012). Thus, the focus in this chapter is primarily on social exclusion.
2 The use of the forward slash between references denotes that the later works represent extensions of the former as original conceptualisations.
3 Despite its empirical grounding, Feng (2012) does not formally acknowledge place-based exclusion.

References

Abbott, P. and Sapsford, R. (2005). Living on the margins: Older people, place and social exclusion. *Policy Studies*, 26(1), 29–46. http://dx.doi.org/10.1080/01442870500041660

Andrews, G., Evans, J. and Wiles, J. (2013). Re-placing and re-spacing gerontology: Relationality and affect. *Ageing & Society*, 33(8), 1339–1373. http://dx.doi.org/10.1017/S0144686X12000621

Atkinson, A. B. (1998). *Social exclusion, poverty and unemployment*. London: Center for Analysis of Social Exclusion (CASE) paper #4. Available from: http://sticerd.lse.ac.uk/dps/case/cp/Paper4.pdf

Barnes, M., Blom, A., Cox, K. and Lessof, C. (2006). *The social exclusion of older people: Evidence from the first wave of the English Longitudinal Study of Ageing (ELSA): Final Report*. London: Office for the Deputy of Prime Minister. Available from: www.ifs.org.uk/docs/odpm_social_exclusion.pdf

Buffel, T. (2012). *Experiences of place and neighbourhood in later life: Developing age-friendly communities*. Brussels:Vrije Universiteit Brussel.

Buffel, T., De Donder, L., Phillipson, C., Dury, S., De Witte, N. andVerte, D. (2013a). Social participation among older adults living in medium-sized cities in Belgium: The role of neighbourhood perceptions. *Health Promotion International*, 29(4), 655–668. All versions of the article are available from: http://heapro.oxfordjournals.org/content/29/4/655.full.pdf+html.

Buffel, T. and Phillipson, C. (2016). Can global cities be 'age-friendly cities'? Urban development and ageing populations. *Cities*, 55, 94–100. http://dx.doi.org/10.1016/j.cities.2016.03.016

Buffel, T., Phillipson, C. and Scharf, T. (2013b). Experiences of neighbourhood exclusion and inclusion among older people living in deprived inner-city areas in Belgium and England. *Ageing & Society*, 33(Special Issue 1), 89–109. http://dx.doi.org/10.1017/S0144686X12000542.

Burns, V. F., Lavoie, J-P. and Rose, D. (2012). Revisiting the role of neighbourhood change in social exclusion and inclusion of older people. *Journal of Aging Research*, 2012, 12pp. http://dx.doi.org/10.1155/2012/148287

Buys, D. R., Howard, V. J., McClure, L. A., Buys, K. C., Sawyer, P., Allman, R. M. and Levitan, E. B. (2015). Association between neighborhood disadvantage and hypertension prevalence, awareness, treatment, and control in older adults: Results from the University of Alabama at Birmingham Study of Aging. *American Journal of Public Health*, 105(6), 1181–1188. http://dx.doi.org/10.2105/AJPH.2014.302048

Cameron, A. (2006). Geographies of welfare and exclusion: Social inclusion and exception. *Progress in Human Geography*, 30, 396–404.

Cutchin, M. P. (2003). The process of mediated aging-in-place: A theoretically and empirically based model. *Social Science &Medicine*, 57(6), 1077–1090. http://dx.doi.org/10.1016/S0277-9536(02)00486-0

Day, R. (2010). Environmental justice and older age: Consideration of a qualitative neighbourhood-based study. *Environment and Planning A*, 42(11), 2658–2673. http://dx.doi.org/10.1068/a43109

De Donder, L., Verté, D. and Messelis, E. (2005). Fear of crime and elderly people: Key factors that determine fear of crime among elderly people in West Flanders. *Ageing International*, 30(4), 363–376. http://dx.doi/org/10.1007/s12126-005-1021-z

Dwyer, P. and Hardill, I. (2011). Promoting social inclusion? The impact of village services on the lives of older people living in rural England. *Ageing & Society*, 31, 243–264. http://dx.doi.org/10.1017/S0144686X10000851

Feng, W. (2012). Social exclusion of the elderly in China: One potential challenge resulting from the rapid population ageing, in C. Martinez-Fernandez, N. Kubo, A. Noya and T. Weyman (eds), *Demographic change and local development: Shrinkage, regeneration and social dynamics* (pp. 221–230). Paris: OECD.

Galčanová, L. and Sýkorová, D. (2015). Socio-spatial aspects of ageing in an urban context: An example from three Czech Republic cities. *Ageing & Society*, 35(6), 1200–1220. http://dx.doi.org/10.1017/S0144686X14000154

Giarchi, G. G. (2006). Older people 'on the edge' in the countrysides of Europe. *Social Policy and Administration*, 40(6), 705–721. http://dx.doi.org/10.1111/j.1467-9515.2006.00528.x

Gjertsen, T., Ryser, L. and Haleth, G. (2016). Gamvik, 'a good place to grow old': The role of voluntary organisations in an ageing resource-dependent municipality in northern Norway, in M. Skinner and N. Hanlon (eds), *Ageing resource communities: New frontiers of rural population change, community development and voluntarism* (pp. 59–73). London and New York: Routledge.

Golant, S. M. (2011). The quest for residential normalcy by older adults: Relocation but one pathway. *Journal of Aging Studies*, 25(3), 193–205. http://dx.doi.org/10.1016j.jaging.2011.03.003

Grant, T. L., Edwards, N., Sveistrup, H., Andrew, C. and Egan, M. (2010). Inequitable walking conditions among older people: Examining the interrelationship of neighbourhood socio-economic status and urban form using a comparative case study. *BMC Public Health*, 10, 16pp. http://dx.doi.org/10.1186/1471-2458-10-677

Grenier, A. M. and Guberman, N. (2009). Creating and sustaining disadvantage: The relevance of a social exclusion framework. *Health and Social Care in the Community*, 17, 116–124. http://dx.doi.org/10.1111/j.1365-2524.2007.00804.x

Guberman, N. and Lavoie, J. P. (2004). *Equipe vies: Framework on social exclusion*. Montréal, QC: Centre de recherche et d'expertise de gérontologie sociale – CAU/CSSS Cavendish.

Hossen, A. and Westhues, A. (2010). A socially excluded space: Restrictions on access to health care for older women in rural Bangladesh. *Qualitative Health Research*, 20(9), 1192–1201. http://dx.doi.org/10.1177/1049732310370695

Jehoel-Gijsbers, G. and Vrooman, C. (2008). *Social exclusion of the elderly: A comparison study of EU member states*. Centre for European Policy Studies, European Network of Policy Research Institutes [ENEPRI]. Available from: http://aei.pitt.edu/9443/2/9443.pdf

Keating, N., Eales, J. and Phillips, J. E. (2013). Age-friendly rural communities: Conceptualizing 'Best-Fit'. *Canadian Journal on Aging*, 32(4), 319–332. http://dx.doi.org/10.1017/S0714980813000408

Keene, D. E. and Ruel, E. (2013). "Everyone called me grandma": Public housing demolition and relocation among older adults in Atlanta. *Cities*, 35(April), 359–364. http://dx.doi.org/10.1016/j.cities.2012.10.011

Kneale, D. (2012). *Is social exclusion still important for older people?* The International Longevity Centre – UK Report. Available from: www.ilcuk.org.uk/index.php/publications/publication_details/is_social_exclusion_still_important_for_older_people

Krause, N. (2006). Neighborhood deterioration, social skills, and social relationships in late life. *International Journal of Aging & Human Development*, 62(3), 185–207. http://dx.doi.org/10.2190/7PVL-3YA2-A3QC-9M0B

Lager, D., Van Hoven, B. and Huigen, P.P.P. (2013). Dealing with change in old age: Negotiating working-class belonging in a neighbourhood in the process of urban renewal in the Netherlands. *Geoforum*, 50(2013), 54–61. http://dx.doi.org/10.1016/j.geoforum.2013.07.012

Marquet, O. and Miralles-Guasch, C. (2015). Neighbourhood vitality and physical activity among the elderly: The role of walkable environments on active ageing in Barcelona, Spain. *Social Science & Medicine*, 135, 24–30. http://dx.doi.org/10.1016/j.socscimed.2015.04.016

Milbourne, P. and Doheny, S. (2012). Older people and poverty in rural Britain: Material hardships, cultural denials and social inclusions. *Journal of Rural Studies*, 28(4), 389–397. http://dx.doi.org/10.1016/j.jrurstud.2012.06.007

Milne, A., Hatzidimitriadou, E. and Wiseman, J. (2007). Heath and quality of life among older people in rural England: Exploring the impact and efficacy of policy. *Journal of Social Policy*, 36(3), 477–495. http://dx.doi.org/10.1017/S0047279407001055

Moffatt, S. and Glasgow, N. (2009). How useful is the concept of social exclusion when applied to rural older people in the United Kingdom and the United States? *Regional Studies*, 43, 1291–1303. http://dx.doi.org/10.1080/00343400903002697

Phillipson, C. (2007). The 'elected' and the 'excluded': Sociological perspectives on the experience of place and community in old age. *Ageing & Society*, 27(3), 321–342. http://dx.doi.org/10.1017/S0144686X06005629

Phillipson, C. (2015). Austerity policies and new forms of solidarity, in K. Walsh, G. Carney and Á. Ní Léime (eds), *Ageing through austerity: Critical perspectives from Ireland* (pp. 145–153). Bristol: Policy Press.

Rogers, M., Winterton, R., Warburton, J. and O'Keefe, S. (2015). Water management and healthy ageing in rural Australia: Economic, social, and cultural considerations. *Environment and Behavior*, 47(5), 551–569. http://dx.doi.org/10.1177/0013916513502355

Rowles, G. D. (1978). *Prisoners of space? Exploring the geographical experience of older people* Boulder, CO: Westview Press.

Rowles, G. D. (1983). Place and personal identity in old age: Observations from Appalachia. *Journal of Environmental Psychology*, 3(4), 299–313. http://dx.doi.org/10.1016/S0272-4944(83)80033-4

Rowles, G. D. (2008). Place in occupational science: A life course perspective on the role of environmental context in the quest for meaning. *Journal of Occupational Science*, 15(3), 127–135. http://dx.doi.org/10.1080/14427591.2008.9686622

Russell, C., Hill, B. and Basser, M. (1998). Older people's lives in the inner city: Hazardous or rewarding? *Australian and New Zealand Journal of Public Health*, 22(1), 98–106. http://dx.doi.org/10.1111/j.1467-842X.1998.tb01152.x

Ryvicker, M., Gallo, W. T. and Fahs, M. C. (2012). Environmental factors associated with primary care access among urban older adults. *Social Science & Medicine*, 75(5), 914–921. http://dx.doi.org/10.1016/j.socscimed.2012.04.029

Scharf, T. and Bartlam, B. (2008). Ageing and social exclusion in rural communities, in N. Keating (ed.), *Rural ageing: A good place to grow old?* (pp. 97–108). Bristol: The Policy Press.

Scharf, T. and Keating, N. (2012). Social exclusion in later life: A global challenge, in T. Scharf and N. Keating (eds), *From exclusion to inclusion in old age: A global challenge* (pp. 1–16). Bristol: The Policy Press.

Scharf, T., Phillipson, C. and Smith, A. (2003). Older people's perceptions of the neighbourhood: Evidence from socially deprived urban areas. *Sociological Research Online*, 8(4). http://dx.doi.org/10.5153/sro.867

Scharf, T., Phillipson, C. and Smith, A. E. (2005). Social exclusion of older people in deprived urban communities of England. *European Journal of Ageing*, 2(2), 76–87. http://dx.doi.org/10.1007/s10433-005-0025-6

Scharf, T., Phillipson, C., Smith, A. and Kingston, P. (2002). *Growing older in socially deprived areas: Social exclusion in later life*. London: Help the Aged, 124pp. Available from: www.ageuk.org.uk/documents/en-gb/for-professionals/communities-and-inclusion/id2255_a_growing_older_in_socially_deprived_areas_social_exclusion_in_later_life_2002_pro.pdf?dtrk=true

Scharlach, A. E. and Lehning, A. J. (2013). Ageing-friendly communities and social inclusion in the United States of America. *Ageing & Society*, 33(1), 110–136. http://dx.doi.org/10.1017/S0144686X12000578

Shergold, I. and Parkhurst, G. (2012). Transport-related social exclusion amongst older people in rural Southwest England and Wales. *Journal of Rural Studies*, 28(4), 412–421. http://dx.doi.org/10.1016/j.jrurstud.2012.01.010

Shiovitz-Ezra, S. (2015). Loneliness in Europe: Do perceived neighbourhood characteristics matter? in A. Börsch-Supan, T. Kneip, H. Litwin, M. Myck and G. Weber (eds), *Ageing in Europe – supporting policies for an inclusive society* (pp. 169–178). Berlin: de Gruyter.

Skinner, M. W., Cloutier, D. and Andrews, G. J. (2015). Geographies of ageing Progress and possibilities after two decades of change. *Progress in Human Geography*, 39(6), 776–799. http://dx.doi.org/10.1177/0309132514558444

Skinner, M. and Hanlon, N. (2016). Introduction to ageing resource communities, in M. Skinner and N. Hanlon (eds), *Ageing resource communities: New frontiers of rural population change, community development and voluntarism* (pp. 106–118). London and New York: Routledge.

Stoeckel, K. J. and Litwin, H. (2015). Social cohesiveness and neighbourhood environmental deprivation: How are they related to life satisfaction in late life? in A. Börsch-Supan, T. Kneip, H. Litwin, M. Myck and G. Weber (eds), *Ageing in Europe – Supporting policies for an inclusive society* (pp. 149–158). Berlin: de Gruyter.

Temelová, J. and Slezáková, A. (2014). The changing environment and neighbourhood satisfaction in socialist high-rise panel housing estates: The time-comparative perceptions of elderly residents in Prague. *Cities*, 37(April), 82–91. http://dx.doi.org/10.1016/j.cities.2013.12.002

Vitman A, Iecovich, E. and Alfasi, N. (2014). Ageism and social integration of older adults in their neighborhoods in Israel. *Gerontologist*, 54(2), 177–189. http://dx.doi.org/10.1093/geront/gnt008

Wahl, H-W., Iwarsson, S. and Oswald, F. (2012). Aging well and the environment: Toward an integrative model and research agenda for the future. *The Gerontologist*, 52(3), 306–316. http://dx.doi.org/10.1093/geront/gnr.154

Walker, J., Orpin, P., Baynes, H., Stratford, E., Boyer, K., Mahjouri, N . . . Carty, J. (2013). Insights and principles for supporting social engagement in rural older people. *Ageing and Society*, 33(6), 938–963. http://dx.doi.org/10.1017/S0144686X12000402

Walker, R. B. and Hiller, J. E. (2007). Places and health: A qualitative study to explore how older women living alone perceive the social and physical dimensions of their neighbourhoods. *Social Science & Medicine*, 65(6), 1154–1165. http://dx.doi.org/10.1016/j.socscimed.2007.04.031

Walsh, K. (2015). Interrogating the 'age-friendly community' in austerity: Myths, realties and the influence of place context, in K. Walsh, G. Carney and Á. Ní Léime (eds), *Ageing through austerity: Critical perspectives from Ireland* (pp. 9–95). Bristol: Policy Press.

Walsh, K. (2016). Creating, maintaining and losing home: Productions of ageing and migration in Ireland, in K. Walsh and L. Näre (eds), *Transnational migration and home in older age* (pp. 153–164). New York and Abingdon: Routledge.

Walsh, K. and Gannon, B. (2011). Perceived neighbourhood context, disability onset and old age. *Journal of Socio-Economics*, 40(5), 631–636. http://dx.doi.org/10.1016/j.socec.2011.05.001

Walsh, K., O'Shea, E. and Scharf, T. (2012). *Social exclusion and ageing in diverse rural communities: Findings of a cross border study in Ireland and Northern Ireland*. Galway: Irish Centre for Social Gerontology [ICSG], National University of Ireland. Available from: http://citeseerx.ist.psu.edu/viewdoc/download?doi=10.1.1.426.9536&rep=rep1&type=pdf

Walsh, K., O'Shea, E. and Scharf, T. (2017). Ageing in farm communities in Ireland: Changing agricultural resources – changing lives, in M. Skinner and N. Hanlon (eds), *Ageing resource communities: New frontiers of rural population change, community development and voluntarism* (pp. 59–73). London and New York: Routledge.

Walsh, K., O'Shea, E., Scharf, T. and Murray, M. (2012b). Ageing in changing community contexts: Cross-border perspectives from rural Ireland and Northern Ireland. *Journal of Rural Studies*, 28(4), 347–357. http://dx.doi.org/10.1016/j.rurstud.2012.01.012

Walsh, K., Scharf, T. and Keating, N. (2017). Social exclusion of older persons: A scoping review and conceptual framework. *European Journal of Ageing*, 14(1), 81–98. Online First. http://dx.doi.org/10.1007/s10433-016-0398-8

Walsh, K., Scharf, T. and Shucksmith, M. (2014). Exploring the impact of informal practices on social exclusion and age-friendliness for older people in rural communities. *Journal of Community & Applied Social Psychology*, 24(1), 37–49. http://dx.doi.org/10.1002/casp.2176

Warburton, J., Scharf, T. and Walsh, K. (2016). Flying under the radar? Risks of social exclusion for older people in rural communities in Australia, Ireland and Northern Ireland. *Sociologia Ruralis*. Online first: 23 May 2016. http://dx.doi.org/10.1111/soru.12129

Wiles, J. L., Leibing, A., Guberman, N., Reeve, J. and Allen, R.E.S. (2012). The meaning of "aging in place" to older people. *The Gerontologist*, 52(3), 357. http://dx.doi.org/10.1093/geront/gnr098

Wood, B. E. (2015). Border spaces: Geographies of youth exclusion, inclusion, and liminality, in P. Kraftl, K. Nairn and T. Skelton (eds), *Space, place and environment: Geographies of children and young people* (vol. 3, pp. 1–18). Singapore: Springer Verlag.

Ziegler, F. (2012). You have to engage with life, or life will go away: An intersectional life course analysis of older women's social participation in a disadvantaged urban area. *Geoforum*, 43(6), 1296–1305. http://dx.doi.org/10.1016/j.geoforum.2012.03.013

21 Employed caregivers in the ageing family

Conceptualising respite as an embodied space of resistance

Gillian M. Joseph and Alun E. Joseph

Introduction

> Each of us needs to withdraw from the cares which will not withdraw from us.
>
> —— Maya Angelou

Nurtured by interdisciplinary collaboration and innovative theorisation, the field of geographical gerontology has grown considerably from its formal, albeit cautious, beginnings in the 1970s (Sylvestre, 1999). This evolution has led to a heightened awareness of the "reflexive, transactional and mutually constitutive" relationships among older people, their families and their communities, while at the same time it has highlighted the need for greater understanding of how space and place shape peoples' lives (Skinner, Cloutier and Andrews, 2015: 778). This 'relational thinking', as Skinner *et al.* (2015) describe it, involves considering the connections between multiple places and spaces and exploring how these connections shape interactions between generations – which, in turn, also shapes the experience of aging itself. A case in point is this chapter. It is not about old age directly but rather indirectly, considering, as it does, the embodied experiences of employed spouses, siblings and adult children who are in need of a break from the unpaid care they provide to elderly family members.

Increasingly, there are calls to recognise that the experience of caregiving can be perceived and acted out differently by different people and that it may even be experienced differently in different places (Geisbrecht *et al.*, 2016). While there has been growing interest in space, place and issues associated with family caregiving in the geographical literature (e.g., Milligan, 2000; Sylvestre, 1999; Wiles, 2005), few studies have focused specifically on employed family caregivers. Nevertheless, it is well documented that all people give meaning to spaces which are not only context-dependent but also shaped by and embedded in political power-based relationships (Petersen *et al.*, 2016). Indeed, space itself, and a human body in it, can be inherently political. A powerful example of this is presented by Sharpe and Gorman-Murray (2013). In their editorial, reference is made to the now famous video taken in New York on 11 September 2001 of a body as it fell from the Twin Towers in an attempt to escape the destruction that was taking place there. The authors suggest that the 'falling man', as the person in the video was named, is more than just a body extending into empty space. Instead, we are invited to consider that the falling body has a presence in that particular space that has meaning that is simultaneously social, political and emotional (Sharpe and Gorman-Murray, 2013).

Focusing on the embodied experiences of employed family caregivers of the elderly addresses important gaps in the geographical gerontology literature. Few studies have explored in depth the nature of spaces where respite takes place and, as Herron notes elsewhere in this volume (Chapter 14), even fewer have explored issues about space through the lens of embodiment. Yet as Merleau-Ponty (2004: 126) asserts, "we are in the world through our body, and ... we perceive that world within our body".

In acknowledging that family caregiving (which is hard work) overlaps boundaries that are somewhat nebulous in nature, we step out of the traditional definitions of 'employment' and 'leisure', which are commonly seen as diametrically opposed concepts, and consider them as one and the same. That is, we argue that when their bodies are situated in the space where paid employment takes place, caregivers can also, at the same time, experience that space as leisure/respite. Moreover, simply being in that respite space is a highly political action. Thus, the topic of respite is an important focal point for geographical gerontology, both now and in the future, because it makes a relational link between different generations – younger caregivers and their older recipients of care. At the same time, it opens up new opportunities for thinking about how being in everyday spaces such as those of work and leisure can have different meanings for different people over the life course.

The remainder of our discussion is organised into four major sections. In the first, we introduce the urgent need for caregiver respite and the importance of using a geographical lens to understand this experience. We then present research showing that employment space can also be a place of respite for caregivers. Following this, we consider three powerful and highly gendered forces that act as barriers for caregivers, making access to employment/respite space precarious. These barriers include the impact of internalized sexism, institutional sexism and family conflict. The discussion of barriers is then complemented by a section on resistance, illustrating how some employed caregivers fight against powerful forces that try to keep them from being in the employment/respite space. Finally, in a short concluding section we draw out major insights and offer suggestions for further research.

Setting the scene

In Canada, a rapidly aging population has made informal caregiving an increasingly important issue (Joseph, Leach and Turner, 2007). Cuts in government funding, the increasing complexity of care and a shortage of health professionals have shifted responsibility for many aspects of eldercare from declining formal services to the 'community' – in other words, to families and friends (Canadian

Healthcare Association, 2012; Joseph and Skinner, 2012). This has created what Schroeder, MacDonald and Shamian (2012: 40) call a "perfect storm". Indeed, in 2012 it was estimated that 8.1 million Canadians aged 15 years and older had provided care to a chronically ill, disabled, or aging family member or friend in the preceding 12 months, and that 60 per cent of family caregivers were also in paid employment (Sinha, 2013). Although research suggests that families are willing to provide such care (Morgan *et al.*, 2016), we also know that for family members who are engaged in more intense forms of caregiving, especially those who care for loved ones with chronic conditions such as dementia, the stress associated with eldercare can profoundly affect their health and well-being (Pew Research Center, 2015; Sinha, 2013).

In other countries the situation is similar and, in some cases, even worse (Pew Research Center, 2015; Rose, Noelker and Kagan, 2015). Indeed, it is estimated that there are over 47 million people worldwide who suffer with dementia and that this figure will double by 2030 and triple by 2050 (World Health Organization [WHO] i-support, n.d.), thereby increasing the pressure on family caregivers. Yet as the WHO (2015: 1) notes, "[a]ll over the world, the family remains the cornerstone of care for older people who have lost the capacity for independent living. In high-income countries, the vital caring role of families, and their need for support, is often overlooked".

In North America, respite has long been the number one item on a family caregiver's 'wish list' of supports that are urgently needed (Boscart, 2016; Pyper, 2006; WHO i-support, n.d.). However, policy responses have been partial at best. Indeed, with the exception of a few donor-funded charitable initiatives, most caregiver respite in Canada and the United States is paid for privately by the families who need it, if they can afford it (Canadian Healthcare Association, 2012; Joseph and Leach, 2016). Where formal respite services and supports are available, eligibility is often tied to the recipient alone and not to the family caregiver (Rose *et al.*, 2015). For example, in a study undertaken for the Ontario Caregiver Coalition that sought to identify Local Health Integration Network (LHIN)–funded initiatives designed specifically for family caregivers, Joseph and Rivera (2010) found that there was often an underlying, but rarely evaluated, assumption expressed by both LHIN and community agency spokespeople that services aimed at the care recipient would also somehow fulfil the needs associated with respite for the caregiver. Yet, again, the literature on the high levels of caregiver stress suggests otherwise.

In moving this discussion forward, a good place to start is to ask the question, 'How does one define caregiver respite?' This is a question that is easily asked but difficult to answer. Definitions of respite vary (Evans and Lee, 2013; Reinhard, Bemis and Huhtala, 2005), and this lack of consistency has been problematic for researchers, practitioners and policy makers alike (Reinhard *et al.*, 2005; Rose *et al.*, 2015). In their study of family eldercare providers, Chappell, Reid and Dow (2001: 202) state that respite is most often defined as "a pause, a temporary cessation, or an interval of rest". Although this definition has permeated much of the Canadian literature on caregiver respite, it is limited intrinsically by a conceptualisation of respite as being temporal in nature, thereby ignoring its non-temporal characteristics. Indeed, some studies suggest that the *space* in which respite takes place may be more important for the caregiver than other factors. For example, when comparing the words that family caregivers used to conceptualise respite space with those deployed in feminist writings to conceptualize leisure, Strang (2001) found a number of similarities. When her participants were in their respite/leisure "world", a space where they were no longer required to maintain their identity as a provider of family eldercare, they reported feeling a sense of freedom and ability to just "be themselves" (Strang, 2001: 77). The *space* where respite occurred then, and their bodies being in that space, facilitated a leisure-type experience for caregivers through which a renewed sense of well-being could emerge (Strang, 2001).

Employment as place of respite

Although there are few studies that focus specifically on the respite experiences of employed family caregivers alone, some researchers have reflected on their findings from mixed samples and speculated that employment may be being used as a space of respite by employed caregivers. In a study of Norwegian caregivers, Hansen and Slagsvold (2015) found that for women providing care to elderly people who resided with them, a demanding caregiving role had a negative impact on their psychological well-being – but not if the caregiver was also employed full-time. Similarly, van Campen, de Boer and Iedema (2012) found that ratings of happiness were higher for caregivers who were also employed compared to those who engaged only in caregiving. Dawid, Ludescher and Trukeschitz's study of Viennese caregivers (as cited in Schneider *et al.*, 2013), not only described the special meaning of the workplace to caregivers as a retreat from caregiving but also pointed out that, particularly for caregivers of elderly people with dementia, employment had an important respite value that actually strengthened caregiver attachment to the labour force. Moreover, in a US study of employed women caregivers of the elderly, Martire, Stephens and Atienza (1997) found that those who worked a greater number of hours were buffered from the negative effects of caregiving stress.

Taken together, these studies suggest that the space where employment takes place can also be a place in which caregiver respite can be experienced. Furthermore, when their bodies are in that space, caregiver well-being can be enhanced (Barnett and Hyde, 2001). Nevertheless, some researchers have uncovered forces that act as barriers for caregivers, making access to employment as a space of respite difficult or impossible. It is to these barriers that we now turn.

The political nature of space: gender and caregiving barriers

From the time our bodies are labelled as female or male we are taught about 'appropriate' gender identity. This can be observed in the clothing we are given to wear, the ways we are influenced to use our bodies and in the spaces which we are encouraged to inhabit or not (Freysinger *et al.*, 2013). Thus, throughout life, gender is not only a biological trait but also an ongoing process of socialisation (Freysinger *et al.*, 2013). More broadly, attitudes and assumptions about gender evolve over time. This can be observed through exploring a set of structured power relations that are enacted by laws, legislation and policies shaping our expectations, activities, interactions and practices (Freysinger *et al.*, 2013).

When it comes to a gender analysis of caregiving, feminist studies have sought to expose these power relations, focusing particularly on how caregiving work is passed from the state to women in families. However, the ways in which caregiving work has been taken away from men is rarely discussed (Geisbrecht *et al.*, 2016). We argue that the experiences of both female and male family caregivers of the elderly are shaped, albeit differently, by gender-based power relations, of which we consider three here – internalised sexism, institutional sexism and family conflict.

Internalised sexism

In her discussion of people with dementia elsewhere in this volume, Herron emphasises the importance of examining body–self relationships as being essential to gaining a deeper understanding of rights, freedoms and injustices (see Chapter 14). Self-expectations are born of powerful forces that can strongly influence behaviour which, in turn, can create self-fulfilling prophecies (Patton and Johns, 2007). Internalised sexism occurs at the individual level when people voluntarily or involuntarily accept the myths and stereotypes about their gender to be accurate and act accordingly. For example, Glenn (2010) asserts that because there is a societal belief that women are naturally better suited to caregiving than are men, a powerful expectation is formed through which women are coerced into accepting the role of unpaid caregiver. Moreover, when it comes to taking a break from caregiving, studies routinely show that women can internalise expectations about their role as caregivers and refuse to engage with spaces of respite because of their beliefs that women are not entitled to leisure and that respite is a form of abandonment (Bedini and Guinan, 1996; de la Cuesta-Benjumea, 2009). Thus, it has been suggested that the growing demand for eldercare will actually increase gender inequality if women continue to feel more obligated than men to provide it (Campbell and Carroll, 2017; Maume, 2015).

Ferber (2012) asserts that stubborn gender essentialism is bolstered by a widespread assumption that because modern women enjoy the same legal rights and opportunities – to a greater extent than in the past – as men, any remaining inequality between them must be the result of their different natures and choices. However, without acknowledging how such *choices* are influenced by broader structural forces such as globalisation, restructuring and sociocultural norms, the assumption that there is choice in role taking is grossly misleading (Gazso, 2004). When women believe that they are natural caregivers who are better suited than men to assume the role and that leisure/respite is inappropriate for them because of their caregiving responsibilities, internalised sexism creates a self-imposed barrier that can prevent caregivers from engaging with employment as a space of respite.

Institutional Sexism

Even if caregivers do not hold internalized sexist beliefs, there are powerful forces in the employment sphere that make employment precarious for both female and male caregivers. In some cases caregivers have even been terminated because of their family commitment (United States Equal Employment Opportunity Commission, 2012). For example, women and men are subject to gender-based stereotyping when employers make and act on internalised sexist assumptions about who should prioritise the demands of family caregiving over those of employment and vice versa. Organisational studies have consistently shown that women are expected to be nurturing and communal, while men are expected to be ambitions and results-oriented (Moss-Racusin, Phelan and Rudman, 2010; Sandberg and Grant, 2015). Indeed, it has also been shown that assumptions about gender-based biological determinism lead to discrimination against women in the workplace, affecting their access to benefits, wages and leadership roles. Moreover, they come under greater scrutiny than do men when the demands of work and family become unbalanced (Bobbitt-Zeher, 2011).

Men are also affected by gender-based assumptions about family caregiving – albeit in different ways from that experienced by women. For example, it has been shown that women perceived as *not actively caring* for family members experience greater mistreatment in the workplace than do those who take on more caregiving (Berdahl and Moon, 2013). In contrast, men who violate traditional gender norms by *actively caring* for family members are shown to experience more workplace mistreatment than do men who *don't actively provide care* (Berdahl and Moon, 2013; Garcia, 2012). Thus, both male and female employees are subordinates of a powerful organisational patriarchal view about who should provide care and who should not. Although research suggests that younger cohorts appear to be more family-centred than their predecessors, there is little evidence that organisations have adjusted to better accommodate men as they attempt to reconcile work and family (Maume, 2015). This may explain why men are less likely than women to tell their supervisors about their eldercare responsibilities (Lero and Joseph, 2007) and why men are less likely than women to utilise organisational caregiver supports (Mitchell, Eby and Lorys, 2015).

There are a number of studies that show how gender and age-based discrimination combines with organisational policies to make employment precarious for caregivers of the elderly (Bobbitt-Zeher, 2011). For example, in a seminal Canadian study, Medjuck, Keefe and Fancey (1998) found that erroneous assumptions about caregiving shaped organisational policy and that this became particularly problematic for caregivers of elderly parents. They noted, for instance, significant differences in the length of allowable absence for employees caring for an ill child compared to those caring for an ill parent – something that has been shown to still exist today (Calvert, 2016). Additionally, such benefits and programmes allowed for both longer-term, hands-on provision and shorter-term management (i.e., making arrangements) for child care but only shorter-term management of eldercare. Since men are more likely to manage care and women to provide hands-on care, Medjuck *et al.* (1998) assert that workplace policies that support care management activities alone are not gender neutral.

When caregivers do not use existing family-friendly policies because they are inappropriate or because of an unsupportive employment environment, employers can conclude that such supports are not needed. Moreover, in the absence of organisational supports, when caregivers, and women in particular, are forced to withdraw from the workforce or to make adjustments because of caregiving responsibilities, this is often considered to be their choice (Lero and Joseph, G., 2007). Yet, as noted earlier, caregiving can be considered 'voluntary' only if there are alternatives that allow true choices to be made. Thus, institutional sexism with its intent to "pull men out of the home and push women into it" (Berdahl and Moon, 2013: 343), makes employment precarious for caregivers of the elderly who desire a non-traditional path, putting access to this space of respite in jeopardy.

Family conflict

Recently, studies have suggested that eldercare is most often shared among friends and family members, taking the form of a dynamic network (Davey, Szinovacz and Bauer, 2017). However, not all relationships between caregivers and family members are positive. Conflict, resentment and other negative interactions between primary caregivers and family members may undermine well-being and add to stress (Davey *et al.*, 2017). There may even be different perceptions of impairment and quality of care among family members, and some may have very different ideas about the amount of care that should be provided to frail parents, who should provide it and where it should be undertaken (Kwak, Ingersoll-Dayton and Kim, 2012). Adult children, especially women, may find themselves trapped in the caregiving role due to pressure from other family members (Family Caregiver Alliance, 2011). However, research suggests that male caregivers receive less assistance from family members than do their female counterparts, putting even more pressure on their ability to balance the demands of work and family care (Friedemann and Buckwalter, 2014). The power of gender-based assumptions in the family can create a barrier for caregivers and make access to employment precarious, particularly if siblings believe that eldercare should be undertaken full-time by a particular family member in the home space.

The extent to which family members can create strain for employed caregivers has not received much attention in the literature (Li, Shaffer and Bagger, 2015). However, research has shown that when family members put strain on caregivers rather than providing support, such behaviours result in a loss of resources that can significantly affect caregivers' abilities to manage the delicate balance between care and work responsibilities effectively – placing employment in jeopardy (Li *et al.*, 2015).

Resistance

Resistance is usually conceptualised in terms of actions or processes that challenge constraints or forms of oppression. Therefore, resistance is really about the struggle for freedom of choice, human agency and empowerment – and it is highly political (Shaw, 2005; Skinner, Joseph, and Herron, 2016). It is within this frame of reference that we consider some of the ways that caregivers of the elderly push back against barriers that affect their access to employment/respite space. This includes asserting their entitlement to employment space, litigating against employers and changing jobs or quitting temporarily.

Shaw (2005) asserts that claiming entitlement to self-determined leisure is a way to push back against powerful forces. This might occur if caregivers decide to remain connected to their space of employment despite pressure to disengage. Carr *et al.* (2016) found that the longer caregivers who provided intense family caregiving stayed employed, the less likely they were to leave their existing jobs or to exit the workplace altogether. This suggests that caregivers can be determined to maintain access to employment as a space of respite despite pressures to do otherwise.

Some caregivers overcome barriers by accessing the legal system. In the United States, most state and local laws prohibit discrimination against employees because of familial or parental status based on caring for children. However, few prohibit discrimination based on caring for elderly parents (Calvert, 2016; Canadian Human Rights Commission, 2016). Interestingly, in the United States, cases brought against employers related to employee discrimination because of eldercare responsibilities have increased in the last decade by a staggering 650 per cent (Calvert, 2016). Moreover, according to Calvert (2016), male employees have brought to court 55 per cent of spousal care cases and 39 per cent of eldercare cases. In Canada, laws against discrimination based on family status have been incorporated into human rights legislation, and recent challenges have highlighted the expensive penalties levied against employers who do not accommodate the family care–related needs of their employees (Joseph, 2013).

Employees may also resort to labour attachment strategies as a form of resistance. For example, they may retire early, quit their jobs or look for another one that better suits their needs as caregivers. Indeed, research shows that employee retention is increasingly becoming associated with the availability of good family-friendly benefits and the existence of a workplace culture that supports their use. This is especially the case for university-educated employees (Hughes, Lowe and Schellenberg, 2003; Schneider *et al.*, 2013). While the impacts of family-friendly policies are difficult to estimate precisely, Morissette,

Schellenberg and Silver (2004) report that 6 per cent of the people who retired between 1992 and 2002 stated that they would have continued working if suitable arrangements for caregiving had been available to them. This might suggest that quitting work was an act of resistance to a lack of organisational support.

In bringing this discussion of resistance to a conclusion, we need to acknowledge again the interdependence of pressures – from society, employers and the family – and of responses, including resistance against powerful forces that prevent caregivers from using employment as a space of respite. This is brought out clearly in a study of caregiver siblings reported by Li *et al.* (2015). The authors found that a positive perception of support from supervisors was an important factor in determining psychological well-being for caregivers, particularly if they were experiencing family conflict (Li *et al.*, 2015). In other words, employed caregivers may resist the pressures of family conflict if occupational support, and, in particular, supervisor support, is present.

Conclusion

Suggesting that a place of employment can simultaneously be a space of *work and leisure/respite* shakes up traditional views of these spaces as distinct and separate spheres. Yet doing so teases out where the overlaps and intersections lie and uncovers both the healthy as well as the difficult aspects of everyday life in those spaces.

Conceptualising employment as a space of leisure/respite opens up new ways of thinking about how bodies can experience the same space in different ways. For us, it reveals a complex reality in which caregivers seek a break from the caregiving space and relocate temporarily in the space where paid work takes place, viewing the latter as a space of respite. Yet this is not an easy task for caregivers given the gender-based barriers that make connecting to employment space so precarious. Caregivers work to resist powerful and deeply embedded assumptions, in society at large, among employers and in families, about how women and men should differentially situate themselves in the spaces of caregiving and employment. Notwithstanding important differences between the challenges faced by men and women, we emphasise that all employed family caregivers are marginalised within the highly politicized spaces that they must navigate to gain respite. Thus, viewing respite as an embodied space of resistance helps to advance the field of geographical gerontology by challenging researchers to step out of their assumptions about the common use of everyday spaces and explore how different people at different stages of life might not only view those same spaces differently but also may be marginalised because they do so.

In moving forward, we highlight the need for case studies that examine, in detail, the unique interpretations of spaces in which employed caregivers are immersed. In particular, we suggest the need to tease out the ways in which caregivers of the elderly challenge the barriers that prevent their access to such spaces. There is also a need to more fully explore how employed caregivers themselves define successful respite, and how the relationship between activities and the spaces where they occur contributes to their well-being. This has potential benefits for both employee and employer, not only in terms of enhanced well-being but also in terms of organisational concerns about employee retention as the workforce ages. Further studies are also needed on the engendered nature of barriers preventing access to particular spaces, while simultaneously exposing the forces that create those barriers. The employment/respite space in which caregivers' bodies are situated has meaning that lies far beyond the boundaries of the four walls of a workplace. Exploring these mysteries through a geographical lens will provide a greater understanding of the aging family, both now and in the future.

References

Angelou, M. (n.d.). Available from: www.azquotes.com/quote/836130

Barnett, R. C. and Hyde, J. S. (2001). Women, men, work and family: An expansionist theory. *American Psychologist*, 56(10), 781–796. http://dx.doi.org/10.1037/0003-066X.56.10.781

Bedini, L. A. and Guinan, D. M. (1996). "If I could just be selfish . . .": Caregivers' perceptions of their entitlement to leisure. *Leisure Sciences*, 18(3), 227–239. http://dx.doi.org/10.1080/01490409609513284

Berdahl, J. L. and Moon, S. H. (2013). Workplace mistreatment of middle class workers based on sex, parenthood, and caregiving. *Journal of Social Issues*, 69(2), 341–366. http://dx.doi.org/10.1111/josi.12018

Bobbitt-Zeher, D. (2011). Connecting gender stereotypes, institutional policies, and gender composition of workplace. *Gender & Society*, 25(6), 764–786. http://dx.doi.org/10.1177/0891243211424741

Boscart, V. (2016). *Dementia in Canada: Presentation to the standing senate committee on social affairs, science and technology.* Toronto: Registered Nursing Association of Ontario, Canadian Gerontological Nursing Association. Available from: http://rnao.ca/sites/rnao-ca/files/Confidential_Joint_Submission_re_Dementia_-_May_24_2016.pdf

Calvert, C. T. (2016). *Caregivers in the workplace: Family responsibilities discrimination litigation update 2016.* San Francisco, CA: University of California, Hastings College of the Law. Available from: http://worklifelaw.org/pubs/FRDupdate2016.pdf

Campbell, L. D. and Carroll, M. P. (2017). Men providing care to aging parents in the internet age: Part of the stalled gender revolution? in G. Joseph (ed.), *Diverse perspectives on aging in a changing world* (pp. 71–85). London: Routledge.

Canadian Healthcare Association. (2012). *Respite care in Canada.* Ottawa: Canadian Healthcare Association. Available from: www.healthcarecan.ca/wp-content/uploads/2012/11/Respite_Care_in_Canada_EN_web.pdf

Canadian Human Rights Commission. (2016). *Beyond recognition – caregiving and human rights in Canada. A policy brief.* Ottawa: Canadian Human Rights Commission. Available from: www.creges.ca/wp-content/uploads/2016/03/PolicyBrief_Eng_Web.pdf

Carr, E., Murray, E. T., Zaninotto, P., Cadar, D., Head, J., Stansfield, S. and Stafford, M. (2016). The association between informal caregiving and exit from employment among older workers: Prospective findings from the UK houshold longitudinal study. *Journal of Gerontology*. Online first. http://dx.doi.org/10.1093/geronb/gbw156

Chappell, N., Reid, R. C. and Dow, E. (2001). Respite reconsidered. A typology of meanings based on the caregiver's point of view. *Journal of Aging Studies*, 15(2), 201–216. http://dx.doi.org/10.1016/S0890-4065(00)00026-8

Davey, A., Szinovacz, M. E. and Bauer, K. W. (2017). Dynamic care networks: The changing face of family caregiving, in G. Joseph (ed.), *Diverse perspectives on aging in a changing world* (pp. 86–103). London: Routledge.

de la Cuesta-Benjumea, C. (2009). The legitimacy of rest: Conditions for the relief of burden in advanced dementia care-giving. *Journal of Advanced Nursing*, 66(5), 988–998. http://dx.doi.org/10.1111/j.1365-2648.2010.05261.x

Evans, D. and Lee, E. (2013). Respite services for older people. *International Journal of Nursing Practice*, 19(4), 431–436. http://dx.doi.org/10.1111/ijn.12096

Family Caregiver Alliance. (2011). *Caregiving with your siblings*. Available from: www.care giver.org/caregiving-with-your-siblings

Ferber, A. L. (2012). The culture of privilege: Color-blindness, postfeminism and christonormativity. *Journal of Social Issues*, 68(1), 63–77. http://dx.doi.org/10.1111/j.1540-4560.2011.01736.x

Freysinger, V. J., Shaw, S. M., Henderson, K. A. and Bialeschki, M. D. (2013), Introduction: Constructing a framework, in V. J. Freysinger, S. M. Shaw, K. A. Henderson and M. D. Bialeschki (eds), *Leisure, women and gender* (pp. 3–20). Urbana, IL: Venture Publishing.

Friedemann, M. and Buckwalter, K. C. (2014). Family caregiver role and burden related to gender and family relationships. *Journal of Family Nursing*, 20(3), 313–336. http://dx.doi.org/10.1177/1074840714532715

Garcia, K. K. (2012). The gender bind: Men as inauthentic caregivers. *Duke Journal of Gender, Law & Policy*, 20(1), 1–43. Available from: http://scholarship.law.duke.edu/cgi/viewcontent.cgi?article=1234&context=djglp

Gazso, A. (2004). Women's inequality in the workplace as framed in news discourse: Refracting from gender ideology. *The Canadian Review of Sociology and Anthropology*, 41(4), 449–473. http://dx.doi.org/10.1111/j.1755-618X.2004.tb00786.x

Geisbrecht, M., Williams, A., Duggleby, W., Ploeg, J. and Markle-Reid, M. (2016). Exploring the daily geographies of diverse men caregiving for family members with multiple chronic conditions. *Gender, Place & Culture*, 23(11), 1586–1598. http://dx.doi.org/10.1080/0966369X.2016.1219329

Glenn, E. N. (2010). *Forced to care: Coercion and caregiving in America*. Cambridge, MA: Harvard University Press.

Hansen, T. and Slagsvold, B. (2015). Feeling the squeeze? The effects of combining work and informal caregiving on psychological well-being. *European Journal of Ageing*, 12(1), 51–60. http://dx.doi.org/10.1007/s10433-014-0315-y

Hughes, K., Lowe, G. S. and Schellenberg, G. (2003). *Men's and women's quality of work in the new Canadian economy*. Ottawa: Canadian Policy Research Networks. Available from: www.cprn.org/documents/17786_en.PDF

Joseph, A. E. and Skinner, M. W. (2012). Voluntarism as a mediator of the experience of growing old in evolving rural spaces and changing rural places. *Journal of Rural Studies*, 28(4), 380–388. http://dx.doi.org/10.1016/j.jrurstud.2012.01.007

Joseph, G. (2013). *Our differences and our sameness: Why do organizations rarely accommodate diversity in family-friendly initiatives? A critical theory approach*. Philadelphia, PA: Work and Family Research Network, 25pp. Available from: https://workfamily.sas.upenn.edu/wfrn-repo/object/t2rk47qa361qu7ae

Joseph, G. and Leach, B. (2016). Neoliberalism, temporary foreign workers and precarious eldercare/eldercare work, in G. Joseph (ed.), *Diverse perspectives on aging in a changing world* (pp. 115–129). London: Routledge.

Joseph, G., Leach, B. and Turner, S. (2007). *Caring at a distance: Working women, rural to urban migration and the compassionate care challenge*. Ottawa: Status of Women Canada. Available

from:www.worklifecanada.ca/cms/resources/files/261/CADfinalreporttoSWC
complete.pdf

Joseph, G. and Rivera, A. (2010). *LHIN initiatives that support caregivers in Ontario: A closer look*. Guelph: Clear Pane Research Services for The Ontario Caregiver Coalition.

Kwak, M., Ingersoll-Dayton, B. and Kim, J. (2012). Family conflict from the perspective of adult child caregivers: The influence of gender. *Journal of Social and Personal Relationships*, 29(4), 470–487. http://dx.doi.org/10.1177/0265407511431188

Lero, D. and Joseph, G. (2007). *A systematic review of the literature on combining work and eldercare in Canada*. Guelph: Centre for Families, Work & Well-being, University of Guelph. Available from: www.familycaregiversbc.ca/wp-content/uploads/2015/04/systemic-review-of-combining-work-and-elder-care-final-report.pdf

Li, A., Shaffer, J. and Bagger, J. (2015). The psychological well-being of disability caregivers: Examining the roles of family strain, family-to-work conflict and perceived supervisor support. *Journal of Occupational Health Pschology*, 20(1), 40–49. http://dx.doi.org/10.1037/a0037878

Martire, L. M., Stephens, M.A.P. and Atienza, A. A. (1997). The interplay of work and caregiving: Relationships between role satisfaction, role involvement and caregivers' well-being. *Journal of Gerontology*, 52B(5), S279–S289. Available from: http://dx.doi.org/10.1093/geronb/52B.5.S279

Maume, D. J. (2015). Can men make time for family? Paid work, care work, work-family reconcilliation policies, and gender equality. *Social Currents*, 3(1), 43–63. http://dx.doi.org/10.1177/2329496515620647

Medjuck, S., Keefe, J. M. and Fancey, P. J. (1998). Available but not accessible: An examination of the use of workplace policies for caregivers of elderly kin. *Journal of Family Issues*, 19(3), 274–299. http://dx.doi.org/10.1177/019251398019003003

Merleau-Ponty, M. (2004). Part II: The world as perceived from introduction, in T. Baldwin (ed.), *Maurice Merleau-Ponty: Basic writings* (pp. 122–136). London: Routledge.

Milligan, C. (2000). 'Bearing the burden': Towards a restructured geography of caring. *Area*, 32(1), 49–58. http://dx.doi.org/10.1111/j.1475.4762.2000.tb00114.x

Milligan, C. and Power, A. (2010). The changing geography of care, in T. Brown, S. McLafferty and G. Moon (eds), *A companion to health and medical geography* (pp. 567–586). Sussex: Blackwell.

Mitchell, M. E., Eby, L. T. and Lorys, A. (2015). Feeling work at home: A transactional model of women and men's negative affective spillover from work to family, in M. Mill (ed.), *Gender and the work-family experience: An intersection of two domains* (pp. 121–140). New York: Springer.

Morgan, T., Williams, L. A., Trussardi, G. and Gott, M. (2016). Gender and family caregiving at the end-of-life in the context of old age: A systematic review. *Palliative Medicine*, 30(7), 616–624. http://dx.doi.org/10.1177/0269216315625857

Morissette, R., Schellenberg, G. and Silver, C. (2004). Retaining older workers. *Perspectives on Labour and Income*, 5(3), 15–24. Available from: www.statcan.gc.ca/studies-etudes/75-001/10304/5018840-eng.pdf

Moss-Racusin, C. A., Phelan, J. E. and Rudman, L. A. (2010). When men break the gender rules: Status incongruity and backlash against modest men. *Psychology of Men & Masculinity*, 11(2), 140–151. http://dx.doi.org/10.1037/a0018093

Patton, E. and Johns, G. (2007). Women's absenteeism in the popular press: Evidence for a gender-specific absence culture. *Human Relations*, 60(11), 1579–1612. http://dx.doi.org/10.1177/0018726707084301

Petersen, M., Wilson, J., Wright, O., Ward, E. and Capra, S. (2016). The space of family care-giving in Australian aged care facilities: Implications for social work. *British Journal of Social Work*, 46, 81–97. http://dx.doi.org/10.1093/bjsw/bcu108

Pew Research Center. (2015). *Family support in graying societies: How Americans, Germans and Italians are coping with an aging population.* Washington, DC: Pew Research Center. Available from: www.pewsocialtrends.org/files/2015/05/2015-05-21_family-support-relations_FINAL.pdf

Pyper, W. (2006). Balancing career and care. *Perspectives on Labour and Income*, 7(2), 5–15. Available from: www.statcan.gc.ca/pub/75-001-x/11106/9520-eng.pdf

Reinhard, S. C., Bemis, A. and Huhtala, N. (2005). *Defining respite care.* New Brunswick, NJ: Rutgers Center for State Health Policy.

Rose, M. S., Noelker, L. S. and Kagan, J. (2015). Improving policies for caregiver respite services. *The Gerontologist*, 55(2), 302–308. http://dx.doi.org/10.1093/geront/gnu120

Sandberg, S. and Grant, A. (2015). Madam C.E.O., get me a coffee. *The New York Times*, 6 February. Available from: www.nytimes.com/2015/02/08/opinion/sunday/sheryl-sandberg-and-adam-grant-on-women-doing-office-housework.html?_r=0

Schneider, U., Trukeschitz, B., Muhlmann, R. and Ponocny, I. (2013). "Do I stay or do I go?" – job change and labor market exit intentions of employees providing informal care to older adults. *Health Economics*, 22(10), 1230–1249. http://dx.doi.org/10.1002/hec.2880

Schroeder, B., MacDonald, J. and Shamian, J. (2012). Older workers with caregiving respon-sibilities: A Canadian perspective on corporate caring. *Ageing International*, 37(1), 39–56. http://dx.doi.org/10.1007/s12126-011-9134-z

Sharpe, S. and Gorman-Murray, A. (2013). Special issue: Bodies in place, bod-ies displaced: A body falls. *Geographical Research*, 51(2), 115–179. http://dx.doi.org/10.1111/1745.5871.12013

Shaw, S. M. (2005). *Leisure as resistance: Academic debates, social action and applications to profes-sional practice.* Paper presented at the eleventh Canadian Congress on Leisure Research, Nanaimo, BC, May. Available from: http://lin.ca/sites/default/files/attachments/CCLR11-139.pdf

Sinha, M. (2013). *Portrait of caregivers, 2012.* 2013(001), 22pp. Ottawa: Statistics Canada: Social and Aboriginal Statistics Division. Available from: www.statcan.gc.ca/pub/89-652-x/89-652-x2013001-eng.pdf

Skinner, M. W., Cloutier, D. and Andrews, G. J. (2015). Geographies of ageing: Pro-gress and possibilities after two decades of change. *Progress in Human Geography*, 39(6), 776–799.

Skinner, M. W., Joseph, A. E. and Herron, R. V. (2016). Voluntarism, defensive localism and spaces of resistance to health care restructuring. *Geoforum*, 72, 67–75. http://dx.doi.org/0.1016/j.geoforum.2016.04.004

Strang, V. R. (2001). Family caregiver respite and leisure: A feminist perspective. *Scandina-vian Journal of Caring Sciences*, 15, 74–81. http://dx.doi.org/10.1046/j.1471-6712.2001.1510074.x

Sylvestre, G. (1999). The geography of aging: A geographical contribution to gerontol-ogy. *Prairie Perspectives*, 2(October), 214–224. Available from: http://pcag.uwinnipeg.ca/Prairie-Perspectives/PP-Vol02/sylvestre.pdf

United States Equal Employment Opportunity Commission. (2012). *Unlawful discrimination against pregnant workers and workers with caregiving responsibilities.* Transcript of the meeting, 15 February. Available from: www.eeoc.gov/eeoc/meetings/2-15-12/williams.cfm

van Campen, C., de Boer, A. H. and Iedema, J. (2012). Are informal caregivers less happy than noncaregivers? Happiness and the intensity of caregiving in combination with paid and voluntary work. *Scandinavian Journal of Caring Sciences*, 27(1), 44–50. http://dx.doi.org/10.1111/j.1471-6712.2012.00998.x

Wiles, J. (2005). Conceptualizing place in the care of older people: The contributions of geographical gerontology. *International Journal of Older People Nursing in Association with Journal of Clinical Nursing*, 14(s2), 100–108. http://dx.doi.org/j.1365-2702.2005.01281.x

World Health Organization. (2015). *Supporting informal caregivers of people living with dementia*, 4pp. Available from: www.who.int/mental_health/neurology/dementia/dementia_thematicbrief_informal_care.pdf

World Health Organization. (n.d.). *WHO i-support: An e-health solution to support caregivers of people living with dementia*, 2pp. Available from: www.who.int/mental_health/neurology/dementia/iSupport_flyer.pdf?ua=1

22 Ageing landscapes

Real and imagined

Robin Kearns and Tara M. Coleman

Introduction

According to Nigerian writer Chimamanda Ngozi Adichie there is a danger in assuming there is a 'single story'. To her, complex experiences and situations are too often simplified into uniform narratives. Using the example of Africans being stereotypically regarded as impoverished victims she makes the point that every person and population embody a heterogeneous compilation of stories. In this chapter we argue that Adichie's observation applies to older people and their geographies of everyday life. Hence, although there may be objectively documented 'real' landscapes of ageing such as retirement communities, these publicly recognisable 'single stories' can overlook and override the varied ways that people imagine and construct their own place-in-the-world. In other words, reducing people's experience to a singular story risks detracting from the diversity of their place experience and ultimately their humanity itself (Brooks, 2016).

We seek to survey ways in which conventional 'landscapes of old age' (Laws, 1993) are challenged by the place preferences and mobilities of older people especially given the diversity of contemporary Western societies. We argue that imagined and virtual landscapes are as important as real ones; the places we associate with older people and the metaphors we use to 'place them', as well as the places older people create for themselves, are all part of the broader landscape of ageing in Western societies. We contend that thinking in terms of real and 'objective' landscapes risks creating singular stories about the experience of old age; rather, we suggest that there are always 'other' stories anchored in the less-visible geographies of everyday life. The rationality of 'real' landscapes can overlook the multiplicity of less immediately visible landscapes of significance to older people – places to which they return in memory or create in response to a sense of loss or closure that might otherwise engulf people in advanced years. Our interest is in both stereotypical 'real' landscapes of old age, as well as examples of imagined (but arguably no less real) creative expressions of ageing. In the remainder of the chapter we survey the changing contexts for 'real' landscapes of ageing; we then consider two examples of imagined landscapes: the role of views as generative of memories and of imagining both a present and future self and the role of place-specific intergenerational musical performance keeping seniors 'young at heart'.

Diverse landscapes, diverse experiences

Geographers regard landscapes as both objectively observable and creatively imagined. To understand the places of ageing and ageing in place, we need to embrace the idea of landscapes – those accumulations of elements, resources and symbols – of which specific places are a necessary but not sufficient part. We are reliant on observable patterns and material evidence in assessing landscapes as 'real'. Imagined landscapes are idiosyncratic and created, therefore they are often dismissed in policy discourse or by social services as ephemeral or of little consequence. Yet, unquestionably, with reference to imagined engagements and relationships, humans "recognize; remember, and memorialize; maintain a sense of belonging; and commemorate their connections to landscapes" (Russell, 2012: 401). In other words, the manner in which people imagine and feel connected to a given landscape is dependent upon empathy, or "the projection of one's own consciousness into another being, thing or place" (Whiston Spirn, 2008: 44). The real consequences associated with human connection to imagined places are clear in the manner in which people experience and describe who they are.

In our work on ageing in place on Waiheke Island (Auckland, New Zealand), for example, several older people described themselves as "put out to pasture" and "past it" in response to living at a distance from the mainland (see Coleman and Kearns, 2015). For these seniors, imagining themselves as set apart from the mainland intensified negative perceptions of ageing as being separated from the mainstream and being of less value than others in society. Others in the study expressed a strong sense of insideness and agency in everyday life through decorating their home spaces with objects that reminded them of previous achievements (e.g., certificates, photographs and other memorabilia). This process allowed people to construct a landscape imbued with affirmations that buffered them from stereotypes associated with ageing. In these examples, imagined landscapes participate in the reproduction of, or resistance to, social life and norms. Furthermore, social identity is produced with reference to landscape and diverse experiences and feelings. Indeed, people commonly desire to assert diverse identities through marking out both individuality and group membership with reference to real and imagined places (Russell, 2005) in order to belong and to fit into a preferred landscape – real or imagined. The act of imagining and feeling connected to a particular place is, therefore, part of the performance of belonging (Bell, 1999), while also giving form to individual selves. Exploring the diversity of human engagements with real and imagined landscapes, including how such engagements inform processes of identity construction and belonging, highlights the complexities associated with peoples' interactions with their environments, as well as their unique human experiences of place and ageing. Whereas 'real' landscapes can be understood as sets of tangible places, those that are 'imagined' are less easy to specify and identify. One immediate distinction involves time and space; the imagination can creatively transport the self to other (frequently past) times and (frequently distant) places. By way of example, Blaikie (1997, 2005) examines representations of ageing with respect to maritime as well as rural heritage through the lens of nostalgia and reminiscence.

Both 'real' and imagined landscapes are, in our estimation, inherently more complex given the diversity of contemporary Western societies. Cities have always been diverse in terms of housing and population groups, but in recent decades the 'restless urban landscape' (Knox, 1993) has seen new levels of diversification: lifestyles, ethnicity, attitudes, consumption patterns and preferred activities. This 'hyper-diversity' (Tasan-Kok *et al.*, 2013) has been fueled by increased rates of immigration and more fine-grained segregation within and among groups in terms of ethnicity, age and socio-economic status, leading to widening sets of opportunities and inequalities. The imperative of ageing in place no longer necessarily implies particular and predictable landscapes of ageing.

A key part of the 'single story' of ageing is the prevailing imaginings of seniors' 'real' landscapes: residential concentrations and environments of high dependency. Hence, images of stylish retirement villages and/or bland pensioner units occupy public imaginings of landscapes of older age. Advertising adds a persuasive sense of inevitability to such destinations: smiling seniors happily embedded in contrived surrounds, maintaining relaxed poses or undertaking passive pursuits. Laws (1993) asked what such readings of the observable 'real' urban landscape can tell us about society's attitudes toward the ageing process as well as aged citizens themselves. Her conclusion was that ageing involved not only separations in time but also generational separations in space. Such spatio-temporal segregation is most graphically seen in mass-market retirement destinations like Florida or Arizona, where, in some instances, opportunities to take up residence is formally codified by age restriction. Master-planned communities which feature a minimum age create, *de facto*, a real landscape of ageing and the localised phenomenon of retirement villages becomes entire landscapes, albeit affluent, of old age.

One of the challenges to stereotypical 'landscapes of old age' has been an embrace of the 'new mobilities paradigm' which has helped researchers recognise that a focus on fixed places risks diminishing the mobile life of older people and the importance of that mobility (Sheller and Urry, 2006). This view tacitly argues against a 'sedentarist' view of ageing that treats stability, meaning and place as normal, and distance, change and movement as the exception. The influence of digital technologies is a further influence that is increasingly important as a means of maintaining relationships at a distance. Thus, the use of cell phones and e-mail can, for some, at least potentially compensate for a lack of social relationships with proximate neighbours.

While the phrase 'ageing in place' has been readily adopted by both researchers and policy-makers, the term ageing has arguably been better understood than place. Place, we contend, has been too easily reduced to location. However, geographical scholarship has offered a rich vein of nuance that has helped extend understanding of the imagined as well as real landscapes of ageing. Place, according to Pred (1984), is always in a state of becoming – a description that helps account for not only the inertia that comes with the built environment but also the accumulation of other less concrete, but no less potent, building blocks: memories. This conceptualisation of place as involving deeply interwoven influences of space and time can be complemented by more recent scholarship acknowledging the enabling possibilities of places.

Building on the ideas of Duff (2011) allows us to see a range of resources embedded in the local landscape – material, social or symbolic – as potentially enabling well-being. The subtlety, if not elusiveness, of these combinations of enabling properties can mean that significant landscapes of old age can lie 'beneath the radar' of policy scans and service organisations yet nonetheless be potent anchors within the geographies of everyday life. In the next two sections we draw on case examples to explore the contention that how older people behave, identify and *feel* within particular places arises from everyday encounters and relations with place itself. In so doing, we claim the benefits of conceptually broadening the idea of place as an enabler of older people's mobility, agency and well-being.

Imagined landscapes I: views, memories and the self

People commonly interpret places of physical beauty as offering emotional, aesthetic and spiritual opportunities, and the feeling of dwelling closer to nature (Kearns and Collins, 2010; Peart, 2009). Landscape views offer opportunities for moments of stillness, reflection and vicarious participation in everyday life, as well as productive engagement with memories. In turn, such views play a role in generating and sustaining affective ties and a sense of well-being (Kaplan, 1995; Kellert, 2005) by facilitating reminiscence, and providing opportunities to imagine both present and future selves (Coleman and Kearns, 2015). In this section, we consider the role of landscape views as generative of memories and supportive of cultivating the self and well-being during ageing.

Living within landscapes that offer physical beauty such as views of water bodies has been shown to have a positive influence on daily life and well-being. For instance, Mahmood *et al.* (2012) suggest that older people who consider their neighbourhoods to be aesthetically pleasing are more engaged in everyday activities and intentional physical exercise. In this sense place is both an enabling resource and generative of activity. Others claim that 'bluespaces' have stress-reducing, mood-enhancing powers (Karmanov and Hamel, 2008). Views, themselves, can be enabling. Troubles can seem less significant, and a settled sea can calm the inner waters of the human spirit.

Views of bluespaces enjoyed from the comfort of home may increase older adults' opportunities to cope with increasing frailties associated with ageing, as well as mitigate wider issues and problems occurring within their broader social contexts (across time and within place) (Coleman and Kearns, 2015; Coleman, Kearns and Wiles, 2016). Waiheke is an island (population 8,000) reachable from downtown Auckland, New Zealand's largest city, by a 35-minute ferry ride. There, views of bluespaces were found to enable people to imagine themselves living in a beautiful, peaceful and healing place, even in the context of dilapidated housing, economic constraints and health issues. As one participant commented in the larger study, despite reduced income and mobility which meant he could not repair a leaking roof in his home,

> I still feel peaceful. The sea moves me, its beauty and rhythm. I feel a deep connection to here and now, and to things years ago, looking at the in and out of the sea [from the kitchen window]. There are a few problems . . . but I look at this view and feel good, I can keep moving.
>
> (Tom, aged 80)

As this narrative illustrates, in addition to providing an alternative literal and metaphoric view in the context of challenges, water views, may be deeply moving. They offer a window on mobilities that may now be beyond reach, allowing connection to where one is in the here and now. They can also facilitate a 'looking back' and a 'moving forward' as the ebb and flow of water is perceived. Witnessing the peace, beauty and movement of water is also an undemanding way for people to imagine themselves as fortunate, content and located in the 'now' with strong connections to the past as well as to possible futures. As another study participant described it, viewing the sea "takes you away from whatever is on your mind, you remember things . . . you imagine things . . . then your life seems blessed" (Rose, aged 80). The opportunity to put challenges in perspective, relax the mind and spirit and engage in remembering and imagining through an undemanding view may be particularly important for those experiencing physical, social and emotional difficulties related to ageing.

Participants would pause when encountering a view from a window at home, describing themselves as deeply moved by the topography beyond: the beaches, the hills, the changing light and the greenery. For some, pausing to consider such views led to imagining the landscape observed as a mirror image of their own bodies, in turn enabling them to cope with their bodily ailments. One participant, Jenny (aged 83), for instance, spent long periods of time gazing at a view from her bedroom window which showed a vast seascape and several distant islands, framed by an expansive blue sky. She explained that as she considered these small islands, the water surrounding them and the sky above, she imagined her body itself as an island. By doing so, Jenny perceived her various ailments as part of the landscape. This allowed her to understand and take control of her health through likening ailments to 'natural' landmarks. Experiences of pain and discomfort became topographical features that could be, in Jenny's own words, "gently drifted away from".

A key quality of water is that, in calmness, it can act as a mirror (Illich, 2000). While it literally reflects light, the sea can also symbolically reflect one's life back, perhaps re-awakening memories of engagement with the sea (Coleman and Kearns, 2015). One study participant, Sam (aged 70), described viewing the sea from his dining room windows where the "light moving across the water" would elicit memories of his childhood sailing trips and successes as a fisherman in his younger years. Sam imagined he was "out there again, advising the captain a wind is coming or saying to the fisherman 'you won't catch anything today'". Views of water and light, then, allow vicarious participation in everyday life in an undemanding manner from the comfort of home. For Sam, such views not only allowed connection to the past and previous achievements (journeys made, fish caught) but also provided opportunities to resolve present challenges. In the context of deteriorating relationships with his adult children and painful memories of a difficult childhood, it was possible for Sam to "pause, think, ask questions and see different possible solutions" through contemplating the reflectiveness of water and being moved to "reflect deeply on life".

Engaging landscape views from the comfort of home may enable seniors and others spending considerable time in home spaces to connect with past relationships. Eileen (aged 80) maintained a relationship with her deceased husband by gazing at the sea, which her husband had loved, from the windows of her home. Engaging with views of the sea enabled Eileen to process her feelings about her husband's death and their relationship. Similarly, Maggie (82) looked out through her lounge windows to a gazebo located in her garden that was built many years earlier by her long-term partner. This view enabled her to connect with memories of her partner who had died 10 years earlier. Consequently, Maggie felt warmed by her partner's imagined presence and recalled shared happy events. By providing positive memories, these reminiscences – generated through recollection triggered by landscape views – facilitated a pleasant daily experience for participants and lessened the weight of their concerns and challenges. By connecting to the world beyond and providing a bounded space within which the past may be memorialised (i.e., via views), people may perceive their home environments as therapeutic, thereby supporting well-being by (re-)affirming positive affective ties to place, relationships and identities. This process may provide security and restoration, as well as a pleasant experience (Curtis *et al.*, 2007; Dyck *et al.*, 2005).

Imagined landscapes II: engaging diversity across intergenerational and intercultural soundscapes

In this section we remain on our field site of Waiheke Island, but turn to consider a case example of boundary-crossing involving performance and music to illustrate the potency of re-imagining ageing identities. Our starting point is that a widespread imagining of older people within a youth-centric Western culture is their placement in landscapes of immobility, nostalgia and conservatism. Through the 'metaphors we live by' (Lakoff and Johnson, 1980), public perceptions can 'place' especially frailer older people in (dis)positions of passivity. Whether it be 'rest' homes or 'nursing' homes, the implication of these prefixes is that passivity prevails and residents possess a distinct lack of agency. Even the term *retirement* bears a connotation of stepping back from useful and productive activity into a liminal zone of 'betweenness' (Entrikin, 1991) involving temporal and locational zones that bridge having a 'worldly' identity and experiencing mental and bodily dissolution. While some have discussed the 'age-less self' as the quest of older people to maintain a veneer of youthful vigour (McHugh, 2000), in this section we examine the case of older people incorporating cultural practices of youth but self-consciously celebrating their age.

While music is widely recognised as a medium for the evocation of memory and a link with the past (Jones, 2011), it can also offer opportunities to embody other images of ageing. In so doing, participation in alternative musical forms can endorse as well as forge new links between music, place and well-being (Andrews, Kingsbury and Kearns, 2014). In this section we explore the potential of music to be a site of resistance to perceptions of confinement and decline among otherwise frail older people. Specifically, we profile the example of a group of middle-class 'white' seniors who have embraced a distinctly 'other' form of musical expression in terms of 'race', age and class.

As a genre, hip hop has its origins in African-American, Caribbean and Latino youth subcultures, especially in New York, in the early 1970s. It incorporates rapping, DJ-ing, dance moves and dress styles, as well as being associated with particular forms of graffiti art. Elements of hip-hop have been associated with an amplification of the virtues of the otherwise disparaged neighbourhood (the 'hood) or 'ghetto' (Hess, 2009). In the case of the 'Hip Op-eration Crew' based on Waiheke Island, participating in this genre is a vehicle for a 'breaking out' of the normative discursive ghetto (ageism) and an act of reclaiming their place (in society, and on the island).

Hip Op-eration is registered in the Guinness World Records as the world's oldest dance group and was founded by Christchurch earthquake survivor Billie Jordan. At the time of writing, the dance crew consists of seven senior citizens aged 71 to 96 years old. The average age is 80, and there are two men and five women in the group. In a challenge to perceptions of passivity in advanced years, the Hip Op-eration website states that

> one member is legally blind and partially deaf, one member uses a mobility aid, one member is profoundly deaf, two members have had major heart surgery . . . many have artificial knees and hips, there is a married couple in the group (aged 83 and 85 years old) and all members have arthritis.
>
> (www.hipop-eration.com)

They started as a flash mob in August 2012, offering spontaneous performances around Auckland but soon specialised in hip-hop and eventually fund-raised their way to the world hip-hop championships in Las Vegas. An explicit aim is to use this form of dance as a medium through which to promote change in societal attitudes towards older people as well as forming intergenerational connections with young people. According to the founder, the group seeks "to also address . . . mis-perceptions and prejudices whilst forming a stronger connection with youth" (www.hipop-eration.com). The result has been described as "one of the most bizarre, improbable and joyful sights you will ever see" (Perrott, 2014: p. 2).

The founder expresses an explicit attempt to cross boundaries and bestow agency on elders:

> When we started I treated them like porcelain, but you have to understand life for old people ... they're treated like they can't do anything, they can't make a cup of tea, they can't go to the bathroom alone. They're incompetent and nobody has any expectations of them. That's demoralizing.
>
> (Perrott, 2014: 9)

Hip-hop is a genre which, for many in New Zealand, is associated with racialised Māori and Pacific youth in the poorer southern suburbs of Auckland, a considerable social and spatial distance from largely white Waiheke Island. In choosing this genre, Jordan was tacitly embracing the diversity of post-millennial Auckland (Murphy, Friesen and Kearns, 1999). To underline the intergenerational connectivity, their motto is RHY, Respect and Honour Youth, and they maintain a programme of exchange visits with the Dziah Dance Academy in Otara, a suburb commonly associated with poverty and deprivation (Friesen and Kearns, 2010).

The 2014 documentary titled *Hip Hop-eration* follows the group as they prepare for, and travel to, the World Hip-Hop Dance Championship in Las Vegas in 2013. The film focuses on three nonagenarian dancers in the ensemble – Maynie (95), Kara (94) and Terri (94), also known as Quicksilver, Kara Bang Bang and Terri 2-Cents, respectively – as they rehearse moves. The film recounts the story of amateurs moving from an ignorance of the genre to performing on the world stage. Participants include nonagenarians who intermittently embody alternate personas and transform whatever landscapes they occupy, subtly contesting the predictable and 'realistic' landscapes of old age (Laws, 1993). The message is that engagement with sound and music can bridge communities and that while elders can learn from younger people, their youthful mentors, in turn. develop respect for elders by cheering on the ageing dancers.

How can we 'read' Hip Op-eration (along with precedents such as the Young at Heart chorus from Massachusetts which similarly involved a younger facilitator enabling the participation of elders in music outside their stereotypical tastes)? As Twigg suggests, "we are aged by culture" (Twigg, 2004: 61). Her argument is that as the body is central to identity in old age, we need to know more about how prevailing cultural perceptions age the body and person. To her, performance of self and identity are critical to one's position, degree of empowerment and care for self (Twigg, 2000). Ultimately, performance and identity literally and metaphorically *take place*. The soundscapes and choreographies of well-being that are produced (Andrews *et al.*, 2014) amount to mobile and, arguably, subversive landscapes of old age. Hence, engagements like Hip Op-eration can be read as the performance of resistance to the passivity of 'rest' homes and an explicit engagement with the otherwise dissonant influence of the hyper-diverse city. Notably, the Hip Op-eration Crew's 'performance' is costumed such that they embody, and are re-named, into hip-hop character (but in other ways do not hide evidence of their age). In this act of simultaneous age-denying and age-embracing agency (Twigg, 2007) the participants play with an audience's perceptions of what is, or is not, age-appropriate. Interestingly, this 'audience' comprises not only those who come to their shows but also fellow residents of the island who are aware of their boundary-crossing activity. Indeed, some Waiheke residents expressed disdain towards the activities of those involved in Hip Op-eration. By implication, the Hip Op-eration crew were being disciplined and told to 'grow up'. Unlike the nostalgic 'saucy' postcards of older people that Blaikie (1997) analysed in coastal England, this contemporary group actively contests rather than fully embraces advanced old age. Indeed, Blaikie (2005) suggests there are two ways of denying the difficulties of the present: to disclaim continuity with the past and instead assert self-sufficiency or to embrace the past so fully that reality is subsumed by nostalgia. We believe the Hip Op-eration Crew epitomise a third possibility: emulating contemporary trends in popular culture while self-consciously parodying one's old age. The net result is embracing a more light-hearted landscape of ageing. In engaging in performances cutting across divides of age, ethnicity and class, these geriatric dancers are therefore, and paradoxically, 'growing up' and growing into an embrace of the hyper-diverse city (Hodkinson, 2013; Tasan-Kok *et al.*, 2013).

Conclusion

With considerable understatement, the introduction to an edited collection on ageing published a quarter century ago claimed that "to present the elderly population as a uniform group is misleading" (Koopman-Boyden, 1993: 2). In this chapter we have considered some of the ways a diversity of experiences and landscapes of old age exist 'beneath the radar' of objectively measured and mapped gerontological patterns and processes. This diversity confounds any assumptions of there being a 'single story' of ageing in place.

As our first case example indicated, views and memories are closely connected in imagined landscapes, whether as actual windows on the world or as evocation through photographs or other visual representations. Views of physical beauty may be a catalyst for memories, dreams and fantasies. Such views may entail connection and isolation, and the performance of emotional and imagined realities while being at an island-like distance (Baldacchino, 2007). We conclude that it is possible for people to imagine views in multiple and contradictory ways. This imagining is facilitated by the island-like boundaries of a given view itself. Views are in a sense akin to a defined landscape with clear boundaries (what can be seen through a window or from a particular position). Yet, like islands, they are connected to in- and outflows (of other times, places, ideas, people, identities, emotions and experiences). While it is possible to picture a particular view in one's mind as if it were a singular landscape, a 'single story' – it is also possible to imagine so much more. Thus, views offer multiple opportunities for metaphorical engagement with everyday life and phenomena such as ageing.

Our second case example, drawn from the same island setting, illustrated a very different story. The evolution of Hip Op-eration amounts to a narrative of independence, well-being and mobility that actively re-imagines landscapes of ageing, subverting the negative stereotypes associated with decline and dependency (Katz and Marshall, 2003). In a sense, then, what may, on first sight, have seemed a camp appropriation of youth culture revealed the ultimate malleability of the hip-hop genre and has highlighted the fact that youth culture itself is increasingly ambiguous and open to imaginative interpretation (Bennett and Hodkinson, 2012). Through this performative boundary-crossing, the presumed and otherwise 'real' landscape of old age is being re-imagined and music is enabling the reproduction of place (Hudson, 2006, Andrews *et al.*, 2014).

We have suggested that the social and cultural diversity of contemporary societies offers opportunities for creativity as well as new forms of social cohesion (Tasan-Kok *et al.*, 2013). Yet, as bodily and mental capabilities change and older people's activity spaces contract, 'outlook' on the world takes on a new meaning. We conclude that in ageing societies, the enumeration of people's needs and locations is an emphatically important task. This engagement with 'real' landscapes of hope as well as despair needs to be complemented by analysis and identification of service delivery solutions. However, the diversity of 'windows' into the human significance of place and performance offered by geographical theory and methods is opening new horizons of understanding both imagined and 'real' landscapes of old age.

References

Andrews, G J., Kingsbury, P. and Kearns, R. A. (eds). (2014). *Soundscapes of wellbeing in popular music.* Farnham: Ashgate Press.

Baldacchino, G. (ed.). (2007). *A world of islands: An island studies reader.* Charlottetown: Institute of Island Studies.

Bell, V. (1999). Performativity and belonging: An introduction, in V. Bell (ed.), *Performativity and belonging* (pp. 1–10). London: Sage. Available from: http://dx.doi.org/10.1177/02632769922050511

Bennett, A. and Hodkinson, P. (eds). (2012). *Ageing and youth cultures: Music, style and identity.* London: Bloomsbury.

Blaikie, A. (1997). Beside the sea: Visual imagery, ageing and heritage. *Ageing and Society,* 17, 629–648. http://dx.doi.org/10.1017/S0144686X97006703

Blaikie, A. (2005). Imagined landscapes of age and identity, in G. J. Andrews and D. R Phillips (eds), *Ageing and place: Perspectives, policy, practice* (pp. 164–175). London: Routledge.

Brooks, D. (2016). The danger of a single story. *New York Times,* 19 April. Available from: www.nytimes.com/2016/04/19/opinion/the-danger-of-a-single-story.html [Accessed 27 July 2016]

Coleman, T. and Kearns, R. A. (2015). The role of blue spaces in experiencing place, aging and wellbeing: Insights from Waiheke Island, New Zealand. *Health & Place,* 35, 206–217. http://dx.doi.org/10.1016/j.healthplace.2014.09.016

Coleman, T., Kearns, R. and Wiles, J. (2016). Older adults' experiences of home maintenance issues and opportunities to maintain aging-in-place. *Housing Studies,* 31(8), 964–983. doi.org/10.1080/02673037.2016.1164834

Curtis, S., Gesler, W., Fabian, K., Francis, S. and Priebe, S. (2007). Therapeutic landscapes in hospital design: A qualitative assessment by staff and service users of the design of a new hospital inpatient unit. *Environment and Planning C*, 25(4), 591–610. http://dx.doi.org/10.1068/c1312r

Duff, C. (2011). Networks, resources and agencies: On the character and production of enabling places. *Health & Place*, (Special Section: *Health Geographies of Voluntarism*), 17(1), 149–156. http://dx.doi.org/10.1016/j.healthplace.2010.09.012

Dyck, I., Kontos, P., Angus, J. and McKeever, P. (2005). The home as a site for long-term care: Meanings and management of bodies and spaces. *Health & Place*, (Special Section: *Geographies of Intellectual Disability*), 11(2), 173–185. http://dx.doi.org/10.1016/j.healthplace.2004.06.001

Entrikin, N. (1991). *The betweenness of place: Towards a geography of modernity*. New York: Palgrave Press.

Friesen, W. and Kearns, R. A. (2010). Otara and Dannemora: Contrasting landscape and ethnic identities in two South Auckland suburbs, in J. Stephenson, M. Abbott and J. Ruru (eds), *Beyond the scene: Landscape and identity in Aotearoa New Zealand* (pp. 93–110). Dunedin: Otago University Press.

Hess, M. (2009). *Hip Hop in America: A regional guide*. Santa Barbara: Greenwood Publishing Group.

Hodkinson, P. (2013). Spectacular youth cultures and ageing: Beyond refusing to grow up. *Sociology Compass*, 7(1), 13–22. http://dx.doi.org/10.1111/soc4.12008

Hudson, R. (2006). Regions and place: Music, identity and place. *Progress in Human Geography*, 30(5), 626–634. http://dx.doi.org/10.1177/0309132506070177

Illich, I. (2000). *H2O and the waters of forgetfulness*. London: Marion Boyars.

Jones, O. (2011). Geography, memory and non-representational geographies. *Geography Compass*, 5(12), 875–885. http://dx.doi.org/10.1111/j.1749-8198.2011.00459.x

Kaplan, S. (1995). The restorative benefits of nature: Toward an integrative framework. *Journal of Environmental Psychology*, 15(3), 169–182. http://dx.doi.org/10.1016/0272-4944(95)90001-2

Karmanov, D. and Hamel, R. (2008). Assessing the restorative potential of contemporary urban environment(s): Beyond the nature versus urban dichotomy. *Landscape and Urban Planning*, 86(2), 115–125. http://dx.doi.org/10.1016/j.landurbplan.2008.01.004

Katz, S. and Marshall, B. (2003). New sex for old: Lifestyle, consumerism, and the ethics of aging well. *Journal of Aging Studies*, 17(1), 3–16. http://dx.doi.org/10.1016/S0890-4065(02)00086-5

Kearns, R. A. and Collins, D.C.A. (2010). Health geography, in T. Brown and G. Moon (eds), *A companion to health and medical geography* (pp. 15–32). Oxford: Oxford University Press

Kellert, S. (2005). Coastal values and a sense of place, in D. Whitelaw and G. Visgilio (eds), *America's changing coasts: Private rights and public trust* (pp. 13–25). Cheltenham: Edward Elgar.

Knox, P. (1993). *The restless urban landscape*. New York: Prentice Hall.

Koopman-Boyden, P. (1993). *New Zealand's ageing society: The implications*. Wellington: Daphne Brasell Associated Press.

Lakoff, G. and Johnson, M. (1980). *Metaphors we live by*. Chicago: University of Chicago Press.

Laws, G. (1993). "The land of old age": Society's changing attitudes towards urban built environments for elderly people. *Annals of the Association of American Geographers*, 83(4), 672–693. http://dx.doi.org/10.1111/j.1467-8306.1993.tb01960.x

Mahmood, A., Chaudhury, H. Michael, Y. L., Campo, M., Hay, K. and Sarte, A. (2012). A photovoice documentation of the role of neighbourhood physical and social environments in older adults' physical activity in two metropolitan areas in North America. *Social Science & Medicine*, 74(8), 1180–1192. http://dx.doi.org/10.1016/j.socscimed.2011.12.039

McHugh, K. E. (2000). The "ageless self"? Emplacement of identities in sunbelt retirement communities. *Journal of Aging Studies*, 14(1), 103–115.

Murphy, L., Friesen, W. and Kearns, R. A. (1999). Transforming the city: People, property and identity in Millennial Auckland. *New Zealand Geographer*, 55(2), 60–65. http://dx.doi.org/10.1111/j.1745-7939.1999.tb00543.x

Peart, R. (2009). *Castles in the sand: What's happening to the New Zealand coast?* Port Nelson: Potton & Burton.

Perrott, A. (2014). Hip Op-eration Crew: Golden oldies. *New Zealand Herald*, 20 September. Available from: www.nzherald.co.nz/canvas-magazine/news/article.cfm?c_id=532&objectid=11326330 [Accessed 6 August 2016]

Pred, A. (1984). Place as historically contingent process: Structuration and the time-geography of becoming places. *Annals of the Association of American Geographers*, 74(2), 279–297. http://dx.doi.org/10.1111/j.1467-8306.1984.tb01453.x

Russell, L. (2005). "Either, or, neither nor": Resisting the production of gender, race and class dichotomies in the pre-colonial period, in E. Cassela and C. Fowler (eds), *The archaeology of plural and changing identities: Beyond identification* (pp. 33–54). New York: Plenum/Kluwer.

Russell, L. (2012). Remembering places never visited: Connections and context in imagined and imaginary landscapes. *International Journal of Historical Archaeology*, 16(2), 401–417. http://dx.doi.org/10.1007/s10761-012-0182-1

Sheller, M. and Urry, J. (2006). The new mobilities paradigm. *Environment and Planning A*, 38(2), 207–226. http://dx.doi.org/10.1068/a37268

Tasan-Kok, T., van Kempen, R., Raco, M. and Bolt, G. (2013). *Towards hyper-diversified European cities: A critical literature review*, 6pp. Utrecht: Utrecht University. Available from: www.urbandivercities.eu/wp-content/uploads/2013/05/DIVERCITIES_Policy_Brief_1.pdf

Twigg, J. (2000). Carework as a form of body work. *Ageing & Society*, 20(4), 389–411. http://dx.doi.org/10.1017/S0144686X99007801

Twigg, J. (2004). The body, gender and age: Feminist insights in social gerontology. *Journal of Aging Studies*, 18(1), 59–73. http://dx.doi.org/10.1016/j.jaging.2003.09.001

Twigg, J. (2007). Clothing, age and the body: A critical review. *Ageing & Society*, 27(2), 285–305. http://dx.doi.org/10.1017/S0144686X06005794

Whiston Spirn, A. (2008). "One with nature": Landscape, language, empathy, and imagination. In J. Elkins and R. De Lue (eds), *Landscape theory* (pp. 43–68). London and New York: Routledge.

23 Therapeutic landscapes of ageing

Rachel Winterton

Introduction

Places act as critical settings for the development and maintenance of population health (Williams, 2002; Wiles, Chapter 3). While this notion is reflected among current emphases on person–environment transactions in active ageing policies (World Health Organization (WHO), 2002), the healing powers of environmental contexts in older age have long been acknowledged within geographical gerontology (Cutchin, Marshall and Aldrich, 2010; Gesler, 1992; Martin *et al.*, 2005; Williams, 1999a). In recent decades, research in this field has sought to elucidate how older people experience and create meaning for themselves within environments and the subsequent impact on health and well-being in older age (Milligan *et al.*, 2015). While concepts and theories from varied fields of inquiry (environmental psychology, health and cultural geography) have been employed to interrogate this phenomenon, one concept of particular relevance is the notion of therapeutic landscapes (Gesler, 1992; Williams, 1999a).

Williams (1999b: 2) defines the therapeutic landscape as the "changing places, settings, situations, locales and milieus that encompass the physical, psychological and social environments associated with treatment or healing." This description was expanded in later years to encompass general well-being (Williams, 2002). This notion of environments as 'therapeutic' signals a discursive shift from a purely medical approach, toward a more holistic, ecological model which emphasises the individual, lived experience of health (Kearns and Andrews, 2005; Williams, 2002). It also encourages a transition from understanding place as a physically located entity, toward the notion of landscape as something that is socially constructed by means of human values and interactions across the life course (Gesler, 1992; Williams, 2002). Consequently, the therapeutic nature of particular settings lies primarily in the meanings they are assigned by diverse individuals (Williams, 1999b).

Although the therapeutic landscape can act as a geographic metaphor for understanding how places facilitate healing (Gesler, 1992), from a gerontological perspective, it also articulates how individuals adapt to both changes in environment and the ageing process (Cheng *et al.*, 2011). Therefore, while health and cultural geographers have long promoted the utility of the therapeutic landscape as a heuristic device for exploring well-being in older age (Cutchin, 2005;

Kearns and Andrews, 2005), to date there has been limited systematic, empirical application of the concept (Andrews *et al.*, 2009; Cutchin, 2005; Finlay *et al.*, 2015). Yet, the benefits of such an approach are significant in the context of increased rates of international population ageing (Cutchin *et al.*, 2010), the increased risk of disease, injury and hospitalisation faced by older adults (Finlay *et al.*, 2015), and the diverse, more nuanced interactions older adults have with places (Milligan, Gatrell and Bingley, 2004). The importance of place-based interactions is more pronounced for older adults living with dementia, because diminished memory and reasoning capacities prompt more intensive responses to, and engagement with, the physical environment (Sloane and Mathew, 1990). By means of a comprehensive review of the literature, this chapter considers how therapeutic landscapes are constructed and experienced in older age.

Locating therapeutic landscapes in older age

Therapeutic landscapes are often physically or spatially located in some way (Gesler, 1992; Kearns and Andrews, 2005), with the physical environment simultaneously influencing, regulating and providing a setting for interaction and behaviour in older age (Edvardsson, 2008). Within the gerontological literature, the therapeutic potential of physically located landscapes has been explored from the perspective of naturally occurring everyday spaces such as the home (Andrews, 2003), nature-based settings such as gardens, open skies or green spaces (Egdell, 2013; Finlay *et al.*, 2015; Milligan *et al.*, 2004), or community public spaces (Brittain *et al.*, 2010). Therapeutic landscapes can also reflect spaces designed specifically to fulfil a therapeutic purpose, with key examples in the literature including community-based activity centres (Milligan *et al.*, 2015), or institutional health and aged-care settings (Andrews, 2003; Cheng *et al.*, 2011; Edvardsson, 2008). However, therapeutic spaces can also exist at the spatial level, with older adults describing towns or regions as conducive to well-being (Andrews, 2003; Winterton and Warburton, 2012). Certain types of places can also be construed as therapeutic, with rural places possessing certain healing qualities (Winterton and Warburton, 2012; also see Skinner and Winterton, Chapter 11). Consequently, these different scales and types of landscapes often interact to produce an overall therapeutic experience for older adults (Andrews, 2003). However, given that the therapeutic nature of landscapes is related to how they are socially constructed (Gesler, 1992; Kearns and Gesler, 1998), therapeutic landscapes can also exist as imagined, rather than physical, spaces. Social gerontologists and health geographers have highlighted the importance of the imagined landscape in facilitating well-being in older age (Blaikie, 2005; Peace, Holland and Kellaher, 2006; Rose and Lonsdale, 2016; Rowles, 1978).

Therapeutic outcomes associated with landscape production in older age

Whether they are physically located, spatially situated or imagined, the therapeutic nature of certain spaces for older adults generally reflects their ability to improve or foster physical, mental or spiritual well-being (Cheng *et al.*, 2011; Hawkins *et al.*, 2013). Therapeutic settings promote feelings of renewal, peace, pleasure, relaxation and spiritual connectedness (Edvardsson, 2008; Finlay *et al.*, 2015; Milligan *et al.*, 2004), with certain spaces perceived to alleviate or improve the ability to cope with stress, improve mood and reinforce health-promoting behaviours (Cutchin *et al.*, 2010; Hawkins *et al.*, 2013; Milligan *et al.*, 2004; Milligan *et al.*, 2015). From the perspective of social well-being, spaces and settings that foster planned or impromptu social connectivity and inclusion (see Walsh, Chapter 20) have also been noted for their therapeutic qualities (Cheng *et al.*, 2011; Cutchin *et al.*, 2010, Finlay *et al.*, 2015; Milligan *et al.*, 2004).

As Gesler (1998) has observed, the therapeutic component of landscapes often reflects their ability to meet certain human needs. If the literature relating to therapeutic landscapes in older age is further examined, common themes are the ability of settings to compensate for losses relating to ageing and the capacity of landscapes to accommodate change (Cutchin *et al.*, 2010). Given that ageing is frequently accompanied by a loss of personal agency associated with reduced mobility and poorer health (Gilleard and Higgs, 2011), landscapes are often construed as therapeutic where they can enable greater independence, autonomy and capacity for self-care (Cheng *et al.*, 2011; Cutchin *et al.*, 2010; Martin *et al.*, 2005). In institutional settings, therapeutic potential is often linked with the ability of that space to take the focus away from declining mobility or illness (Edvardsson, 2008). If this is explored in relation to the importance of control and mastery over individual health outcomes for older adults (Glendinning, 2008), landscapes that enable choice and individualisation within the health care experience and that enable older adults to understand their care trajectory are also therapeutic (Andrews, 2003; Edvardsson, 2008; Verbeek *et al.*, 2009). In the event of acute, health-related events that challenge independence, spaces that facilitate functional improvement and rehabilitation are also significant in enabling healing (Finlay *et al.*, 2015; Martin *et al.*, 2005), particularly where they provide opportunities and support for participation (Andrews, 2003; Cheng *et al.*, 2011). Often, opportunities for participation are related to use and occupancy of physical space, with walkability and increased use of public space commonly linked with well-being (Andrews, 2003; Finlay *et al.*, 2015). In this context, spaces that are perceived as 'safe' by means of their physical or social features, are also deemed to have therapeutic qualities in older age, particularly for older adults living with dementia (Day, Carreon and Stump, 2000; Sloane and Mathew, 1990).

The therapeutic value of certain settings for older adults also includes their ability to reinforce certain identity processes (Brittain *et al.*, 2010; Wiersma, 2008). To some extent, benefits associated with identity reinforcement reflect the challenges that ageing-related processes such as retirement, reduced social roles and reduced mobility can pose to personal identity (Hay, 1998; Peace *et al.*, 2006; Rubinstein and Parmelee, 1992). Consequently, therapeutic spaces reflect both the ability of older people to be themselves (Edvardsson, 2008), and to re-situate their 'selves', by means of building competence, establishing new skills, developing alternate social roles or through fostering a sense of achievement

(Milligan *et al.*, 2004; Milligan *et al.*, 2015). These outcomes are associated with the ability of particular landscapes to create feelings of value, or lessen the perception of being a burden (Edvardsson, 2008), and assume greater significance for those experiencing substantial challenges to identity, such as older adults living with dementia (Milligan *et al.*, 2015).

In response to some of these threats to the self, place consequently assumes a crucial role in self-identity in older age and acts as a source of great meaning (Blaikie, 2005). Williams (1999c), suggests that where a fit exists between one's identity and their place location, that place can be considered therapeutic (see Rowles, Chapter 16). This use of environmental meaning to symbolise or situate the self has been referred to as place identity (Cuba and Hummon, 1993), and many geographical gerontologists have sought to explicate the nature of place identity in older age (e.g., Burholt, 2006; Peace *et al.*, 2006; Rowles, 1983; Winterton and Warburton, 2012). While it is beyond the scope of this chapter to review this literature, discussions of the therapeutic landscape in older age have nonetheless noted the importance of place identity and other accompanying concepts, such as place attachment and sense of place (Andrews, 2003; Blaikie, 1997; Cheng *et al.*, 2011; Martin *et al.*, 2005; Wiersma, 2008). As Gesler (1992) has noted, landscapes act as cultural texts to be read, and for older adults, therapeutic landscapes are instilled with symbols (such as cultural values and socio-economic status) that affirm the self (Cutchin *et al.*, 2010; Milligan *et al.*, 2004). Consequently, for many older adults, the therapeutic properties of landscapes lie in their ability to provide ontological security, by providing either memories and reminders of better times (Blaikie, 2005) or a sense of security or continuity (Brittain *et al.*, 2010; Cheng *et al.*, 2011; Cutchin *et al.*, 2010; Milligan *et al.*, 2004). For older adults, being 'at home' in a place – in terms of being able to locate themselves within a foreign setting – is associated with its therapeutic potential (Edvardsson, 2008). Similarly, for those living in institutional aged-care settings, the therapeutic nature of landscape reflects the ability of a space to facilitate a 'homelike' environment that normalises daily life (Day *et al.*, 2000; Verbeek *et al.*, 2009).

Processes of therapeutic landscape production in older age

The literature suggests that therapeutic landscapes are produced through interactions between physical environments, older adults and diverse social, symbolic, community, institutional and policy contexts (Martin *et al.*, 2005). To some extent, certain elements of therapeutic settings are 'produced' by external actors in response to societal, community and institutional trends. For example, certain places such as popular retirement migration locations and retirement villages are touted as therapeutic in older age in response to historical and social trends (Blaikie, 1997; Kearns and Andrews, 2005). Similarly, gender discourses can reinforce the construction of certain spaces as therapeutic, and gender discourses also are influenced by social and political discourses (Milligan *et al.*, 2015). Other settings are physically constructed using environmental design principles to fulfil therapeutic outcomes for older adults (Day *et al.*, 2000; Sloane and Mathew, 1990; Verbeek *et al.*, 2009). As Edvardsson (2008) has described, the design, maintenance and decoration of physical spaces within institutional settings can represent symbols of care or uncaring, and those symbols can influence social interaction, provision of care, visitation of family and behaviour of staff.

Within these physically or socially constructed spaces, the actions of institutional and policy actors can further enhance or negate their therapeutic potential for older adults. Within institutional health and aged-care settings, social and institutional processes such as management mechanisms, the ambiguous role of families within care settings, and philosophies of care can reinforce or contradict their perceived therapeutic benefits (Cheng *et al.*, 2011; Day *et al.*, 2000; Martin *et al.*, 2005). Edvardsson (2008) notes that the actions of staff within institutional settings can influence their perceived healing power, with a willingness to care and the ability of staff to facilitate welcoming, calm and safe spaces important to older adults. In this vein, certain policy directives can also delimit the capacity of certain spaces to be therapeutic. Martin and colleagues (2005) indicate that while older adults may view spaces of service provision as therapeutic, increasing emphasis on independent living and decreasing reliance on services will discourage interaction within these spaces.

However, older people themselves play an important role in the production and reproduction of therapeutic environments by means of their varied interactions with the landscape (Cheng *et al.*, 2011; Cutchin, 2005; Martin *et al.*, 2005). As such, historical, social and institutional trends interact with individual experiences, characteristics, intentions and actions (Andrews, 2003; Milligan *et al.*, 2004; Williams, 1999b), all of which influence the therapeutic potential of settings in older age. These interactions can also reinforce prominent stereotypes of certain places as therapeutic, as Blaikie (1997) has noted in reference to the continued movement of older adults to popular retirement locations (see Rosenberg and Wilson, Chapter 5). At the individual level, the perception of certain places as therapeutic in older age has been linked to personal characteristics, histories and embodied identities, including gender, race, ethnicity, cultural background, mobility level, self-confidence and sense of safety (Cheng *et al.*, 2011; Finlay *et al.*, 2015). These factors dictate the landscape preferences, needs and uses of older adults, and consequently, the construction of certain places as therapeutic differs considerably among older adults (Finlay *et al.*, 2015).

Construction and experience of therapeutic landscapes in older age can be achieved through active engagement, or by means of more passive interactions. From the perspective of active engagement, construction of therapeutic spaces is often achieved through occupation and activity engagement such as exercise, gardening, crafts, generative acts or constructing goods or objects (Cheng *et al.*, 2011; Finlay *et al.*, 2015; Hawkins *et al.*, 2013; Milligan *et al.*, 2004; Milligan *et al.*, 2015). Cutchin and colleagues (2010) have described these as 'everyday activities of meaning'. Consequently, social practices within physically bounded spaces are also significant in facilitating therapeutic landscapes for older adults (Cheng *et al.*, 2011; Milligan *et al.*, 2004). Older adults' active engagement within landscapes can also expand the scope of particular therapeutic encounters beyond the original space, or even create new therapeutic spaces. Andrews (2003) has observed that in the context of health care, following practitioner advice and employing self-help principles can extend the therapeutic space beyond formal places of care to the home setting. Similarly, social interactions can provide a bridge between public space (community) and private (home) space (Milligan *et al.*, 2004).

Older people also experience therapeutic landscapes in a more passive manner (Martin *et al.*, 2005), and this is achieved in two primary ways. For older adults, therapeutic landscapes are often constructed in the context of receiving care, and the symbolic meaning that older adults assign to the care they receive (Edvardsson, 2008). Landscapes are also often passively constructed and experienced by means of a 'scenic gaze' that shifts the focus from self to the environment (Edvardsson, 2008) and that positively impacts older adults' mental health (Milligan *et al.*, 2004). Often, this passive construction of therapeutic landscapes is linked with processes of reminiscence, which create an imagined landscape (see Kearns and Coleman, Chapter 22). For older adults, these imagined landscapes promote well-being by both providing consolation and compensation for perceived deficiencies in the current landscape (Blaikie, 2005, Wiersma, 2008). Consequently, the construction of certain settings as therapeutic in older age is often relational, that is, where places are deemed as therapeutic in comparison to other, less therapeutic places. For example, the threat of urban, built environments (see Phillips, Chapter 6) to personal safety and perceived competence in older age can render homes and natural environments more therapeutic to older adults (Milligan *et al.*, 2004). Therefore, as the subsequent section outlines, the therapeutic nature of places in older age is context-dependent and subject to changes in both people and place (Martin *et al.*, 2005).

Renegotiating the changing therapeutic landscape in older age

The contextual nature of interactions between older adults and their environments suggests that therapeutic landscapes are dynamic and subject to change throughout the life course (Cutchin *et al.*, 2010). From this perspective, they often exist as contested spaces (Foley, Wheeler and Kearns, 2011) that must be negotiated both at the individual and the collective level. At the individual level, landscapes can be simultaneously therapeutic and non-therapeutic (Cutchin, 2007; Williams, 2010), depending on the scale at which they are examined. While the natural landscape of a bounded place can contribute positively to well-being for older adults, the built environment can simultaneously be construed as threatening and unsafe (Milligan *et al.*, 2004). At the collective level, the negotiation of shared space by diverse individuals in the pursuit of therapeutic outcomes can also reduce the healing power of that setting for particular groups. For example, research has noted that the occupation of certain spaces by older people living with dementia can limit the therapeutic potential of that space for other older users (Day *et al.*, 2000; Milligan *et al.*, 2015).

Similarly, the in-migration of older adults into rural communities in pursuit of therapeutic outcomes can potentially reduce the therapeutic experience of place for longer-term, older residents (Winterton and Warburton, 2012). Therefore, these processes of negotiation and contestation are often prompted by the dynamic nature of places themselves. As Kearns and Gesler (1998) have described, landscapes frequently change in response to alterations in societal institutions, which can impact older people's perceptions of these places as therapeutic. Change in the physical and social characteristics of environments can negatively impact older people's ontological security within a place, which has a negative impact on place identities and subsequently adversely affects their well-being (Blaikie, 2005; Milligan *et al.*, 2004; Winterton and Warburton, 2012).

However, these alterations in physical and social landscapes are often inextricably linked to changes in older people, which can also impact the therapeutic nature of particular settings. For example, changing physical capacity and increasing frailty associated with ageing can impact older people's ability to engage actively with landscapes, leading to negative affect and challenges to personal identity, in turn reducing the healing properties of that particular setting (Blaikie, 2005; Milligan *et al.*, 2004). Cognitive changes, such as those related to dementia, can also destablise older adults' relationships with, and use of, environments, to the extent that these settings become threatening (Brittain *et al.*, 2010). Furthermore, while spaces can remain therapeutic over time for older adults living with dementia, the aspects of that place that produce healing outcomes may change. As Milligan and colleagues (2015) have noted, older people living with dementia often place more value on the ability of spaces to foster occupation and self-worth rather than social space and interactions. In this context, it is not necessarily the physical or social setting that changes but the older person's experience of the landscape.

While these examples refer to active engagement with landscape, changing health care needs in older age can also impact the passive experience of therapeutic landscapes, through deconstructing existing boundaries relating to public and private space (Cutchin, 2007). As Egdell (2013) notes, while the home is often a source of ontological security for older adults, care provision in the home can limit, or alter, the therapeutic nature of the home for older care recipients. This is particularly pronounced where caregiving practices within the homespace are construed as 'institutional' in nature (Day *et al.*, 2000). Consequently, while the home space may still be viewed as therapeutic from the perspective of its ability to facilitate health care and support, its therapeutic qualities as a 'retreat' from the external lifeworld may be reduced. Furthermore, while increasing advocacy for home-based care within contemporary policy structures is challenging the therapeutic potential of the home for older adults, this will also have implications for family carers. Where carers are co-resident with their older care recipient, provision of care in the home may consequently impact the therapeutic qualities of this setting for the carer. This will significantly impact carers' well-being, which may have an impact on their ability to facilitate a therapeutic home space for the care recipient (Williams, 2002).

While relocation to new environments in older age is often not conducive to well-being (Day *et al.*, 2000), older people also frequently experience change in their living environments due to movements across and within homes, communities and institutional care spaces (Cutchin, 2005). Consequently, whether these moves are voluntary or involuntary, the need for this cohort to seamlessly negotiate the therapeutic potential of new settings will become increasingly important to well-being and quality of life (see Cutchin, Chapter 17). This is particularly relevant where sites of relocation are generally stereotyped as non-therapeutic, such as residential aged-care facilities (Blaikie, 2005; Cutchin *et al.*, 2010). Consequently, the therapeutic potential of landscapes is a product of continued negotiation, struggle and resistance (Blaikie, 2005) and contingent on both the ability of people and settings to respond. Research suggests that the success of these responses, and subsequently the maintenance of therapeutic spaces, is contingent on processes of accommodation, incorporation, adaptation and compensation (Blaikie, 2005, Milligan *et al.*, 2004). However, support is often needed for older adults to undertake compensatory actions that allow them to maintain the therapeutic potential of particular settings or to locate new therapeutic spaces (Milligan *et al.*, 2004). This is particularly relevant where the therapeutic nature of landscape is predicated on active engagement and is influenced by individual- and macro-level factors such as cost and access (Cutchin *et al.*, 2010; Milligan *et al.*, 2004). The ability of landscapes to be therapeutic for certain groups, such as older people living with dementia, has also been questioned (Wahl, 2001). Certainly, the therapeutic outcomes of particular spaces are often gauged by the views of relatives, carers and staff rather than older people themselves (Cioffi *et al.*, 2007). In this vein, there is a pressing need for additional critical research to explore how the diverse life-course trajectories and living environments of older adults can impact capacity to construct, maintain and adapt therapeutic landscapes in response to change.

Towards a new research agenda for therapeutic landscapes and ageing

This critical review of the literature has highlighted the complex, dynamic and interactional nature of therapeutic landscape construction among older adults. As such, the value of this concept lies primarily in its ability to facilitate a much more nuanced, ecological understanding of health and well-being for diverse ageing populations. However, a major issue in tracking the development of the therapeutic landscapes paradigm within geographical gerontology is the lack of systematic use of the concept (Kearns and Andrews, 2005). While there has been much work within the field in relation to affective relationships with place in older age, little of this work has drawn on the therapeutic landscape as a heuristic device, nor has this work sought to explicate its relationship to broader person–place concepts (i.e., place identity, place attachment) in older age. This is problematic, as continued anecdotal, inconsistent use of the concept will impact its comprehensiveness (Williams, 2010). Consequently, there is a concerted need for geographical gerontologists to engage more closely with the therapeutic landscapes paradigm, utilising a critical lens. To date, there has been little systematic attention to how the construction of therapeutic landscapes in older age is influenced by factors relating to class, gender, culture and the life course. There is considerable scope for a more critical exploration of how particular spaces hold different therapeutic potentials for diverse older adults and how these therapeutic spaces are contested and renegotiated over time.

References

Andrews, G. (2003). Placing the consumption of private complementary medicine: Everyday geographies of older peoples' use. *Health & Place*, 9(4), 337–349. http://dx.doi.org/10.10 16S1353-8292(02)00068-0

Andrews, G., Milligan, C., Phillips, D. and Skinner, M. (2009). Geographical gerontology: Mapping a disciplinary intersection. *Geography Compass*, 3(5), 1641–1659. http://dx.doi.org/10.1111/j.1749-8198.2009.00270.x

Blaikie, A. (1997). Beside the sea: Visual imagery, ageing and heritage. *Ageing & Society*, 17(6), 629–648. http://dx.doi.org/10.1017/S0144686X97006703

Blaikie, A. (2005). Imagined landscapes of age and identity, in G. Andrews and D. Phillips (eds), *Ageing and place: Perspectives, policy, practice*. London: Routledge.

Brittain, K., Corner, L., Robinson, L. and Bond, J. (2010). Ageing in place and technologies of place: The lived experience of people with dementia in changing social, physical and technological environments. *Sociology of Health and Illness*, 32(2), 272–287. http://dx.doi.org/10.1111/j.1467-9566.2009.01203.x

Burholt, V. (2006). Adref: Theoretical contexts of attachment to place for mature and older people in rural North Wales. *Environment and Planning A*, 38(6), 1095–1114. http://dx.doi.org/10.1068/a3767

Cheng, Y., Rosenberg, M. W., Wang, W., Yang, L. and Li, H. (2011). Aging, health and place in residential care facilities in Beijing, China. *Social Science & Medicine*, 72(3), 365–372. http://dx.doi.org/10.1016/j.socscimed.2010.10.008

Cioffi, J. M., Fleming, A., Wilkes, L., Sinfield, M. and Le Miere, J. (2007). The effect of environmental change on residents with dementia: The perceptions of relatives and staff. *Dementia*, 6(2), 215–231. http://dx.doi.org/10.1177/1471301207080364

Cuba, L. and Hummon, D. (1993). A place to call home: Identification with dwelling, community and region. *Sociological Quarterly*, 34(1), 111–131. http://dx.doi.org/10.1111/j.1533-8525.1993.tb00133.x

Cutchin, M. (2005). Spaces for inquiry into the role of place for older people's care. *International Journal of Older People Nursing*, 14(8B), 121–129. http://dx.doi.org/10.1111/j.1365-2702.2005.01280.x

Cutchin, M. (2007). Therapeutic landscapes for older people: Care with commodification, liminality, and ambiguity, in A. Williams (ed.), *Therapeutic landscapes: Advances and applications* (pp. 181–198). Aldershot: Ashgate.

Cutchin, M., Marshall, V. and Aldrich, R. (2010). Moving to a continuing care retirement community: Occupations in the therapeutic landscape process. *Journal of Cross-Cultural Gerontology*, 25(2), 117–132. http://dx.doi.org/10.1007/s10823-010-9113-y

Day, K., Carreon, D. and Stump, C. (2000). The therapeutic design of environments for people with dementia: A review of the empirical research. *The Gerontologist*, 40(4), 397–416. http://dx.doi.org/10.1093/geront/40.4.397

Edvardsson, D. (2008). Therapeutic environments for older adults: Constituents and meanings *Journal of Gerontological Nursing*, 34(6), 32–40. http://dx.doi.org/10.3928/00989134-20080601-05

Egdell, V. (2013). Who cares? Managing obligation and responsibility across the changing landscapes of informal dementia care. *Ageing & Society*, 33(5), 888–907. http://dx.doi.org/10.1017/S0144686X12000311

Finlay, J., Franke, T., Mckay, H. and Sims-Gould, J. (2015). Therapeutic landscapes and well-being in later life: Impacts of blue and green spaces for older adults. *Health & Place*, 34, 97–106. http://dx.doi.org/10.1016/j.healthplace.2015.05.001

Foley, R., Wheeler, A. and Kearns, R. (2011). Selling the colonial spa town: The contested therapeutic landscapes of Lisdoonvarna and Te Aroha. *Irish Geography*, 44(2–3), 151–172. http://dx.doi.org/10.1080/00750778.2011.616059

Gesler, W. M. (1992). Therapeutic landscapes: Medical issues in light of the new cultural geography. *Social Science & Medicine*, 34(7), 735–746. http://dx.doi.org/10.1016/0277-9536(92)90360-3

Gesler, W. M. (1998). Bath's reputation as a healing place, in R. Kearns and W. M. Gesler (eds), *Putting health into place: Landscape, identity and well-being*. New York: Syracuse University Press.

Gilleard, C. and Higgs, P. (2011). Ageing abjection and embodiment in the fourth age. *Journal of Aging Studies*, 25(2), 135–142. http://dx.doi.org/10.1016/j.jaging.2010.08.018

Glendinning, C. (2008). Increasing choice and control for older and disabled people: A critical review of new developments in England. *Social Policy & Administration*, 42(5), 451–469. http://dx.doi.org/10.1111/j.1467-9515.2008.00617.x

Hawkins, J., Mercer, J., Thirlaway, K. and Clayton, D. (2013). "Doing" gardening and "being" at the allotment site: Exploring the benefits of allotment gardening for stress reduction and healthy aging. *Ecopsychology*, 5(2), 110–125. http://dx.doi.org/10.1089/eco.2012.0084

Hay, R. (1998). Sense of place in developmental context. *Journal of Environmental Psychology*, 18(1), 5–29. http://dx.doi.org/10.1006/jevp.1997.0060

Kearns, R. and Andrews, G. (2005). Placing ageing, in G. Andrews and D. Phillips (eds), *Ageing and place: Perspectives, policy and practice*(pp. 13–23). London: Routledge.

Kearns, R. and Gesler, W. M. (1998). Introduction, in R. Kearns and W. M. Gesler (eds), *Putting health into place: Landscape, identity and well-being* (pp. 1–14). New York: Syracuse University Press.

Martin, G., Nancarrow, S., Parker, H., Phelps, K. and Regen, E. (2005). Place, policy and practitioners: On rehabilitation, independence and the therapeutic landscape in the changing geography of care provision to older people in the UK. *Social Science & Medicine*, 61(9), 1893–1904. http://dx.doi.org/10.1016/j.socscimed.2005.04.001

Milligan, C., Gatrell, A. and Bingley, A. (2004). "Cultivating health": Therapeutic landscapes and older people in northern England. *Social Science & Medicine*, 58(9), 1781–1793. http://dx.doi.org/10.1016/S0277-9536(03)00397-6

Milligan, C., Payne, S., Bingley, A. and Cockshott, Z. (2015). Place and wellbeing: Shedding light on activity interventions for older men. *Ageing & Society*, 35(1), 124–149. http://dx.doi.org/10.1017/S0144686X13000494

Peace, S., Holland, C. and Kellaher, L. (2006). *Environment and identity in later life*. Berkshire: Open University Press.

Rose, E. and Lonsdale, S. (2016). Painting place: Re-imagining landscapes for older people's subjective wellbeing. *Health & Place*, 40, 58–65. http://dx.doi.org/10.1016/j.healthplace.2016.05.002

Rowles, G. (1978). *Prisoners of space? Exploring the geographical experience of older people.* Boulder, CO: Westview Press.

Rowles, G. (1983). Place and personal identity in old age: Observations from Appalachia. *Journal of Environmental Psychology*, 3(4), 299–313. http://dx.doi.org/1.1016/s0272-4944(83)80033-4

Rubinstein, R. and Parmelee, P. (1992). Attachment to place and the representation of the life course by the elderly. In I. Altman and S. Low (eds), *Place attachment* (pp. 139–163). New York: Plenum Press.

Sloane, P. D. and Mathew, L. J. (1990). The therapeutic environment screening scale: An observational screening instrument to assess the quality of nursing home environments for residents with dementia. *American Journal of Alzheimer's Disease and Other Dementias*, 5(5), 22–26. http://dx.doi.org/10.1177/153331759000500606

Verbeek, H., Van Rossum, E., Zwakhalen, S.M.G., Kempen, G.I J.M. and Hamers, J.P.H. (2009). Small, homelike care environments for older people with dementia: A literature review. *International Psychogeriatrics*, 21(2), 252–264. http://dx.doi.org/10.1017/S104161020800820X.

Wahl, H. 2001). Environmental influences on aging and behavior, in J. Birren and K. W. Schaie (eds), *Handbook of the psychology of aging* (pp. 215–237). San Diego, CA: Academic Press.

Wiersma, E. C. (2008). The experiences of place: Veterans with dementia making meaning of their environments. *Health & Place*, 14(4), 779–794. http://dx.doi.org/10.1016/j.healthplace.2008.01.001

Williams, A. (1999a). *Therapeutic landscapes: The dynamic between place and wellness.* Lanham, MD: University Press of America.

Williams, A. (1999b). Introduction, in A. Williams (ed.), *Therapeutic landscapes: The dynamic between place and wellness* (pp. 1–14). Lanham, MD: University Press of America.

Williams, A. (1999c). Place identity and therapeutic landscapes: The case of home care workers in a medically underserviced area, in A. Williams (ed.), *Therapeutic landscapes: The dynamic between place and wellness* (pp. 71–96). Lanham, MD: University Press of America.

Williams, A. (2002). Changing geographies of care: Employing the concept of therapeutic landscapes as a framework in examining home space. *Social Science & Medicine*, 55(1), 141–154.

Williams, A. (2010). Therapeutic landscapes as health promoting places, in T. Brown, S. Mclafferty and G. Moon (eds), *A companion to medical and health geography* (pp. 205–223). Malden, MA: Wiley-Blackwell.

Winterton, R. and Warburton, J. (2012). Ageing in the bush: The role of rural places in maintaining identity for long term rural residents and retirement migrants in North-East Victoria, Australia. *Journal of Rural Studies*, 28(4), 329–337. http://dx.doi.org/10.1016/jjrurstud.2012.01.005

World Health Organization (WHO). (2002). *Active ageing, a policy framework.* WHO. Available from: http://apps.who.int/iris/bitstream/10665/67215/1/WHO_NMH_NPH_02.8.pdf

.

Part V

Discussion

24 On the interdisciplinary challenge of geographical gerontology

Sheila Peace

Introduction

In reviewing the perspectives, concepts and approaches that underpin geographical gerontology, the collection of scholarship in this book provides the reader with a comprehensive account of the current state of the art in a field of study where social, urban, population, health, economic, rural and cultural geographers have come to study age in relation to later life. They are joined in the earlier chapters by geographically sensitive scholars with sociological, architectural and historical backgrounds, among others. In seeking to contribute to the discussion on the evolution and importance of geography for the study of ageing, this chapter provides a viewpoint from this interdisciplinary position.

To start this discussion chapter, I situate my academic self. I am by first discipline a human geographer (Herbert and Peace, 1980; Peace, 1977, 1982) whose doctoral research focused on the interface between environment, context and the experience of later life. In 2005, a comment by Andrews and Phillips highlighted the disciplinary dilemma inherent in this dialogue: "*a wider tension and predicament for many geographical gerontologists regarding what discipline to speak from and to – to be a geographer who studies older people or a gerontologist who takes a geographical perspective*" (2005: 10). My own concern with this interface led to a move into the interdisciplinary field of social gerontology. On reflection, it is possible to address both these disciplinary positions simultaneously, however. For while space, place, territory, environment, dwelling and landscape – in light of population ageing at both individual and collective levels – are the anchors of our work, they gain through collaboration that may occur within or among disciplines, including geography and gerontology (Peace, 2017).

A framework for advancing geographical gerontology

To consider an agenda and future direction for this field of study, I have reread positioning papers emerging during the late twentieth century, particularly those from *Progress in Human Geography* (Warnes, 1981; Rowles, 1986; Harper and Laws, 1995; also Skinner, Cloutier and Andrews, 2015), alongside more recent discussions of how geographies of ageing can be captured and developed within this post-modern, neo-liberal period when the need for a critical approach to gerontology has already been embraced, and environmental gerontology continues to question its theoretical base and application (Andrews, Evans and Wiles, 2013; Andrews and Phillips, 2005; Bernard and Scharf, 2007; Kendig, 2003; Rowles and Bernard, 2013; Scheidt and Schwarz, 2013; Wahl and Weisman, 2003; also Andrews, Cutchin and Skinner, Chapter 2). During this review, a framework has emerged that offers human geographers the unique opportunity to both contribute to gerontology through a holistic view of the spatiality of population ageing and, in alignment with the foundational aims of this book, to further develop geographical gerontology within their own sub-disciplines and beyond. Here I define the spatiality of ageing as encompassing the synergy of space and place at all levels. The framework for discussion includes six themes which interact with each other resulting in a construct that

- considers age as a major factor in intersectional analysis;
- operates at the different levels of spatial scale that capture person-environment interactions;
- enables greater recognition of the relationship between space, place and temporality;
- demands continuing methodological innovation;
- opens up the possibility of examining theoretical perspectives among disciplines; and
- recognises the importance of scholarly, policy and programme debates.

In a time of growing life expectancy through ongoing changes in mortality, fertility and morbidity the framework starts by foregrounding age and the ageing of populations – the essence of gerontology. In doing so, it recognises that within geography the interest in age as a distinguishing variable has come first through studies with children and young people and more recently with older people where policy-related research concerning health, long-term care and community living are major drivers (see Hanlon, Chapter 18; Milligan and Tarrant, Chapter 4; Skinner, Andrews and Cutchin, Chapter 1; Wiles, Chapter 3). Discussion of 'successful' ageing, and the experiences of the third and fourth ages, identifies diversity which needs to be considered through intersectional analyses bringing greater depth to our understanding of ageing alongside gender, ethnicity, social class, sexuality, disability and culture (see Gilleard and Higgs, 2002; 2010). In addition, a life course perspective could aid further development encompassing generational difference(s) (Hopkins and Pain, 2007; see also Walsh, Chapter 20). There is a need to understand ageing and later life beyond legitimation as a key variable.

Second, the detail of population ageing has to be addressed from different levels of spatial scale that are both locational and social. As illustrated in Part III of this book, it is here that geographers have the demographic expertise to analyse and communicate the knowledge base at global, regional, national and local levels. Patterns of demographic change have to be located within an understanding of levels of economic development and ongoing trends in urbanisation, migration, conflict, climate change and technological development already part of geographical study (see Phillips and Feng, Chapter 8). Such diverse trends can also be seen as all-embracing and *macro* impacting on *meso-* and *micro*-level environments – there is an interrelationship among these that has to be acknowledged (see Buffel and Phillipson, Chapter 10; Peace, 2019; Phillipson and Scharf, 2005). The value of this part of the framework is essential to the contextualisation of issues ranging, for example, from the development of national transportation systems to patterns of neighbourhood interaction to the provision of home care within mainstream dwellings – all topics where a basic recognition of spatial scale and the diversity of the population is necessary. The recent spatial turn within the social sciences has recognised that viewing levels of spatial scale as constant and unconnected is simplistic (Andrews *et al.*, 2013).

The third aspect for elaboration situates place alongside space in association with temporality (time). While the concept of space has been debated across disciplines from the philosophical to the architectural, geographical thinking is central. The social construction of space and importance of power within both the natural and human landscapes has been heralded across the years by a number of key geographical thinkers from Anne Buttimer to Yi-Fu Tuan, David Harvey to Doreen Massey, Gillian Rose to Nigel Thrift, alongside many others. The interface between space and place is fluid, with place defining different levels of meaning and attachment of emotional value in what may be defined as public or private spaces. The presence of attachment introduces temporality at a personal level, which, alongside the collective experience of different generations, is informative regarding issues such as ageing in place (see Golant, Chapter 15; Rowles, Chapter 16). We must also acknowledge the impact of diurnal, seasonal, annual and historic time as elements of change. As geographers, the issue of temporality may be the opportunity to confront our own ageism and how this may influence our understanding of the social construction of space. For instance, even though gender differences are central to conceptual development for feminist authors within geography (see Joseph and Joseph, Chapter 21; Laws, 1997; Rose, 1993), later life has been largely overlooked in discussions of gender. More recent feminist contemplation of place through home shows that the impact of age and time on place are beginning to be recognised (see Peace, 2015; Young, 2005), and again, geographers may need to draw on the work of other disciplines (e.g., Krekula, 2007).

The complexity of these associations demands appropriate methodologies utilising more reflexive post-modern approaches. Advancing this fourth area is important in multi-/interdisciplinary research and expertise in both quantitative and qualitative methods (see Cutchin, Chapter 17). The rigorous analysis of big data from both the developed and developing world is necessary for monitoring population ageing, with lobbying for new and ongoing data collection essential. Geographers are leaders when mapping data, using geographical information systems (GIS) and visual methods. Additionally, qualitative methods utilising ethnographic, biographical and oral history skills enable a more detailed life-course approach to personal history of value to both cross-sectional and longitudinal research (see Andrews *et al.*, 2006; Skinner *et al.*, 2015). Last, within contextual studies mixed-method approaches that involve participatory research with older people have been welcomed especially given the applied nature of research in relation to policy development (Peace, 2002).

The fifth part of the framework – which reinforces the aspirations of this book – brings theoretical perspectives to the fore, encouraging geographical gerontologists to demonstrate how contemporary perspectives from geography can be adapted, developed and tested both within and among disciplines. The evolution of environmental gerontology over more than 50 years has moved from a primarily psychosocial understanding of the person-environment fit/congruence model suggested by the ecological theory of ageing (Kleemeier, 1959; Lawton and Nahemow, 1973, etc.), to a broader recognition of the complexities involved when material, social, psychological and biological facets are considered central to the experience and exploration of belonging and agency that impact on identity and autonomy in later life (Wahl and Oswald, 2010). Intervention research leading to improved physical-social environments and better quality of life remains central. While some geographers have followed this environmental gerontological trajectory, others are currently exploring contemporary relational and non-representational theories to consider space and place within the geographies of ageing (see Andrews *et al.*, 2013; Hopkins and Pain, 2007). While relational theory may continue to develop the 'inter-related' nature of place, for example, through intergenerationality, non-representational theory moves to a different level where space and place are experienced in the moment, embodied and witnessed through 'atmosphere' or emotional impact (see Herron, Chapter 14). Further acceptance of these innovations alongside the development of methodologies that capture the fluidity of minute everyday experience in later life is to be valued. It is here that interdisciplinary debate is important. To see theoretical development emerging within geographical gerontology, as outlined by Andrews *et al.* (Chapter 2) and illustrated in various chapters in this book, is central to this agenda and very welcome.

Which brings us to the sixth (and final) aspect addressed here: the centring of geographical gerontology within academic life where it should be recognised through both teaching and research as an important area of scholarship. The geographic community has a language of landscape and mapping that is fundamental to this disciplinary perspective (see Kearns and Coleman, Chapter 22; Winterton, Chapter 23). It is an area of humanistic geography where research often has the capacity to make a direct impact on people's lives across a range of policy areas from health to housing, transport to leisure, financial arrangements to age-friendly cities and communities.

Concluding comments

While Skinner *et al.* (2015), Andrews *et al.* (2013) and Cutchin (2009), as well as the contributors they have gathered together in this volume, have demonstrated how the growth of geographical research contributes to gerontology, geographical gerontologists must continue to be assertive and demonstrate the importance that their skills and expertise regarding the spatiality of ageing can bring to wider debates. There is a need to recognise and acknowledge the geographical view within gerontology alongside psychology, sociology, economics, anthropology, architecture and design, as well as the practice of nursing, social work and occupational therapy, among others. This book provides an essential brief for those within, and those sensitive to, the geographical discipline. In setting an agenda to broaden its scope, it is my hope that the framework of areas for development discussed here offers an exciting starting point. How geographers contribute to and meet the challenge of collaborative interdisciplinary study – particularly at a time when global ageing is a crucial factor in our changing world – is imperative for the future of geographical gerontology.

References

Andrews, G. J., Evans, J. and Wiles, J. (2013). Re-spacing and re-placing gerontology: Relationality and affect. *Ageing & Society*, 33(8), 1339–1373. http://dx.doi.org/10.1017/S0144686X12000621

Andrews, G. J., Kearns, R. A., Kontos, P. and Wilson, V. (2006). "Their Finest hour": Older people, oral histories, and the historical geography of social life. *Social and Cultural Geography*, 7(2), 153–179. http://dx.doi.org/10.1080/1464936000600338

Andrews, G. J. and Phillips, D. R. (eds). (2005). *Ageing and place: Perspectives, policy, practice*. London: Routledge.

Bernard, M. and Scharf, T. (eds). (2007). *Critical perspectives on ageing societies*. Bristol: Policy Press.

Cutchin, M. P. (2009). Geographical gerontology: New contributions and spaces for development. *The Gerontologist*, 49(3), 440–444. http://dx.doi.org/10.1093/geront/gnp095

Gilleard, C. and Higgs, P. (2002). Concept forum – the third age: Class, cohort or generation? *Ageing & Society*, 22(3), 369–382. http://dx.doi.org/10.1017/S0144686X0200870X

Gilleard, C. and Higgs, P. (2010). Aging without agency: Theorizing the fourth age. *Aging & Mental Health*, 14(2), 121–128. http://dx.doi.org/10.1080/13607860903228762

Harper, S. and Laws, G. (1995). Rethinking the geography of ageing. *Progress in Human Geography*, 19(2), 199–221. http://dx.doi.org/10.1177/030913259501900203

Herbert, D. T. and Peace, S. M. (1980). The elderly in an urban environment: A study in Swansea, in D. T. Herbert (ed.), *Geography and the urban environment* (pp. 223–256). Chichester: John Wiley & Sons.

Hopkins, P. E. and Pain, R. (2007). Geographies of age: Thinking relationally. *Area*, 39(3), 287–294. http://dx.doi.org/10.1111/j.1475-4762.2007.00750.x

Kendig, H. (2003). Directions in environmental gerontology: A multidisciplinary field. *The Gerontologist*, 43(5), 611–615. http://dx.doi.org/10.1093/geront/43.5.611

Kleemeier, R. W. (1959). Behavior and the organization of the bodily and external environment, in J. E. Birren (ed.), *Handbook of aging and the individual* (pp. 400–451). Chicago: University of Chicago Press.

Krekula, C. (2007). The intersection of age and gender: Reworking gender theory and social gerontology. *Current Sociology*, 55, 155–171.

Laws, G. (1997). Spatiality and age relations, in A. Jamieson, S. Harper and C. Victor (eds), *Critical approaches to ageing and later life* (pp. 90–100). Maidenhead: Open University Press.

Lawton, M. P. and Nahemow, L. (1973). Ecology and the aging process, in C. Eisdorfer and M. P. Lawton (eds), *The psychology of adult development and aging* (pp. 619–674). Washington, DC: American Psychological Association.

Peace, S. (1977). *The elderly in an urban environment: A study of spatial mobility in Swansea*. PhD Thesis, Dept. of Geography, University College of Swansea.

Peace, S. (1982). The activity patterns of elderly people in differing environments, in A. Warnes (ed.), *Geographical perspectives on the elderly* (pp. 281–302). Chichester: John Wiley & Sons.

Peace, S. (2002). The role of older people in research, in A. Jamieson and C. Victor (eds), *Researching ageing and later life: The practice of social gerontology* (pp. 226–242). London: Sage Publications.

Peace, S. (2015). Meaning of home and age, in J. Twigg and W. Martin (eds), *Routledge handbook of cultural gerontology* (pp. 447–454). Oxford: Routledge.

Peace, S. (2017). Home thoughts, in H. Chaudhury and F. Oswald (eds), *Autobiographical perspectives in environmental gerontology: Understanding the field through personal reflections*, 2018 Annual Review of Gerontology and Geriatrics, 38. New York: Springer Publications.

Peace, S. (2019). *Environment and ageing: Space, place and materiality*. Bristol: Policy Press.

Phillipson, C. and Scharf, T. (2005). Rural and urban perspectives on growing old: Developing a new research agenda. *European Journal of Aging*, 2(1), 67–75. http://dx.doi.org/10.1007/s10433-005-0024-7

Rose, G. (1993). *Feminism and geography: The limits of geographical knowledge*. Minneapolis, MN: University of Minnesota Press.

Rowles, G. (1986). The geography of ageing and the aged: Towards an integrated perspective. *Progress in Human Geography*, 10(4), 511–539. http://dx.doi.org/10.1177/030913258601000403

Rowles, G. and Bernard, M. (eds). (2013). *Environmental gerontology: Making meaningful places in old age*. New York: Springer Publications.

Scheidt, R. J. and Schwarz, B. (2013). Last words, in R. J. Scheidt and B. Schwarz (eds), *Environmental gerontology: What now?* (pp. 326–334). London and New York: Routledge.

Skinner, M. W., Cloutier, D. and Andrews, G. J. (2015). Geographies of ageing: Progress and possibilities after two decades of change. *Progress in Human Geography*, 39(6), 776–799. http://dx.doi.org/10.1177/0309132514558444

Wahl, H-W. and Oswald, F. (2010). Environmental perspectives on ageing, in D. Dannefer and C. Phillipson (eds), *The Sage handbook of social gerontology* (pp. 111–124). Los Angeles, CA and London: Sage. Available from: http://sk.sagepub.com/reference/download/hdbk_socialgerontology/n8.pdf

Wahl, H-W. and Weisman, G. (2003). Environmental gerontology at the beginning of the new millennium: Reflections on its historical, empirical, and theoretical development. *The Gerontologist*, 43(5), 616–627. http://dx.doi.org/10.1093/geront/43.5.616

Warnes, A. (1981). Towards a geographical contribution to gerontology. *Progress in Human Geography*, 5(3), 317–341. http://dx.doi.org/10.1177/030913258100500301

Young, I. M. (2005). *On female body experience: "Throwing like a girl" and other essays*. New York: Oxford University Press. Available from: https://biblioteca-alternativa.noblogs.org/files/2011/11/On_Female_Body_Experience___quot_Throwing_Like_a_Girl_quot__and_Other_Essays__Studies_in_Feminist_Philosophy_.pdf

25 Geographical gerontology

Progress and possibilities

Malcolm P. Cutchin, Mark W. Skinner
and Gavin J. Andrews

> Geographical imagination: A sensitivity towards the significance of place and space, landscape and nature, in the constitution and conduct of life on Earth. As such, a geographical imagination is by no means the exclusive preserve of the academic discipline of geography.
>
> —Gregory (2009: 282)

Introduction

As noted at the outset, the aim of this book is to advance the field of study called 'geographical gerontology', a growing body of interdisciplinary scholarship that uses geographical perspectives, concepts and approaches to study ageing, old age and older populations. Although the field emerged in the late 1970s, to date no volume has systematically covered the major perspectives, concepts and approaches that define the field. This volume provides a representative scope of coverage of the field as a whole, along with insightful depth within each chapter to help readers grapple with the nuances and complexities of ageing and geography. As geographers, we are eager to note that all human experience, including all of it related to ageing, happens *someplace* – together with some geography. And that geography matters for ageing and our understanding of ageing, whether it be considered as place, landscape, spatial relationship or some other conceptualisation. We believe this book will help others understand that foundational knowledge better and apply the knowledge in their own gerontological work.

This book makes the case for geographical gerontology through a planned organisation of thought and research into the subject matter. After our introduction and a review of the evolving collection of theoretical traditions in geographical gerontology inquiry (Part I), some foundational perspectives on geographical gerontology were introduced and assessed by leaders in the field (Part II). This background and orientation to core geographical perspectives on ageing illustrated the diversity of the field from a bird's-eye view, but it also helped to trace the development within those core perspectives and suggested where further development is needed. Part III offered more grounded assessments of the state of the field; each chapter focused on a different scale of inquiry, moving from global ageing to issues of embodiment and emotion in later life. Part IV provided insights into key issues and dynamics involved in geographical gerontology. Although there may be argument with what has not been covered fully enough in this book, the volume as a whole and its constituent chapters present the current state of the geographical imagination in gerontological inquiry.

The purpose of this concluding chapter is to take stock of that geographical gerontological imagination and consider what the most likely and most important future opportunities and challenges exist for the field over the temporal horizon. As geographical gerontology continues to emerge as a truly transdisciplinary field, what will enhance its ability to contribute to our understanding and make positive impacts on an ageing world – theoretically, conceptually, methodologically and practically? What about the future of this ageing world is likely to need geographical gerontology's attention and expertise most? While we expect much of the evolution of the field will continue without any course charting, we offer some suggestions for advancing future research that would simultaneously address gaps in the current state of knowledge and broaden the global reach and critical depth of geographical gerontology.

Hallmarks of, and gaps in, geographical gerontology

We imagine that readers less familiar with geographical gerontology may be surprised by some aspects of the field as depicted in this book. A common concept that helps define geographical gerontology as distinct from other fields is *scale* as a unit of analysis and perspectival frame. While Part III illustrated this for the book, scale also is an important orienting principle in many other chapters. A key part of the uniquely geographical gerontological imagination, scale allows one to zoom in and zoom out on a problem to discern different patterns and processes inherent in the subject matter of ageing, old age and older populations. Geographical gerontologists understand that such use of scale as an orienting principle is fundamental for question formulation, methodology selection and inference making. Such scales range from the body to the globe, as Part III indicated, and other scales not featured explicitly in those chapters would include neighbourhoods and regions (e.g., Appalachia, Scandinavia, sub-Saharan Africa, etc.). The geographic scale of analysis utilised shapes what one sees and understands about a problem. We note that such sensitivity to scale is important regardless of the subject matter or the conceptual and theoretical orientation.

The diversity of orientations and problems studied in the field is remarkably illustrated by the preceding chapters. Among its various attributes, geographical gerontology is informed by a broad set of theoretical traditions, models and concepts – many of which are reviewed or reflected in the chapters. Although space and place are the most central ideas in geographical gerontology, environment, landscape and movement (migration and mobility) are additional core ideas. Those concepts often carry multiple meanings, each informed by the theoretical tradition in which the geographical gerontologist locates him- or herself. Geography is the field of origin for many of those theories and concepts, and geography also is integrative because of its frequent use of theories and concepts from other disciplines. Geography is by its integrative character, therefore, predisposed to inter-/transdisciplinary endeavours. The beauty of geographical gerontology is due, in part, to its openness to new perspectives, concepts and approaches to understanding and solving diverse, complex, and important problems of ageing. Sensitivity and attention to the diversity and complexity of people and places and the ideas needed to understand them are other hallmarks of the geographical gerontological imagination – and this is apparent in many of the preceding chapters. As such, we should be clear: this book reflects a field rich in imaginative pluralism. While that pluralism might make the scope of the field more diffuse, we argue that the more important outcome is a collective intellectual toolbox ready for many purposes and problems.

We recognise that the story of geographical gerontology could be told in different ways (for instance, see the interdisciplinary viewpoint offered by Peace, Chapter 24 in this volume) and that there are gaps in the story told in this book. For example, social policy is not fully addressed, nor are the issues of indigenous peoples (although see Rosenberg and Wilson, Chapter 5) or political geographies of older adults (e.g., LGBTQ issues, activism by older groups and individuals). The types of geographical gerontology included here also tend toward the more critical – pointing to the social, economic, political and other causes of inequalities for those in old age. While this reflects our disciplinary backgrounds and our personal orientations as much as it does the field itself, we suggest that there are many examples of excellent geographical gerontology that are not overtly critical in the political economy or post-structural sense of the term (including some of the chapters herein).

Yet perhaps the most important aspect of geographical gerontology that defines its current state is a dynamism in thinking and a concomitant evolution of its theories, concepts and approaches. Many of the chapters trace the evolution of their subject matter in a way that such dynamism is evident. For example, in Chapter 2, Andrews, Cutchin and Skinner illustrated the constant drive toward new (and hopefully complimentary or better) ways of theorising gerontological issues, with non-representational theory being the latest contribution to the field. The use of landscapes as imagined or therapeutic is another more recent conceptual addition to gerontological inquiry, although the concept has deep roots in geography. Ageing also is a territorialised experience (Cutchin, 2014; Rosenberg and Everitt, 2001), and the concepts of territory and territorial power and behavior (e.g., Sack, 1986; Storey, 2012) are waiting to be employed more in geographical gerontology.

Opportunities and challenges for the field

As with other fields of inquiry, geographical gerontology is – and will be – shaped by a combination of its academic and societal contexts. On the academic side, there is a need for the discipline of geography to more fully embrace the importance of ageing societies and associated issues – and for other disciplines active in gerontological inquiry to recognise the discipline of geography and its theoretical and conceptual contributions to ageing studies. These dual challenges for geographical gerontology will likely remain in the future. A more specific opportunity exists for the field to contribute directly to a slightly more mature part of the academy. The life-course perspective is a distinct, interdisciplinary orientation (often focused on gerontological issues) that has both entered its mature phase and has many promises and opportunities left unfulfilled (Mayer, 2009). While geographical gerontology has contributed little to the life-course perspective to date, there is a great opportunity for it to do so. As Hendricks (2012: 231) argued about the life-course perspective, "[l]ike historical time, where we live affects how lives unfold . . . we need to consider whole lives in the contexts in which they unfold". There is a tremendous opportunity for the concepts and approaches discussed in the preceding chapters to inform the life course perspective – and not just in the sense of geography as a context. As many of the chapters in this volume have argued, the dynamic, relational interplay of places/landscapes/environments/spaces and individuals/groups/communities is more than context to use for comparison, as in cross-national studies. The interplay of geography and the social is a nuanced force that helps shape the other central dimensions of the life-course perspective. At the same time, geographical gerontology faces an ongoing challenge to keep up with the rapid evolution of theory and practice in human geography and the social sciences more generally (e.g., new materialisms, post-humanism, public geography/sociology, emotional turns, etc.). Although geographical gerontology may never be a leader in that evolution, it must continue to keep up with new ways of viewing its various subject matters.

From our vantage point, more opportunities for the contribution of geographical gerontology will be related to societal issues stemming from ageing populations in the twenty-first century; they will demand a greater role for geographical gerontology's theories, concepts and approaches. The 'age-friendly (and dementia-friendly) communities' movement is currently informing urban planning around the world – and is discussed in numerous chapters here. Yet geographical gerontologists and their thinking have not figured as prominently in the movement and related planning efforts as we would expect. Understanding and assessment of emerging models of community-based care, for example, the Village model (Lehning, Scharlach and Davitt, 2017), also would benefit from more geographical gerontological imagination. We hope the work featured in this book will encourage geographical gerontologists to be more engaged in those efforts. At the same time, we also believe that those leading such movements will further recognise the inherent benefits that geographical gerontology offers their efforts.

Broader social changes, however, are driving many of the questions that will challenge gerontologists for the coming decades – and they will provide opportunities for geographical gerontology to contribute. There are many that could be discussed here, but we note just a few examples. One seminal change is an ever-faster world, where movement and sensation take precedence over narrative and meaning, where we live simultaneously across multiple dimensions and places through mobile devices, where technology will help us live longer and look younger, where we will work and play longer. The impacts of that ever-faster world on ageing and on older populations and their well-being will need geographical gerontology to understand the problems and help solve them. Another important change relates to the first, and specifically to working later in life. The political and economic shifts in every part of the world intersect with ageing populations in ways we cannot foresee completely. Yet the trend toward later 'retirement' – if it is an option at all – is likely to remain and to shift the experience of ageing dramatically for many, in different ways in different places. This issue also will be better addressed by the inclusion of perspectives and concepts of geographical gerontology. An additional issue looming on the horizon, climate change and its associated upheavals, will be a huge challenge to older populations and those who serve them. Yet those challenges will be both geographically differentiated as well as influenced by physical geography, population dynamics and sociocultural, economic, and political geographies in varied spaces and places. It would seem that geographical gerontology should be well positioned to serve society on that front as well. Finally, ongoing migration and refugee crises will include the needs of older people more than ever. As the European Union continues to struggle with ongoing waves of refugees and migrants, we presume that climate change and other forms of disaster and conflict will continue to affect large numbers of older people – those left behind and those who move. Not only are individuals and groups who move affected by such upheaval, but sending and receiving places are changed as well. All affected people and places will need the understanding and knowledge we can provide. They also will need multiple types of planning and support.

The vision we have for the field is tied to its development so far and to our expectation of future needs, and that vision includes a greater volume and sophistication of empirical work in geographical gerontology. The concepts and perspectives discussed on these pages are tools for many forms of inquiry. Empirical research has not been the focus of this book, but there is a need for additional exemplary studies to be developed and collected – partly as demonstrations of how to operationalise such concepts in various types of geographies – and by way of refining theories and concepts in geographical gerontology and continuing the development of the field. Any field must demonstrate validity and utility of its ideas and their relevance to real-world problems. That continuing project lies in the future as well.

Towards a truly transdisciplinary field of study

The field of geographical gerontology can certainly lay claim to being multidisciplinary and in many cases interdisciplinary in nature. Becoming more fully transdisciplinary, where researchers integrate and transcend disciplinary approaches to develop new frameworks, theories, models or applications (Cooke and Hilton, 2015) should be a goal for this field. The problems of ageing and older populations are becoming increasingly complex, and increasingly nuanced perspectives and approaches are necessary for their study. While geographical gerontology has grown from geography, the geographical imagination, as noted by Gregory (2009) at the outset of this conclusion, is not an idea located exclusively in the province of geography, for geographers. And, as alluded to earlier and made clear by the many contributors in this volume, there are multiple geographic imaginations – there is a healthy pluralism of perspectives, theories, concepts and approaches to the subject matter of geographical gerontology. To build on that imaginative pluralism, the field needs more contributors from outside geography as well as from within geography. So-called team science, or other notions of transdisciplinary inquiry that are becoming the norm in scholarly practice, requires a shared language and shared concepts to unite scholars in vision, collaboration and the production of useful insights about ageing. This book offers a foundation for the maturation of geographical gerontology into a more fully transdisciplinary enterprise. We sincerely hope that readers agree, and we hope to see that enterprise flourish.

In closing, we have been inspired by those who have come before us in the field of geographical gerontology, but we are equally inspired by the contributors to this volume. Moreover, although the academic and material challenges to geographical gerontology are significant, we are ever-hopeful that the field will continue to develop its intellectual toolbox for inquiry as well as the cadre of scholars who will use it. By doing so, we expect geographical gerontology will provide important insights and contribute to making the world a better place to age. We look forward to meeting that challenge with you.

References

Cooke, N. J. and Hilton, M. L. (eds). (2015). *Enhancing the effectiveness of team science*. Committee on the Science of Team Science, Behavioral, Cognitive, and Sensory Sciences, Division of Behavioral and Social Sciences and Education, National Research Council. Washington, DC: The National Academies Press.

Cutchin, M. P. (2014). *Beyond environmental gerontology: The many geographies of aging*. Invited plenary address to the 2014 Annual Scientific and Educational Meeting of the Canadian Association on Gerontology. Available from: mpc@wayne.edu.

Gregory, D. (2009). Geographical imagination, in D. Gregory, R. Johnston, G. Pratt, M. J. Watts and S. Whatmore (eds), *The dictionary of human geography* (5th edn., pp. 282–285). Chichester: Wiley-Blackwell. Available from: www.univpgri-palembang.ac.id/perpus-fkip/Perpustakaan/Geography/Kamus%20Geografi/Kamus%20Geografi%20Manusia.pdf

Hendricks, J. (2012). Considering life course concepts. *The Journals of Gerontology, Series B: Psychological and Social Sciences*, 67(2), 226–231. http://dx.doi.org/10.1093/geronb/gbr147

Lehning, A. J., Scharlach, A. E. and Davitt, J. K. (2017). Variations on the village model: An emerging typology of a consumer-driven community-based initiative for older adults. *Journal of Applied Gerontology*, 36(2), 234–246. http://dx.doi.org/10.11177/0733464817690676

Mayer, K. U. (2009). New directions in life course research. *Annual Review of Sociology*, 35, 413–433. http://dx.doi.org/10.1146/annurev.soc.34.040507.134619

Rosenberg, M. and Everitt, J. (2001). Planning for aging populations: Inside or outside the walls. *Progress in Planning*, 56(3), 119–168. http://dx.doi.org/10.1016/S0305-9006(01)00014-9

Sack, R. (1986). *Human territoriality: Its theory and history*. Cambridge: Cambridge University Press.

Storey, D. (2012). *Territories: The claiming of space*. Hoboken, NJ: Taylor and Francis.

Index

Page numbers in italics indicate a figure.

Printed in the United States
by Baker & Taylor Publisher Services